RANDY Stevenson

BASIC MATHEMATICS FOR ECONOMISTS

While economics students are not always expected to be mathematical geniuses, it is generally accepted that some basic mathematical skills are necessary. *Basic Mathematics For Economists* recognizes that not everyone is comfortable with figures and aims to develop mathematical skills and build the confidence of mature students and those without A-level maths, to the level required for a general economics degree course.

The author provides a gentle introduction, concentrating on revision of arithmetical and algebraic methods that students have probably learned but forgotten. Here, as throughout the book, the information is set out, where possible, in the context of applications in economics. As the book progresses, so the pace increases, as new information is gradually introduced. However, the techniques are kept as simple and relevant to economic use as possible. Thus, familiarizing students with practical usage as quickly as possible, while avoiding abstract techniques. The book concentrates on those techniques which are likely to be useful to all students and avoids complex proofs and special cases. Particular attention is given to the increasingly important field of business economics. In recognition of the increased accessibility to computers the book also includes sections on the use of the Lotus 1–2–3 spreadsheet programme in solving very difficult or time-consuming problems.

Mike Rosser is Principal Lecturer in Economics in the Business School at Coventry University. He has wide teaching experience, including courses in quantitative methods for part-time mature students with little or no formal mathematical qualifications.

BASIC MATHEMATICS FOR ECONOMISTS

Mike Rosser

London and New York

First published 1993
by Routledge
11 New Fetter Lane, London EC4P 4EE

Simultaneously published in the USA and Canada
by Routledge
29 West 35th Street, New York, NY 10001

Reprinted 1995, 1996, 1998, 1999

Routledge is an imprint of the Taylor & Francis Group

Typeset in Times by Pure Tech Corporation, Pondicherry, India
Printed and bound in Great Britain by
TJ International Ltd, Padstow, Cornwall

British Library Cataloguing in Publication Data
A catalogue record for this book is available from the British Library

Library of Congress Cataloging in Publication Data
A catalog record for this book is available from the Library of Congress

ISBN 0–415–08424–5 (hbk)
ISBN 0–415–08425–3 (pbk)

Contents

Preface   ix
Acknowledgements   xi

1 Introduction   1
  1.1 Why Study Mathematics?   1
  1.2 Calculators and Computers   3
  1.3 Using the Book   6

2 Arithmetic   9
  2.1 Revision of Basic Concepts   9
  2.2 Multiple Operations   10
  2.3 Brackets   12
  2.4 Fractions   13
  2.5 Elasticity of Demand   16
  2.6 Decimals   19
  2.7 Negative Numbers   22
  2.8 Powers   23
  2.9 Roots and Fractional Powers   27
  2.10 Logarithms   29

3 Introduction to Algebra   35
  3.1 Representation   35
  3.2 Evaluation   38
  3.3 Simplification: Addition and Subtraction   39
  3.4 Simplification: Multiplication   41
  3.5 Simplification: Factorizing   46
  3.6 Simplification: Division   50
  3.7 Solving Simple Equations   52
  3.8 The Summation Sign Σ 57
  3.9 Inequality Signs   61

4 Graphs and Functions   65
  4.1 Functions   65
  4.2 Inverse Functions   67

Contents

	4.3 Graphs of Linear Functions	70
	4.4 Fitting Linear Functions	76
	4.5 Slope	79
	4.6 Budget Constraints	85
	4.7 Non-linear Functions	90
	4.8 Composite Functions	94
	4.9 Functions with Two Independent Variables	100
	4.10 Summing Functions Horizontally	104
5	**Linear Equations**	**110**
	5.1 Simultaneous Linear Equation Systems	110
	5.2 Solving Simultaneous Linear Equations	110
	5.3 Graphical Solution	111
	5.4 Equating to Same Variable	113
	5.5 Substitution	116
	5.6 Row Operations	118
	5.7 More than Two Unknowns	120
	5.8 Which Method?	123
	5.9 Price Discrimination and Multiplant Monopoly	128
	Appendix: Linear Programming	144
6	**Quadratic Equations**	**166**
	6.1 Solving Quadratic Equations	166
	6.2 Graphical Solution	167
	6.3 Factorization	172
	6.4 The Quadratic Formula	174
	6.5 Quadratic Simultaneous Equations	176
	6.6 Polynomials	180
7	**Series, Time and Investment**	**186**
	7.1 Growth and Decline	186
	7.2 Discrete and Continuous Functions	186
	7.3 Interest	188
	7.4 Time Periods, Initial Amounts and Interest Rates	195
	7.5 Investment Appraisal: Net Present Value	202
	7.6 The Internal Rate of Return	215
	7.7 Geometric Series and Annuities	224
	7.8 Perpetual Annuities	232
	7.9 Loan Repayments	236
	7.10 Other Applications of Growth and Decline	243
8	**Introduction to Calculus**	**251**
	8.1 The Differential Calculus	251
	8.2 Rules for Differentiation	253
	8.3 Marginal Revenue and Total Revenue	256
	8.4 Marginal Cost and Total Cost	262

Contents

8.5	Profit Maximization	265
8.6	Respecifying Functions	267
8.7	Point Elasticity of Demand	269
8.8	Tax Yield	271
8.9	The Keynesian Multiplier	274
9	**Unconstrained Optimization**	**277**
9.1	Maximization and Zero Slope	277
9.2	Second-order Condition for a Maximum	278
9.3	Second-order Condition for a Minimum	281
9.4	Summary of Second-order Conditions	282
9.5	Profit Maximization	285
9.6	Inventory Control	288
10	**Partial Differentiation**	**292**
10.1	Partial Differentiation and the Marginal Product	292
10.2	Further Applications of Partial Differentiation	298
10.3	Second-order Partial Derivatives	309
10.4	Unconstrained Optimization: Functions with Two Variables	315
10.5	Total Differentials and Total Derivatives	329
11	**Constrained Optimization**	**338**
11.1	Constrained Optimization and Resource Allocation	338
11.2	Constrained Optimization by Substitution	338
11.3	The Lagrange Multiplier: Constrained Maximization with Two Variables	346
11.4	The Lagrange Multiplier: Second-order Conditions	352
11.5	Constrained Minimization Using the Lagrange Multiplier	355
11.6	Constrained Optimization with more than Two Variables	361
12	**Further Topics in Calculus**	**370**
12.1	Overview	370
12.2	The Chain Rule	370
12.3	The Product Rule	377
12.4	The Quotient Rule	382
12.5	Integration	389
12.6	Definite Integrals	395
13	**Dynamics and Difference Equations**	**402**
13.1	Dynamic Economic Analysis	402
13.2	The Cobweb: Iterative Solutions	403
13.3	The Cobweb: Difference Equation Solutions	415
13.4	The Lagged Keynesian Macroeconomic Model	425
13.5	Duopoly Price Adjustment	438
14	**Exponential Functions and Continuous Growth**	**444**
14.1	Continuous Growth	444

Contents

14.2 Exponential Functions   444
14.3 Natural Logarithms   451
14.4 Differentiation of Logarithmic and Exponential Functions   454
14.5 Discrete and Continuous Growth Compared   455

Answers   459
Index   468

Preface

Over half of the students who enrol on economics degree courses have not studied mathematics beyond GCSE or an equivalent level. These include many mature students whose last encounter with algebra, or any other mathematics beyond basic arithmetic, is now a dim and distant memory. It is for these students that this book is intended. It aims to develop their mathematical ability up to the level required for a general economics degree course (i.e. one not specializing in mathematical economics) or for a modular degree course in economics and related subjects, such as business studies. To achieve this aim it has several objectives.

First, it provides a revision of arithmetical and algebraic methods that students probably studied at school but have now largely forgotten. It is a misconception to assume that, just because a GCSE mathematics syllabus includes certain topics, students who passed examinations on that syllabus two or more years ago are all still familiar with the material. They usually require some revision exercises to jog their memories and to get into the habit of using the different mathematical techniques again. The first few chapters are mainly devoted to this revision, set out where possible in the context of applications in economics.

Second, this book introduces mathematical techniques that will be new to most students through examples of their application to economic concepts. It also tries to get students tackling problems in economics using these techniques as soon as possible so that they can see how useful they are. Students are not required to work through unnecessary proofs, or wrestle with complicated special cases that they are unlikely ever to encounter again. For example, when covering the topic of calculus, some other textbooks require students to plough through abstract theoretical applications of the technique of differentiation to every conceivable type of function and special case before any mention of its uses in economics is made. In this book, however, we introduce the basic concept of differentiation followed by examples of economic applications in Chapter 8. Further developments of the topic, such as the second-order conditions for optimization, partial differentiation and the rules for differentiation of composite functions, are then gradually brought in over the next few chapters, again in the context of economics application.

Third, this book tries to cover those mathematical techniques that will be most relevant to economics students in the 1990s. Most applications are in the field of microeconomics, rather than macroeconomics, given the increased emphasis on business economics within many degree courses. In particular, Chapter 7 concentrates on a number of mathematical techniques that are relevant to finance and investment decision-making.

Given that most students now have access to computing facilities, ways of using a spreadsheet package to solve certain problems that are extremely difficult or time-consuming to solve manually are also explained.

Although it starts at a gentle pace through fairly elementary material, so that the students who gave up mathematics some years ago because they thought that they could not cope with A-level maths are able to build up their confidence, this is *not* a watered-down 'mathematics without tears or effort' type of textbook. As the book progresses the pace is increased and students are expected to put in a serious amount of time and effort to master the material. However, given the way in which this material is developed, it is hoped that students will be motivated to do so. Not everyone finds mathematics easy, but at least it helps if you can see the reason for having to study it.

Mike Rosser
Coventry

Acknowledgements

I am grateful to the Lotus Development Corporation for kindly granting authorization to include material based on the Lotus 1–2–3* computer package.

I would like to thank Joy Warren for her efficiency in typing the final manuscript and coping with the numerous mathematical formulations that this required, Mrs M. Fyvie for typing earlier drafts, Chandrika Chauhan for her help in completing the diagrams and tables, and Mick Hayes for checking the proofs.

I would also like to acknowledge the help of my students in shaping the way that this book developed, particularly the mature students on the BA Applied Economics Course at Coventry University. I, of course, am responsible for any remaining errors or omissions.

---

* Lotus and 1–2–3 are Registered Trademarks of Lotus Development Corporation

1

Introduction

1.1 WHY STUDY MATHEMATICS?

Economics is a social science. It does not just describe what goes on in the economy. It attempts to explain how the economy operates and to make predictions about what may happen to specified economic variables if certain changes take place, e.g. what effect a crop failure will have on crop prices, what effect a given increase in sales tax will have on the price of finished goods, what will happen to unemployment if government expenditure is increased. It also suggests some guidelines that firms, governments or other economic agents might follow if they wished to allocate resources efficiently. Mathematics is fundamental to any serious application of economics to these areas.

Quantification

In introductory economic analysis predictions are often explained with the aid of sketch diagrams. For example, supply and demand analysis predicts that in a competitive market if supply is restricted then the price of a good will rise. However, this is really only common sense, as any market trader will tell you. An economist also needs to be able to say by how much price is expected to rise if supply contracts by a specified amount. This quantification of economic predictions requires the use of mathematics.

Although non-mathematical economic analysis may sometimes be useful for making qualitative predictions (i.e. predicting the direction of any expected changes), it cannot by itself provide the quantification that users of economic predictions require. A firm needs to know how much quantity sold is expected to change in response to a price increase. The government wants to know how much consumer demand will change if it increases a sales tax.

Simplification

Sometimes students believe that mathematics makes economics more complicated. Algebraic notation, which is essentially a form of shorthand, can, however, make certain concepts much clearer to understand than if they were set

1

out in words. It can also save a great deal of time and effort in writing out tedious verbal explanations.

For example, the relationship between the quantity of apples consumers wish to buy and the price of apples might be expressed as: 'the quantity of apples demanded in a given time period is 1,200 kg when price is zero and then decreases by 10 kg for every 1p rise in the price of a kilo of apples'. It is much easier, however, to express this mathematically as: $q = 1,200 - 10p$ where q is the quantity of apples demanded in kilograms and p is the price in pence per kilogram of apples.

This is a very simple example. The relationships between economic variables can be much more complex and mathematical formulation then becomes the only feasible method for dealing with the analysis.

Scarcity and choice

Many problems dealt with in economics are concerned with the most efficient way of allocating limited resources. These are known as 'optimization' problems. For example, a firm may wish to maximize the output it can produce within a fixed budget for expenditure on inputs. Mathematics must be used to obtain answers to these problems.

Many economics graduates will enter employment in industry, commerce or the public sector where very real resource allocation decisions have to be made. Mathematical methods are used as a basis for many of these decisions. Even if students do not go on to specialize in subjects such as managerial economics or operational research where the applications of these decision-making techniques are studied in more depth, it is essential that they gain an understanding of the sort of resource allocation problems that can be tackled and the information that is needed to enable them to be solved.

Economic statistics and estimating relationships

Not only is mathematics needed to work out predictions from an economic model where the relationships are already quantified, it is also needed in order to estimate the parameters of the model in the first place.

If the economic model $q = 1,200 - 10p$ described the demand relationship in an actual market then this would mean that the parameters (i.e. the numbers 1,200 and 10) had been estimated from statistical data.

The study of how the parameters of economic models can be estimated from statistical data is known as econometrics. Although this is not one of the topics covered in this book, you will find that a knowledge of several of the mathematical techniques that are covered is necessary to understand the methods used in econometrics. Students using this book will probably also study an introductory statistics course as a prerequisite for econometrics, and here again certain basic mathematical tools will come in useful.

Mathematics and business

Some students using this book may be on courses that have more emphasis on business studies than pure economics. Two criticisms of the material covered that these students sometimes make are as follows.

1 These simple models do not bear any resemblance to the real-world business decisions that have to be made in practice.
2 Even if the models are relevant to business decisions there is not always enough actual data available on the relevant variables to make use of these mathematical techniques.

Criticism (1) should be answered in the first few lectures of your economics course when the methodology of economic theory is explained. In summary, one needs to start with a simplified model that can explain how firms (and other economic agents) behave in general before looking at more complex situations only relevant to specific firms.

Criticism (2) may be partially true, but a lack of complete data does not mean that one should not try to make the best decision using the information that is available. Just because some mathematical methods can be difficult to understand to the uninitiated, this does not mean that efficient decision-making should be abandoned in favour of guesswork, rule of thumb and intuition.

1.2 CALCULATORS AND COMPUTERS

Some students may ask, 'what's the point in spending a great deal of time and effort studying mathematics when nowadays everyone uses calculators and computers for calculations?' There are several answers to this question.

Rubbish in, rubbish out

Perhaps the most important point which has to be made is that calculators and computers can only calculate what they are told to. They are machines that can perform arithmetic computations much faster than you can do by hand, and this speed does indeed make them very useful tools. However, if you feed in useless information you will get useless information back – hence the well-known phrase 'rubbish in, rubbish out'.

At a very basic level, consider what happens when you use a pocket calculator to perform some simple operations. Get out your pocket calculator and use it to answer the problem

$$16 - 3 \times 4 - 1 = ?$$

What answer did you get? 3? 7? 51? 39? It all depends on which order you perform the calculations and the type of calculator you use.

3

There are set rules for the order in which basic arithmetic operations should be performed, which are explained in Chapter 2. Nowadays, these are programmed into most calculators but not some older basic calculators. If you only have a basic calculator then it cannot help you. It is you who must tell the calculator in which order to perform the calculations. (The correct answer is 3, by the way.)

For another example, consider the demand relationship

$$q = 1,200 - 10p$$

referred to earlier. What would quantity demanded be if price was 150? A computer would give the answer -300, but this is clearly nonsense as you cannot have a negative quantity of apples. It only makes sense for the above mathematical relationship to apply to positive values of p and q. Therefore if price is 120, quantity sold will be zero, and if any price higher than 120 is charged, such as 130, quantity sold will still be zero. This case illustrates why you must take care to interpret mathematical answers sensibly and not blindly assume that any numbers produced by a computer will always be correct even if the 'correct' numbers have been fed into it.

Algebra

Much economic analysis involves algebraic notation, with letters representing concepts that are capable of taking on different values (see Chapter 3). The manipulation of these algebraic expressions cannot usually be carried out by calculators and computers.

Rounding errors

Despite the speed of operation of calculators and computers it can sometimes be quicker and more accurate to solve a problem manually. To illustrate this point, if you have a basic calculator, use it to answer the problem

$$\frac{10}{3} \times 3 = ?$$

You will probably get an answer something like 9.9999999. However, if you use a modern mathematical calculator you will have obtained the correct answer of 10. So why do some calculators give a slightly inaccurate answer?

All calculators and computers have a limited memory capacity. This means that numbers have to be rounded off after a certain number of digits. Given that 10 divided by 3 is 3.3333333 recurring, it is difficult for basic calculators to store this number accurately in decimal form. Although modern computers have a vast memory they still perform many computations through a series of algorithms, which are essentially a series of arithmetic operations. At various stages numbers can be rounded off and so the final answer can be slightly inaccurate. More accuracy can often be obtained by using simple 'vulgar fractions' and by limiting the number of calculator operations that round off the answers. In fact the functions available on modern calculators, and on most computer programs, are now designed to try to minimize inaccuracies due to rounding errors.

When should you use calculators and computers?

Obviously pocket calculators are useful for basic arithmetic operations that take a long time to do manually, such as long division or finding square roots. If you only use a basic calculator, care needs to be taken to ensure that individual calculations are done in the correct order so that the fundamental rules of mathematics are satisfied and needless inaccuracies through rounding are avoided.

However, the level of mathematics in this book requires more than these basic arithmetic functions. It is recommended that all students obtain a mathematical calculator that has at least the following function keys:

$$[y^x] \qquad [\sqrt[x]{y}] \qquad [\text{LOG}] \qquad [10^x] \qquad [\text{LN}] \qquad [e^x]$$

The meaning and use of these functions will be explained in the following chapters.

Most of you who have recently left school will probably have already used this type of calculator for GCSE mathematics but mature students may only currently possess an older basic calculator with only the basic square root $[\sqrt{}]$ function. The modern mathematical calculators, in addition to having more mathematical functions, are a great advance on these basic calculators and can cope with most rounding errors and sequences of operations in multiple calculations. In some sections of the book, however, calculations that could be done on a mathematical calculator are still explained from first principles to ensure that all students fully understand the mathematical method employed.

Most students on economics degree courses will have access to computing facilities and be taught how to use various computer program packages. Most of these will probably be used for data analysis as part of the statistics component of your course. The facilities and programs available to students will vary from institution to institution. Your lecturer will advise whether or not you have access to computer program packages that can be used to tackle specific types of mathematical problems. For example, you may have access to a graphics package that tells you when certain lines intersect or solves linear programming problems (see Chapter 5). Spreadsheet programs, such as Lotus 1–2–3, can be useful particularly for the sort of financial problems covered in Chapter 7.

However, even if you do have access to computer program packages that can solve specific types of problem you will still need to understand the method of solution so that you will understand the answer that the computer gives you. Also, many economic problems have to be set up in the form of a mathematical problem before they can be fed into a computer program package for solution.

In the main, the problems and exercises in this book can be tackled without using computers although in some cases solution only using a calculator would be very time consuming. Some students may not have easy access to computing facilities. In particular, part-time students who only attend evening classes may find it difficult to get into computer laboratories. These students may find it worthwhile to invest a few more pounds in a more advanced calculator. Many of the problems requiring a large number of calculations are in Chapter 7 where methods of solution using the Lotus 1–2–3 spreadsheet program are suggested. However, financial calculators are now available that have most of the functions and formulae necessary to cope with these problems. These usually cost approximately double the price of a normal calculator, but this is still a small fraction of the price you would have to pay to purchase your own personal computer and the relevant programs.

As Lotus 1–2–3 is probably the spreadsheet program most commonly used by economics students, the spreadsheet suggested solutions to certain problems are given in Lotus 1–2–3 format. It is assumed that students will be familiar with the basic operational functions of this program (e.g. saving files, using the copy command etc.), and the solutions in this book only suggest a set of commands necessary to solve the set problems.

1.3 USING THE BOOK

Most students using this book will be on the first year of an economics degree course and will not have studied A-level mathematics. Some of you will be following a mathematics course specifically designed for people without A-level mathematics whilst others will be mixed in with more mathematically experienced students on a general quantitative methods course. The book starts from some very basic mathematical principles. Most of these you will already have covered at GCSE (or O-level or CSE for mature students). Only you can judge whether or not you are sufficiently competent in a technique to be able to skip some of the sections.

It would be advisable, however, to start at the beginning of the book and work through all the set problems. Many of you will have had at least a two-year break since last studying mathematics and will benefit from some revision. If you cannot easily answer all the questions in a section then you obviously need to work through the topic. You should find that a lot of material is familiar to you although more applications of mathematics to economics are introduced as the book progresses.

It is assumed that students using this book will also be studying an economic analysis course. The examples in the first few chapters only use some basic economic theory, such as supply and demand analysis. By the time you get to the later chapters it will be assumed that you have covered additional topics in economic analysis, such as production and cost theory. If you come across problems that assume a knowledge of economics topics that you have not yet covered then you should leave them until you understand these topics, or consult your lecturer.

In some instances the basic analysis of certain economic concepts is explained before the mathematical application of these concepts but this should not be considered a complete coverage of the topic.

Practise, practise

You will not learn mathematics by reading this book, or any other book for that matter. The only way you will learn mathematics is by practising working through problems. It may be more hard work than just reading through the pages of a book, but your effort will be rewarded when you master the different techniques. As with many other skills that people acquire, such as riding a bike or driving a car, a book can help you to understand how something is supposed to be done but you will only be able to do it yourself if you spend time and effort practising.

You cannot acquire a skill by sitting down in front of a book and hoping that you can 'memorize' what you read.

Group working

Your lecturer will make it clear to you which problems you must do by yourself as part of your course assessment and which problems you may confer with others over. Asking others for help makes sense if you are absolutely stuck and just cannot understand a topic. However, you should make every effort to work through all the problems that you are set by yourself before asking your lecturer or fellow students for help. When you do ask for help it should be to find out *how* to tackle a problem.

Some students who have difficulty with mathematics tend to copy answers off other students without really understanding what they are doing, or when a lecturer runs through an answer in class they just write down a verbatim copy of the answer given without asking for clarification of points they do not follow.

They are only fooling themselves, however. The point of studying mathematics in the first year of an economics degree course is to learn how to be able to apply it to various economics topics. Students who pretend that they have no difficulty with something they do not properly understand will obviously not get very far.

What is important is that you understand the *method* of solving different types of problems. There is no point in having a set of answers to problems if you do not understand how these answers were obtained.

Don't give up!

Do not get disheartened if you do not understand a topic the first time it is explained to you. Mathematics can be a difficult subject and you will need to read through some sections several times before they become clear to you. If you make the effort to try all the set problems and consult your lecturer if you really get stuck then you will eventually master the subject.

Because the topics follow on from each other, each chapter assumes that students are familiar with material covered in previous chapters. It is therefore very important that you keep up to date with your work. You cannot 'skip' a topic that you find difficult and hope to get through without answering examination questions on it, as it is sometimes possible to do in other subjects.

About half of all students on economics degree courses gave up mathematics at school at the age of 16, many of them because they thought that they were not good enough at mathematics to take it for A-level. However, most of them usually manage to complete their first-year mathematics for economics course successfully and go on to achieve an honours degree. There is no reason why you should not do likewise if you are prepared to put in the effort.

2

Arithmetic

2.1 REVISION OF BASIC CONCEPTS

Most students will have previously covered all, or nearly all, of the topics in this chapter. They are included here for revision purposes and to ensure that everyone is familiar with basic arithmetical processes before going on to further mathematical topics. Only a fairly brief explanation is given for most of the arithmetical rules set out in this chapter. It is assumed that students will have learned these rules at school and now just require something to jog their memory so that they can begin to use them again.

As a starting point it will be assumed that all students are familiar with the basic operations of addition, subtraction, multiplication and division, as applied to whole numbers (or integers) at least. The notation for these operations can vary but the usual ways of expressing them are as follows.

Example 2.1

Addition (+):	$24 + 204 = 228$
Subtraction (−):	$9,089 - 393 = 8,696$
Multiplication (× or .):	$12 \times 24 = 288$
Division (÷ or /):	$4,448 \div 16 = 278$

The sign '.' is sometimes used for multiplication when using algebraic notation but, as you will see from Chapter 2 onwards, there is usually no need to use any multiplication sign to signify that two algebraic variables are being multiplied together, e.g. A times B is simply written AB.

Most students will have learned at school how to perform these operations with a pen and paper, even if their long multiplication and long division may now be a bit rusty. However, apart from simple addition and subtraction problems, it is usually quicker to use a pocket calculator for basic arithmetical operations. If you cannot answer the questions below then you need to refer to an elementary arithmetic text or to see your lecturer for advice.

QUESTIONS 2.1

1. $323 + 3{,}232 =$
2. $1{,}012 - 147 =$
3. $460 \times 202 =$
4. $1{,}288/56 =$

2.2 MULTIPLE OPERATIONS

Consider the following problem involving only addition and subtraction.

Example 2.2

A bus leaves its terminus with 22 passengers aboard. At the first stop 7 passengers get off and 12 get on. At the second stop 18 get off and 4 get on. How many passengers remain on the bus?

Most of you would probably answer this by saying $22 - 7 = 15$, $15 + 12 = 27$, $27 - 18 = 9$, $9 + 4 = 13$ passengers remaining, which is the correct answer.
    If you were faced with the abstract mathematical problem

$$22 - 7 + 12 - 18 + 4 = ?$$

you should answer it in the same way, i.e. working from left to right.
    If you performed the addition operations first then you would get $22 - 19 - 22 = -19$ which is clearly not the correct answer to the bus passenger problem!
    If we now consider an example involving only multiplication and division we can see that the same rule applies.

Example 2.3

A restaurant catering for a large party sits 6 people to a table. Each table requires 2 dishes of vegetables. How many dishes of vegetables are required for a party of 60?

Most people would answer this by saying $60 \div 6 = 10$ tables, $10 \times 2 = 20$ dishes, which is correct.
    If this is set out as the calculation

$$60 \div 6 \times 2 = ?$$

then the left to right rule must be used. If you did not use this rule then you might get

$$60 \div 6 \times 2 = 60 \div 12 = 5$$

which is incorrect.

Thus the general rule to use when a calculation involves several arithmetical operations and

1 only addition and subtraction are involved or
2 only multiplication and division are involved

is that the operations should be performed by working from left to right.

Examples 2.4

   (i) $48 - 18 + 6 = 30 + 6 = 36$
  (ii) $6 + 16 - 7 = 22 - 7 = 15$
 (iii) $68 + 5 - 32 - 6 + 14 = 73 - 32 - 6 + 14$
$$= 41 - 6 + 14$$
$$= 35 + 14 = 49$$
 (iv) $22 \times 8 \div 4 = 176 \div 4 = 44$
  (v) $460 \div 5 \times 4 = 92 \times 4 = 368$
 (vi) $200 \div 25 \times 8 \times 3 \div 4 = 8 \times 8 \times 3 \div 4$
$$= 64 \times 3 \div 4$$
$$= 192 \div 4 = 48$$

When a calculation involves both addition/subtraction and multiplication/division then the rule is: multiplication and division calculations must be done before addition and subtraction calculations (except when brackets are involved – see Section 2.3 below). To illustrate the rationale for this rule consider the following simple example.

Example 2.5

How much change do you get from £5 if you buy 6 oranges at 40p each?

Solution

$$\text{change} = 500 - 6 \times 40 = 500 - 240 = 260\text{p} = £2.60.$$

Clearly the multiplication must be done before the subtraction in order to arrive at the correct answer. (Note that all calculations must be done using the same units, in this case pence.)

QUESTIONS 2.2

1. $962 - 88 + 312 - 267 =$
2. $240 - 20 \times 3 \div 4 =$
3. $300 \times 82 \div 6 \div 25 =$
4. $360 \div 4 \times 7 - 3 =$
5. $6 \times 12 \times 4 + 48 \times 3 + 8 =$
6. $420 \div 6 \times 2 - 64 + 25 =$

2.3 BRACKETS

If a calculation involves brackets then the operations within the brackets must be done first. Thus brackets take precedence over the rule for multiple operations set out in Section 2.2 above.

Example 2.6

A firm produces 220 units of a good which cost an average of £8.25 each to produce and sells them at a price of £9.95. What is its profit?

Solution

$$\text{profit per unit} = £9.95 - £8.25$$
$$\text{total profit} = 220 \times (£9.95 - £8.25)$$
$$= 220 \times £1.70$$
$$= £374$$

In a calculation that only involves addition or subtraction the brackets can be removed. However, you must remember that if there is a minus sign before a set of brackets then all the terms within the brackets must be multiplied by − 1 if the brackets are removed, i.e. all + and − signs are reversed. (See Section 2.7 if you are not familiar with the concept of negative numbers.)

Example 2.7

$$(92 - 24) - (20 - 2) = ?$$

Solution

$$68 - 18 = 50 \text{ using brackets}$$

or

$$92 - 24 - 20 + 2 = 50 \text{ removing brackets}$$

QUESTIONS 2.3

1. $(12 \times 3 - 8) \times (44 - 14) =$
2. $(68 - 32) - (100 - 84 + 3) =$
3. $60 + (36 - 8) \times 4 =$
4. $4 \times (62 \div 2) - 8 \div (12 \div 3) =$
5. If a firm produces 600 units of a good at an average cost of £76 and sells them all at a price of £99 what is its total profit?
6. $(124 + 6 \times 81) - (42 - 2 \times 15) =$

7. How much net (i.e. after tax) profit does a firm make if it produces 440 units of a good at an average cost of £3.40 each, and pays 15p tax to the government on each unit sold at the market price of £3.95, assuming it sells everything it produces?

2.4 FRACTIONS

If computers and calculators use decimals when dealing with portions of whole numbers why bother with fractions? There are several reasons.

1 Certain operations, particularly multiplication and division, can sometimes be done more quickly by fractions if one can cancel out numbers.
2 When using algebraic notation instead of actual numbers one cannot use calculators and operations on formulae have to be performed using the basic principles for operations on fractions.
3 In some cases fractions can give a more accurate answer than a calculator owing to rounding error (see Example 2.15 below).

A fraction is written as

$$\frac{\text{numerator}}{\text{denominator}}$$

and is just another way of saying that the numerator is divided by the denominator. Thus

$$\frac{120}{960} = 120 \div 960$$

Before carrying out any arithmetical operations with fractions it is best to simplify individual fractions. Both numerator and denominator can be divided by any whole number that they are both a multiple of.

Example 2.8

$$\frac{168}{104} = \frac{21 \times 8}{13 \times 8} = \frac{21}{13}$$

In this example it is obvious that the 8s cancel out top and bottom, i.e. the numerator and denominator can both be divided by 8.

Example 2.9

$$\frac{120}{960} = \frac{12 \times 10}{12 \times 8 \times 10} = \frac{1}{8}$$

Addition and subtraction of fractions is carried out by converting all fractions so that they have a common denominator (usually the largest one) and then adding or subtracting the different quantities with this common denominator. To convert fractions to the common (largest) denominator, one multiplies both top and bottom of the fraction by whatever number it is necessary to get the required denominator. For example, to convert $\frac{1}{6}$ to a fraction with 12 as its denominator, one simply multiplies top and bottom by 2. Thus $\frac{1}{6} = \frac{2}{12}$.

Example 2.10

$$\frac{1}{6} + \frac{5}{12} = \frac{2}{12} + \frac{5}{12} = \frac{2+5}{12} = \frac{7}{12}$$

It is necessary to convert any numbers that have an integer (i.e. a whole number) in them into fractions with the same denominator before carrying out addition or subtraction operations involving fractions. This is done by multiplying the integer by the denominator of the fraction and then adding.

Example 2.11

$$1\frac{3}{5} = \frac{1 \times 5}{5} + \frac{3}{5} = \frac{5}{5} + \frac{3}{5} = \frac{8}{5}$$

Example 2.12

$$2\frac{3}{7} - \frac{24}{63} = \frac{17}{7} - \frac{8}{21} = \frac{51-8}{21} = \frac{43}{21} = 2\frac{1}{21}$$

Multiplication of fractions is carried out by multiplying the numerators of the different fractions and then multiplying the denominators.

Example 2.13

$$\frac{3}{8} \times \frac{5}{7} = \frac{15}{56}$$

The exercise can be simplified if one first cancels out any whole numbers that can be divided into a numerator and a denominator.

Example 2.14

$$\frac{20}{3} \times \frac{12}{35} \times \frac{4}{5} = \frac{(4 \times 5) \times (4 \times 3) \times 4}{3 \times 35 \times 5} = \frac{4 \times 4 \times 4}{35} = \frac{64}{35}$$

The usual way of performing this operation is simply to cross through numbers that cancel

$$\frac{\overset{4}{\cancel{20}}}{\underset{1}{\cancel{3}}}\times\frac{\overset{4}{\cancel{11}}}{\underset{7}{\cancel{33}}}\times\frac{4}{5}=\frac{64}{35}$$

Multiplying out fractions may provide a more accurate answer than the one you would get by working out the decimal value of a fraction with a calculator before multiplying. However, nowadays if you use a modern mathematical calculator and store the answer to each part you should avoid rounding errors.

Example 2.15

$$\frac{4}{7}\times\frac{7}{2}=?$$

Solution

$$\frac{4}{7}\times\frac{7}{2}=\frac{4}{2}=2 \text{ using fractions}$$

$$0.5714285\times 3.5 = 1.9999997 \text{ using a basic calculator}$$

Using a modern calculator, if you enter the numbers and commands

$$4\,[\div]\,7\,[\times]\,7\,[\div]\,2\,[=]$$

you should get the correct answer of 2.

However, if you were to perform the operation 4 [÷] 7, note the answer of 0.5714286 and then re-enter this number and multiply by 3.5, you would get the slightly inaccurate answer of 2.0000001.

To divide by a fraction one simply multiplies by its inverse.

Example 2.16

$$3\div\frac{1}{6}=3\times\frac{6}{1}=18$$

Example 2.17

$$\frac{44}{7}\div\frac{8}{49}=\frac{44}{7}\times\frac{49}{8}=\frac{11}{1}\times\frac{7}{2}=\frac{77}{2}=38\frac{1}{2}$$

QUESTIONS 2.4

1. $\frac{1}{6}+\frac{1}{7}+\frac{1}{8}=$

2. $\frac{3}{7}+\frac{2}{9}-\frac{1}{4}=$

---

3. $\dfrac{2}{5} \times \dfrac{60}{7} \times \dfrac{21}{15} =$

4. $\dfrac{4}{5} \div \dfrac{24}{19} =$

5. $4\dfrac{2}{7} - 1\dfrac{2}{3}$

6. $2\dfrac{1}{6} + 3\dfrac{1}{4} - \dfrac{4}{5} =$

7. $3\dfrac{1}{4} \times 4\dfrac{1}{3} =$

8. $8\dfrac{1}{2} \div 2\dfrac{1}{6} =$

9. $20\dfrac{1}{4} - \dfrac{3}{5} \times 2\dfrac{1}{8} =$

10. $6 - \dfrac{2}{3} \div \dfrac{1}{12} + 3\dfrac{1}{2} =$

2.5 ELASTICITY OF DEMAND

The arithmetic operation of dividing a fraction by a fraction is usually the first technique that students on an economics course need to brush up on if their mathematics is a bit rusty. It is needed to calculate 'elasticity' of demand, which is a concept you should encounter fairly early in your microeconomics course where its uses will be explained. Price elasticity of demand is a measure of the responsiveness of demand to changes in price. It is usually defined as

$$e = (-1)\,\frac{\%\text{ change in quantity demanded}}{\%\text{ change in price}}$$

The (-1) in this definition ensures a positive value for elasticity as either the change in price or the change in quantity will be negative. When there are relatively large changes in price and quantity it is best to use the concept of 'arc elasticity' to measure elasticity along a section of a demand schedule. This takes the changes in quantity and price as percentages of the averages of their values before and after the change. Thus arc elasticity is usually defined as

$$\text{arc } e = (-1)\,\frac{\dfrac{\text{change in quantity}}{0.5(1\text{st quantity} + 2\text{nd quantity})} \times 100}{\dfrac{\text{change in price}}{0.5(1\text{st price} + 2\text{nd price})} \times 100}$$

The (-1) can be dropped if the changes in both price and quantity are treated as positive quantities even though a positive change usually corresponds to a negative quantity change, and vice versa. The 0.5 and the 100 will always cancel top and bottom in arc elasticity calculations. Thus we are left with

Arithmetic

$$e = \frac{\dfrac{\text{change in quantity}}{\text{1st quantity} + \text{2nd quantity}}}{\dfrac{\text{change in price}}{\text{1st price} + \text{2nd price}}}$$

as the formula for actually calculating price arc elasticity of demand.

Example 2.18

Calculate the arc elasticity of demand between points A and B on the demand schedule shown in Figure 2.1.

Solution

Between points A and B price falls by 5 from 20 to 15 and quantity rises by 20 from 40 to 60. Using the formula defined above

$$\text{arc } e = \frac{\dfrac{20}{40+60}}{\dfrac{5}{20+15}} = \frac{\dfrac{20}{100}}{\dfrac{5}{35}} = \frac{20}{100} \times \frac{35}{5} = \frac{1}{5} \times \frac{7}{1} = \frac{7}{5}$$

Example 2.19

When the price of a product is lowered from £350 to £200 quantity demanded increases from 600 to 750 units. Calculate the elasticity of demand over this section of its demand schedule.

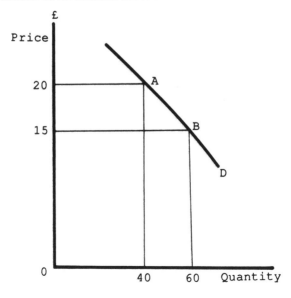

Figure 2.1

17

Solution

Price fall is £150 and quantity rise is 150. Therefore using the concept of arc elasticity

$$e = \frac{\dfrac{150}{600+750}}{\dfrac{150}{350+200}} = \frac{\dfrac{150}{1,350}}{\dfrac{150}{550}} = \frac{150}{1,350} \times \frac{550}{150} = \frac{1}{27} \times \frac{11}{1} = \frac{11}{27}$$

QUESTIONS 2.5

1. With reference to the demand schedule in Figure 2.2 calculate the arc elasticity of demand between the prices of (a) £3 and £6, (b) £6 and £9, (c) £9 and £12, (d) £12 and £15, and (e) £15 and £18.
2. A city bus service charges a uniform fare for every journey made. When this fare is increased from 50p to £1 the number of journeys made drops from 80,000 a day to 40,000. Calculate the arc elasticity of demand over this section of the demand schedule for bus journeys.

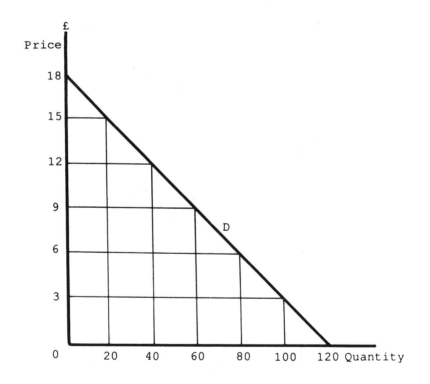

Figure 2.2

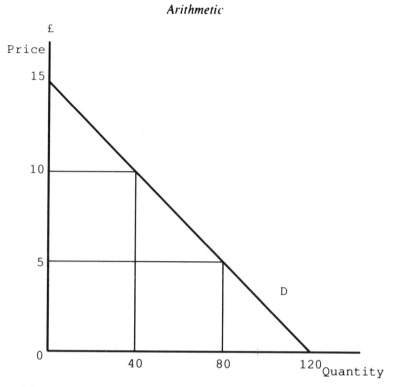

Figure 2.3

3. Calculate the arc elasticity of demand between (a) £5 and £10, and (b) between £10 and £15, for the demand schedule shown in Figure 2.3.
4. The table below shows the quantity demanded of a good at various prices. Calculate the arc elasticity of demand for each £5 increment along the demand schedule.

Price	£40	£35	£30	£25	£20	£15	£10	£5	£0
Quantity	0	50	100	150	200	250	300	350	400

2.6 DECIMALS

Decimals are just another way of expressing fractions.

$$0.1 = 1/10$$
$$0.01 = 1/100$$
$$0.001 = 1/1,000 \text{ etc.}$$

Thus 0.234 is equivalent to 234/1,000.

Most of the time you will be able to perform operations involving decimals by using a calculator and so only a very brief summary of the manual methods of performing arithmetic operations using decimals is given here.

19

Addition and subtraction

When adding or subtracting decimals only like terms must be added or sub-tracted in a similar fashion to the way that you were probably taught in primary school to add whole numbers by putting them in columns for hundreds, tens and units.

Example 2.20

$$1.345 + 0.00041 + 0.20023 = ?$$

Solution

$$
\begin{array}{r}
1.345 \ + \\
0.00041 + \\
0.20023 + \\
\hline
1.54564
\end{array}
$$

Multiplication

To multiply two numbers involving decimal fractions one can ignore the decimal points, multiply the two numbers in the usual fashion, and then insert the decimal point in the answer by counting the total number of digits to the right of the decimal point in both the numbers that were multiplied.

Example 2.21

$$2.463 \times 0.38 = ?$$

Solution

$$
\begin{array}{r}
2{,}463 \ \times \\
38 \\
\hline
19{,}704 \\
73{,}890 \\
\hline
93{,}594
\end{array}
$$

There were a total of 5 digits to the right of the decimal place in the two numbers to be multiplied and so the answer is 0.93594.

Division

When dividing by a decimal fraction one first multiplies the fraction by the multiple of 10 that will convert it into a whole number. Then the number that is being divided is multiplied by the same multiple of 10 and the normal division operation is applied.

Example 2.22

$$360.54 \div 0.04 = ?$$

Solution

Multiplying both terms by 100 the problem becomes

$$36,054 \div 4 = 9,013.5$$

Given that actual arithmetic operations involving decimals can usually be performed with a calculator, perhaps one of the most common problems you are likely to face is how to express quantities as decimals before setting up a calculation.

Example 2.23

Express 0.01p as a decimal fraction of £1.

Solution

$$1p = £0.01$$

Therefore

$$0.01p = £0.0001$$

In mathematics a decimal format is often required for a value that is usually specified as a percentage in everyday usage. For example, interest rates are usually specified as percentages. A percentage format is really just another way of specifying a decimal fraction, e.g.

$$62\% = \frac{62}{100} = 0.62$$

and so percentages can easily be converted into decimal fractions by dividing by 100.

Examples 2.24

$22\% = 0.22$	$0.24\% = 0.0024$
$24.56\% = 0.2456$	$0.02\% = 0.0002$
$2.4\% = 0.024$	

We shall return to this topic in Chapter 7 when investment appraisal methods are explained.

Because some fractions cannot be expressed exactly in decimals, one may need to 'round off' an answer for convenience. In many of the economic problems in this book there is not much point in taking answers beyond two decimal places. Where this is done then the note '(to 2 dp)' is put after the answer. For example, $\frac{1}{7}$ as a percentage is 14.29% (to 2 dp).

QUESTIONS 2.6

(Try to answer these without using a calculator.)
1. $53.024 - 16.11 =$
2. $44.2 \times 17 =$
3. $602.025 + 34.1006 - 201.016 =$
4. $432.984 \div 0.012 =$
5. $64.5 \times 0.0015 =$
6. $18.3 \div 0.03 =$
7. How many pencils costing 12.5p each can be bought for £37.50?
8. What is 1 millimetre as a decimal fraction of
   (a) 1 centimetre   (b) 1 metre   (c) 1 kilometre?
9. Specify the following percentages as decimal fractions:
   (a) 45.2%   (b) 243.1%
   (c) 7.5%   (d) 0.2%

2.7 NEGATIVE NUMBERS

There are numerous instances where one comes across negative quantities, such as temperatures below zero or bank overdrafts. For example, if you have £35 in your bank account and withdraw £60 then your bank balance becomes − £25.

There are instances, however, where it is not usually possible to have negative quantities. For example, a firm's production level cannot be negative.

We have already mentioned negative numbers when removing brackets and calculating elasticity of demand. To add negative numbers one simply subtracts the number after the negative sign. (This is known as the absolute value of a number.)

Example 2.25

$$45 + (-32) + (-6) = 45 - 32 - 6 = 7$$

If it is required to subtract a negative number then the two negatives will cancel out and one adds the absolute value of the number.

Example 2.26

$$0.5 - (-0.45) - (-0.1) = 0.5 + 0.45 + 0.1 = 1.05$$

Arithmetic

The rules for multiplication and division of negative numbers are as follows.

A negative multiplied (or divided) by a positive gives a negative.
A negative multiplied (or divided) by a negative gives a positive.

Example 2.27

Eight students each have an overdraft of £210. What is their total bank balance?

Solution

$$\text{total balance} = 8 \times (-210) = -£1,680$$

Example 2.28

$$\frac{24}{-5} \div \frac{-32}{-10} = \frac{24}{-5} \times \frac{-10}{-32} = \frac{3}{1} \times \frac{2}{-4} = \frac{6}{-4} = -\frac{3}{2}$$

QUESTIONS 2.7

1. Subtract -4 from -6.
2. Multiply -4 by 6.
3. $-48 + 6 - 21 + 30 =$
4. $-0.55 + 1.0 =$
5. $1.2 + (-0.65) - 0.2 =$
6. $-26 \times 4.5 =$
7. $30 \times (4 - 15) =$
8. $(-60) \times (-60) =$
9. $\frac{-1}{4} \times \frac{9}{7} - \frac{4}{5} =$
10. $(-1) \dfrac{\dfrac{4}{30 + 34}}{\dfrac{-2}{16 + 18}} =$

2.8 POWERS

We have all come across terms such as 'square feet' or 'cubic capacity'. A square foot is a rectangular area with each side equal to 1 foot. If a square room had all walls 12 feet long then its area would be $12 \times 12 = 144$ square feet.

When we multiply a number by itself in this fashion then we say we are 'squaring' it. The mathematical notation for this operation is the superscript 2. Thus '12 squared' is written 12^2.

Example 2.29

$$2.5^2 = 2.5 \times 2.5 = 6.25$$

We find the cubic capacity of a room, in cubic feet, by multiplying length $\times$ width $\times$ height. If all these distances are equal, at 12 feet say (i.e. the room is a perfect cube) then cubic capacity is $12 \times 12 \times 12 = 1{,}728$ cubic feet.

When a number is cubed in this fashion the notation used is the superscript 3, e.g. 12^3.

These superscripts are known as 'powers' and denote the number of times a number is multiplied by itself. Although there are no physical analogies for powers other than 2 and 3, in mathematics one can encounter powers of any value.

Example 2.30

$$12^4 = 12 \times 12 \times 12 \times 12 = 20{,}736$$
$$12^5 = 12 \times 12 \times 12 \times 12 \times 12 = 248{,}832 \text{ etc.}$$

To multiply numbers which are expressed as powers of the same number one can add all the powers together.

Example 2.31

$$3^3 \times 3^5 = (3 \times 3 \times 3) \times (3 \times 3 \times 3 \times 3 \times 3)$$
$$= 3 \times 3 \times 3 \times 3 \times 3 \times 3 \times 3 \times 3 = 3^8 = 6{,}561$$

To divide, one subtracts the superscript of the denominator from the numerator.

Example 2.32

$$\frac{6^6}{6^3} = \frac{6 \times 6 \times 6 \times 6 \times 6 \times 6}{6 \times 6 \times 6} = 6 \times 6 \times 6 = 6^3 = 216$$

In the two examples above the multiplication and division processes are set out in full to illustrate how these processes work with exponents. In practice, of course, one need not do this and it is just necessary to add or subtract the indices.

Example 2.33

$$14.23 \times 14.23^3 \times 14.23^2 = 14.23^6 = 8{,}302{,}892.2 \quad (\text{to 1 dp})$$

(Remember that we do not normally write in the index 1.)

Any number to the power of 1 is simply the number itself and any number to the power of 0 is equal to 1. For example, using the above rule for multiplication $8^2 \times 8^0 = 8^{(2+0)} = 8^2$ so 8^0 must be 1.

Powers can take negative values or can be fractions (see Section 2.9). A negative superscript indicates the number of times that one is dividing by the given number.

Example 2.34

$$3^6 \times 3^{-4} = \frac{3^6}{3^4} = \frac{3 \times 3 \times 3 \times 3 \times 3 \times 3}{3 \times 3 \times 3 \times 3} = 3^2$$

Thus multiplying by a number with a negative power when both quantities are expressed as powers of the same number simply involves adding the (negative) power to the power of the number being multiplied.

Example 2.35

$$8^4 \times 8^{-2} = 8^2 = 64$$

Example 2.36

$$14^7 \times 14^{-9} \times 14^6 = 14^4 = 38,416$$

The evaluation of numbers expressed as exponents can be time-consuming without a calculator with the function $[y^x]$, although you could, of course, use a basic calculator and put the number to be multiplied in memory and then multiply it by itself the required number of times. (This method would only work for whole number exponents though.)

To evaluate a number using the $[y^x]$ function on your calculator you should read the instruction booklet, if you have not lost it. The usual procedure can be explained using Example 2.36 above where the answer was 14^4. In this case y, the number to be multiplied, is equal to 14 and x, the exponent, is equal to 4. On your calculator, enter 14 $[y^x]$ 4 $[=]$ where the brackets $[\ ]$ denote the function keys. You should get 38,416 as your answer.

If you do not, then you have either pressed the wrong keys or your calculator works in a slightly different fashion. To check which of these it is, try to evaluate the simpler answer to Example 2.35 (8^2 which is obviously 64) by entering 8$[y^x]$ 2 $[=]$. If you do not get 64 you need to find your calculator instructions.

Most calculators will not allow you to use the $[y^x]$ function to evaluate powers of negative numbers directly. Remembering that a negative multiplied by a positive gives a negative number, and a negative multiplied by a negative gives a positive, we can work out that if a negative number has an even whole number exponent then the whole term will be positive.

Example 2.37

$$(-3)^4 = (-3)^2 \times (-3)^2 = 9 \times 9 = 81$$

Similarly, if the exponent is an odd number the term will be negative.

Example 2.38

$$(-3)^5 = (-3)^4 \times (-3) = 81 \times (-3) = -243$$

Therefore, when using a calculator to find the values of negative numbers taken to powers, one works with the absolute value and then puts in the negative sign if the power value is an odd number.

Example 2.39

$$(-19)^6 = 19^6 = 47,045,881$$

Example 2.40

$$(-26)^5 = -(26^5) = -11,881,376$$

Example 2.41

$$(-2)^{-2} \times (-2)^{-1} = (-2)^{-3} = \frac{1}{(-2)^3} = \frac{1}{-8} = -0.125$$

QUESTIONS 2.8

1. $4^2 \div 4^3 =$
2. $123^7 \times 123^{-6} =$
3. $6^4 \div (6^2 \times 6) =$
4. $(-2)^3 \times (-2)^3 =$
5. $1.42^4 \times 1.42^3 =$
6. $9^5 \times 9^{-3} \times 9^4 =$
7. $8.673^3 \div 8.673^6 =$
8. $(-6)^5 \times (-6)^{-3} =$
9. $(-8.52)^4 \times (-8.52)^{-1} =$
10. $(-2.5)^{-8} + (0.2)^6 \times (0.2)^{-8} =$

2.9 ROOTS AND FRACTIONAL POWERS

The square root of a number is the quantity which when squared gives the original number. There are different forms of notation. The square root of 16 can be written

$$\sqrt{(16)} = 4 \quad \text{or} \quad 16^{0.5} = 4$$

We can check this exponential format of $16^{0.5}$ using the rule for multiplying powers explained in the previous section:

$$(16^{0.5})^2 = 16^{0.5} \times 16^{0.5} = 16^{0.5 + 0.5} = 16^{1.0} = 16$$

Even most basic calculators have a square root function and so it is not normally worth bothering with the rather tedious manual method of calculating square roots when the square root is not obvious, as it is in the above example.

Example 2.42

$$2{,}246^{0.5} = \sqrt{(2{,}246)} = 47.391982 \quad \text{(using a calculator)}$$

There are other roots. For example, $\sqrt[3]{27}$ or $27^{1/3}$ is the number which when multiplied by itself three times equals 27. This is easily checked as

$$(27^{1/3})^3 = 27^{1/3} \times 27^{1/3} \times 27^{1/3} = 27^{1.0} = 27$$

When multiplying roots they can be expressed in the form with a superscript, e.g. $6^{0.5}$, and then the rules for multiplying powers can be applied.

Example 2.43

$$47^{0.5} \times 47^{0.5} = 47$$

Example 2.44

$$15 \times 9^{0.75} \times 9^{0.75} = 15 \times 9^{1.5}$$
$$= 15 \times 9^{1.0} \times 9^{0.5}$$
$$= 15 \times 9 \times 3 = 405$$

These basic rules for multiplying numbers with powers as fractions will prove very useful when we get to algebra in Chapter 3.

Roots other than square roots can be evaluated using the $[\sqrt[x]{y}]$ function key on a calculator.

Example 2.45

To evaluate $\sqrt[5]{(261)}$ enter

$$261\,[\sqrt[x]{y}\,]\,5\,[=]$$

which should give 3.0431832.

Not all fractional powers correspond to an exact root in this sense, e.g. $6^{0.625}$ is not any particular root. To evaluate these other fractional powers you can use the $[\,y^x\,]$ function key on a calculator.

Example 2.46

To evaluate $452^{0.85}$ most calculators require you to enter

$$452\,[\,y^x\,]\,0.85\,[=]$$

which should give you the answer 180.66236.

If you do not have this function key then you can use logarithms to evaluate these fractional powers (see Section 2.10). Roots and other powers less than one cannot be evaluated for negative numbers on calculators. A negative number cannot be the product of two positive or two negative numbers, and so the square root of a negative number cannot exist. Some other roots for negative numbers do exist, e.g. $\sqrt[3]{(-1)} = -1$, but you are not likely to need to find them.

In Chapter 7 some applications of these rules to financial problems are explained. For the time being we shall just work through a few more simple mathematical examples to ensure that the rules for working with powers are fully understood.

Example 2.47

$$24^{0.45} \times 24^{-1} = 24^{-0.55} = 0.1741341$$

Note that you must use the $[+/-]$ key on your calculator after entering 0.55 when evaluating this power. Alternatively you could have calculated

$$\frac{1}{24^{0.55}} = \frac{1}{5.7427007} = 0.1741341$$

Example 2.48

$$20 \times 8^{0.3} \times 8^{0.25} = 20 \times 8^{0.55} = 20 \times 3.1383364 = 62.766728$$

Sometimes it may help to multiply together two numbers with a common power. Both numbers can be put inside brackets with the common power outside the brackets.

Example 2.49

$$18^{0.5} \times 2^{0.5} = (18 \times 2)^{0.5} = 36^{0.5} = 6$$

QUESTIONS 2.9

Put the answers to the questions below as powers and then evaluate.

1. $\sqrt{(625)} =$
2. $\sqrt[3]{8}$
3. $5^{0.5} \times 5^{-1.5} =$
4. $(7)^{0.5} \times (7)^{0.5} =$
5. $6^{0.3} \times 6^{-0.2} \times 6^{0.4} =$
6. $12 \times 4^{0.8} \times 4^{0.7} =$
7. $20^{0.5} \times 5^{0.5} =$
8. $16^{0.4} \times 16^{0.2} =$
9. $462^{-0.83} \times 462^{0.48} \div 462^{0.2} =$
10. $76^{0.62} \times 18^{0.62} =$

2.10 LOGARITHMS

Many people thought that logarithms went out of the window when pocket calculators became widely available. In the author's schooldays logarithms were used as a short-cut method for awkward long multiplication and long division calculations. Although pocket calculators have indeed now made log tables redundant for this purpose they may still be useful for some problems that you may encounter. So, for those of you who have never studied log tables, or have forgotten what they are for, what are these mysterious logarithms?

The most commonly encountered logarithm is the base 10 logarithm. What this means is that the logarithm of any number is the power to which 10 must be raised to equal that number. The usual notation for logarithms to base 10 is 'log'.

Thus the logarithm of 100 is 2 since $100 = 10^2$. This is written as log $100 = 2$. Similarly

$$\log 10 = 1$$

and

$$\log 1,000 = 3$$

The square root of 10 is $3.1622777 = 10^{0.5}$ and so we know that log $3.1622777 = 0.5$.

Basic mathematics for economists

The above logarithms are obvious. For the logarithms of other numbers you can refer to a printed set of log tables or use a calculator with a [LOG] function key.

If two numbers expressed as powers of 10 are multiplied together then we know that the indices are added, e.g.

$$10^{0.5} \times 10^{1.5} = 10^2$$

Therefore, to use logs to multiply numbers, one simply adds the logs, as they are just the powers to which 10 is taken.

To divide, one index is subtracted from the other, e.g.

$$10^{2.5} \div 10^{1.5} = 10^{2.5-1.5} = 10^1 = 10$$

and so logs are subtracted.

The resulting log answer is a power of 10. To transform it back to a normal number one can use the [10^x] function on a calculator or 'Antilog' tables if the answer is not obvious, as it is above.

Although you can obviously do the calculations more quickly by using the relevant function keys on a calculator, the following examples illustrate how logarithms can solve some multiplication, division and power evaluation problems so that you can see how they work. You will then be able to understand how logarithms can be applied to some problems encountered in economics.

Example 2.50

Evaluate $4,632.71 \times 251.07$ using logs.

Solution

Using the [LOG] function key on a calculator

$$\log 4,632.71 = 3.6658351$$
$$\log 251.07 = 2.3997948$$

Thus

$$4,632.71 \times 251.07 = 10^{3.6658351} \times 10^{2.3997948}$$
$$= 10^{6.0656299}$$
$$= 1,163,134.5$$

using the [10^x] function key.

The principle is therefore to put all numbers to be multiplied together in log form, add the logs, and then evaluate.

The above example was solved using the [LOG] function key on a calculator. The following example uses log and antilog tables. These usually only

30

give logs to four decimal places and so there is some slight loss of accuracy but this will not usually be significant.

Example 2.51

Evaluate $56,200 \div 3,484$ using logs.

Solution

From log tables

$$\log 56,200 = 4.7497$$
$$\log 3,484 = 3.5421$$

Note that the log tables only give the logs of numbers between 0 and 10. The number to the left of the decimal point then has to be entered depending on the power to which 10 is taken to get to the actual figure, e.g.

$$\log 5.620 = 0.7497 \text{ from tables}$$

Since

$$56,200 = 5.602 \times 10^4$$

then

$$\log 56,200 = 4 + 0.7497 = 4.7497$$

To divide, we subtract the log of the denominator since

$$56,200 \div 3,484 = 10^{4.7497} \div 10^{3.5421}$$
$$= 10^{4.7497 - 3.5421}$$
$$= 10^{1.2076}$$

From antilog tables the value of antilog 0.2076 can be read off as 1.613. Therefore

$$10^{1.2706} = 10^1 \times 10^{0.2706}$$
$$= 10 \times 1.613 = 16.13$$

Note that when you use the [LOG] function key on a calculator to obtain the logs of numbers less than 1 you get a negative sign, e.g.

$$\log 0.31 = -0.5086383$$

However, if you use log tables to find the log of 0.31 you will have to obtain the value 0.4914 for the log 3.1 and then subtract 1 as $0.31 = 3.1 \times 10^{-1}$. As

31

you will probably be using a calculator to obtain logs it is not worth spending time trying to understand the special methods for dealing with logs of numbers less than 1 using tables.

Logarithms can also be used to work out powers.

Example 2.52

Calculate $1,242.67^6$ using logs.

Solution

$$\log 1,242.67 = 3.0943558$$

This means

$$1,242.67 = 10^{3.0943558}$$

If this is taken to the power of 6, it means that this index of 10 is multiplied 6 times. Therefore

$$\log 1,242.67^6 = 6 \log 1,242.67 = 18.566135$$

Using the $[10^x]$ function to evaluate this number gives

$$3.6824 \times 10^{18} = 3,682,400,000,000,000,000$$

Example 2.53

Use logs to find $\sqrt[8]{(226.34)}$.

Solution

Log 226.34 must be divided by 8 to find the log of the number which when multiplied by itself 8 times gives 226.34, i.e. the eighth root. Thus

$$\log 226.34 = 2.3547613$$
$$\frac{1}{8} \log 226.34 = 0.2943452$$

Therefore $\sqrt[8]{(226.34)} = 10^{0.2943452} = 1.9694509$.

To summarize, the rules for using logs are as follows.

Multiplication:	add logs
Division:	subtract logs
Powers:	multiply log by power
Roots:	divide log by root

The answer is then evaluated by finding 10^x where x is the resulting value of the log.

Apart from allowing you to do some awkward calculations which you could probably do more quickly on a calculator, what is the point of learning how to use logarithms? One application you will come across at some stage in your statistics or economics courses is the estimation of the parameters of non-linear functions. (If you are not familiar with algebraic notation, return to this section after you have covered Chapter 3 on algebra.)

Linear regression analysis allows you to estimate the parameters a and b in linear functions such as the supply schedule

$$p = a + bq$$

using data on p and q.

However, if you have a non-linear function, logs can be used to transform it into a linear form. For example, the Cobb–Douglas production function

$$Q = AK^a L^b$$

can be put into the log form

$$\log Q = \log A + a \log K + b \log L$$

so that a and b can be estimated by linear regression analysis.

There are also some types of problems where logarithms must be used to obtain a solution.

Example 2.54

If $460(1.08)^n = 925$, what is n?

Solution

$$460(1.08)^n = 925$$

$$(1.08)^n = 2.0108696$$

Putting in log form

$$n \log 1.08 = \log 2.0108696$$

$$n = \frac{\log 2.0108696}{\log 1.08}$$

$$= \frac{0.3033839}{0.0334238}$$

$$= 9.0768945$$

We shall return to this type of problem in Chapter 7 when we consider for how long a sum of money needs to be invested at any given rate of interest to accumulate to a specified sum.

Basic mathematics for economists

Although logarithms to the base 10 are perhaps the most commonly encountered ones, logarithms can be based on any number. In Chapter 14 the use of logarithms to the base e = 2.7183, known as natural logarithms, is explained (and also why such an odd base is used.)

QUESTIONS 2.10

Use logs to answer the following.
1. $424 \times 638.724 =$
2. $6,434 \div 29.12 =$
3. $22.43^7 =$
4. $9.612^{8.34} =$
5. $\sqrt[36]{(5,200)} =$
6. $14^{3.2} \times 6.2^4 \times 81^{0.2} =$
7. Use logs to put the production function $Q = AK^{\alpha}L^{\beta}R^{\gamma}$ into a linear format.
8. If $(1.06)^n = 235$ what is n?
9. If $825(1.22)^n = 1,972$ what is n?
10. If $4,350(1.14)^n = 8,523$ what is n?

3

Introduction to algebra

3.1 REPRESENTATION

Algebra is basically a system of shorthand. Symbols are used to represent concepts and variables that are capable of taking different values.

For example, suppose that a biscuit manufacturer uses the following ingredients for each packet of biscuits produced: 0.2 kg of flour, 0.05 kg of sugar and 0.1 kg of butter. One way that we could specify the total amount of flour that the company uses is: '0.2 kg times the number of packets of biscuits produced'. However, it is much simpler if we let the letter q represent the number of packets of biscuits produced. The amount of flour required in kilograms will then be 0.2 times q, written as $0.2q$.

Thus we can also say

$$\text{amount of sugar required} = 0.05q$$

$$\text{amount of butter required} = 0.1q$$

Sometimes an algebraic expression will have several terms in it with different algebraic symbols representing the unknown quantities of different variables. Consider the total expenditure on inputs by the firm in the example above.

Let the price in pounds per kilogram of flour be denoted by the letter a. Let the price in pounds per kilogram of sugar be denoted by the letter b. Let the price in pounds per kilogram of butter be denoted by the letter c. The total cost of the amount of flour the firm uses will therefore be $0.2q$ times a, written as $0.2qa$. Thus total expenditure on inputs is (in £)

$$0.2qa + 0.05qb + 0.1qc$$

When two algebraic symbols are multiplied together it does not matter in which order they are written, e.g. $xy = yx$. This of course, is, the same rule that applies when multiplying numbers. For example:

$$5 \times 7 = 7 \times 5$$

Any operation that can be carried out with numbers (such as division or deriving the square root) can be carried out with algebraic symbols. The difference is that the answer will also be in terms of algebraic symbols rather than numbers. An algebraic expression cannot be evaluated until values have been given to the variables that the symbols represent (see Section 3.2).

For example, an expression for the length of fencing (in metres) needed to enclose a square plot of land of as yet unknown size is as follows.

$$\text{length of fencing} = 4 \times (\text{length of one side})$$

Therefore

$$L = 4\sqrt{A}$$

where L is the total length of fencing and A represents the area of the land in square metres.

Without information on the value of A we cannot say any more. (In some cases, however, algebraic expressions can be simplified, as is explained in section 3.3 onwards.) Once the value of A is specified then one can simply work out the value of the expression using basic arithmetic. For example, if the area is 100 square metres, then

$$\text{length of fencing } L = 4\sqrt{(100)} = 4 \times 10 = 40 \text{ metres}$$

One of the uses of writing an expression in an algebraic form is that it is not necessary to work out a method of solving a problem for every different value of the unknown variable that one is faced with. The different values can just be substituted into the algebraic expression. In this section we start with some fairly simple expressions but later, as more complex relationships are dealt with, the usefulness of algebraic representation will become more obvious.

Example 3.1

You are tiling a bathroom with 4 inch square tiles. The number of square feet to be tiled is as yet unknown and is represented by q. Because you may break some tiles and will have to cut some to fit around corners etc. you work to the rule of thumb that you should buy enough tiles to cover the specified area plus 10%.

Derive an expression for the number of tiles to be bought in terms of q.

Solution

Nine 4 inch square tiles will cover 1 square foot and 110% written as a decimal is 1.1. Therefore the number of tiles required is

$$\frac{q}{9} \times 1.1 = \frac{1.1q}{9}$$

One could write this as the decimal number $0.122222q$, but the above fraction is simpler and more accurate. As one cannot buy fractions of a tile then in practice you would need to round your answer to the nearest whole number.

Introduction to algebra

QUESTIONS 3.1

1. An engineering firm makes a metal component. Each component requires 0.01 tonnes of steel, 0.5 hours of labour plus 0.5 hours of machine time. Let the number of components produced be denoted by x. Derive algebraic expressions for:

   (a) the amount of steel required;
   (b) the amount of labour required;
   (c) the amount of machine time required.

2. If the price per tonne of steel is given by r, the price per hour of labour is given by w and the price per hour of machine time is given by m, then derive an expression for the total production costs of the firm in question 1 above.

3. The petrol consumption of your car is 40 miles per gallon. Let x be the distance you travel in miles and p the price per gallon of petrol in pence. Write expressions for (a) the amount of petrol you use and (b) your expenditure on petrol.

4. Suppose that you are cooking a dinner for a number of people. You only know how to cook one dish, and this requires you to buy $\frac{1}{4}$ lb of meat plus $\frac{1}{2}$ lb of potatoes for each person. (Assume you already have a plentiful supply of any other ingredients.) Define your own algebraic symbols for relevant unknown quantities and then write expressions for:

   (a) the amount of meat you need to buy;
   (b) the amount of potatoes you need to buy;
   (c) your total shopping bill.

5. You are cooking again! This time it's a turkey. The cookery book recommends a cooking time of 15 minutes for every pound weight of the turkey plus another quarter of an hour. Write an expression for the total cooking time (in hours) for your turkey in terms of its weight (in pounds).

6. Make up your own algebraic expression for the total profit of a firm in terms of the amount of output sold, the price of its product and the average cost of production per unit.

7. Someone is booking a meal in a restaurant for a group of people. They are told that there is a set menu that costs £9.50 per adult and £5 per child, and there is also a fixed charge of £1 per head for each meal served. Derive an expression for the total cost of the meal, in pounds, if there are x adults and y children.

8. A firm produces a good which it can sell any amount of at £12 per unit. Its costs are a fixed outlay of £6,000 plus £9 in variable costs for each unit produced. Write an expression for the firm's profit in terms of the number of units produced.

3.2 EVALUATION

An expression can be evaluated when the variables represented by algebraic symbols are given specific numerical values.

Example 3.2

Evaluate the expression $6.5x$ when $x = 8$.

Solution

$$6.5x = 6.5(8) = 52$$

Example 3.3

A firm's total costs are given by the expression

$$0.2qa + 0.05qb + 0.1qc$$

Evaluate these costs if $q = 1,000$, $a = 0.6$, $b = 1.3$ and $c = 2.1$, where a, b and c are measured in pounds.

Solution

$$0.2qa + 0.05qb + 0.1qc = 0.2(600) + 0.05(1,300) + 0.1(2,100)$$
$$= 120 + 65 + 210 = 395$$

Therefore the total cost is £395.

Example 3.4

Evaluate the expression $(3^x + 4)y$ when $x = 2$ and $y = 6$.

Solution

$$(3^x + 4)y = (3^2 + 4)6 = (9 + 4)6 = 13 \times 6 = 78$$

Example 3.5

A Bureau de Change will sell you French francs at an exchange rate of 10 francs to the pound and charges a flat rate commission of £2 on all transactions.

(i) Write an expression for the number of francs that can be bought for £x (any given quantity of sterling), and
(ii) evaluate it for $x = 80$.

Solution

(i) Number of francs bought for £x = 10(x – 2).
(ii) £80 will therefore buy

$$10(80 - 2) = 10(78) = 780 \text{ francs}$$

QUESTIONS 3.2

1. Evaluate the expression $2x^3 + 4x$ when $x = 6$.
2. Evaluate the expression $(6x + 2y)y^2$ when $x = 4.5$ and $y = 1.6$.
3. When the UK government privatized the Water Authorities in 1989 it decided that annual percentage price increases for water would be limited to the rate of inflation plus z, where z was a figure to be determined by the government. Write an algebraic expression for the maximum annual percentage price increase for water and evaluate it for an inflation rate of 6% and a z factor of 3.
4. Make up your own values for the unknown variables in the expressions you have written for questions 3.1 above and then evaluate.
5. Evaluate the expression $1.02^x + x^{-3.2}$ when $x = 2.8$.
6. A firm's average production costs (AC) are given by the expression

$$AC = 450q^{-1} + 0.2q^{1.5}$$

where q is output. What will AC be when output is 175? (AC is measured in pounds.)
7. A firm's profit (in pounds) is given by the expression $7.5q - 1650$. What profit will it make when q is 500?
8. If income tax is levied at a rate of 25% on annual income over £4,000 then:

   (a) write an expression for net monthly salary in terms of gross monthly salary (assumed to be greater than £334), and
   (b) evaluate it if gross monthly salary is £1,650.

3.3 SIMPLIFICATION: ADDITION AND SUBTRACTION

Simplifying an expression means rearranging the terms in it so that the expression becomes easier to work with. Before setting out the different rules for simplification let us work through an example.

Example 3.6

A businesswoman driving her own car on her employer's business gets paid a set fee per mile travelled for travelling expenses. During one week she records one journey of 234 miles, one of 166 miles and one of 90 miles. Derive an expression for total travelling expenses.

Solution

If the rate per mile is denoted by the letter M then her expenses will be $234M$ for the first journey and $166M$ and $90M$ for the second and third journeys respectively. Total travelling expenses for the week will thus be $234M + 166M + 90M$.

We could, instead, simply add up the total number of miles travelled during the week and then multiply by the rate per mile. This would give

$$(234 + 166 + 90) = 490 \text{ miles} \times \text{rate per mile} = 490M$$

It is therefore obvious that, as both methods should give the same answer,

$$234M + 166M + 90M = 490M$$

In other words, in an expression with different terms all in the same format of

$$(\text{a number}) \times M$$

all the terms can be added together.

The general rule is that like terms can be added or subtracted to simplify an expression. 'Like terms' have the same algebraic symbol or symbols, usually multiplied by a number.

Example 3.7

$$3x + 14x + 7x = 24x$$

Example 3.8

$$45A - 32A = 13A$$

It is important to note that only terms that have exactly the same algebraic notation can be added or subtracted in this way. For example, the terms x, y^2 and xy are all different and cannot be added together or subtracted from each other.

Example 3.9

Simplify the expression $5x^2 + 6xy - 32x + 3yx - x^2 + 4x$.

Solution

Adding/subtracting all the terms in x^2 gives $4x^2$.
Adding/subtracting all the terms in x gives $-28x$.
Adding/subtracting all the terms in xy gives $9xy$.

(Note that the terms in xy and yx can be added together since $xy = yx$.)
Putting all these terms together gives the simplified expression

$$4x^2 - 28x + 9xy$$

In fact all the basic rules of arithmetic (as set out in Chapter 2) apply when algebraic symbols are used instead of actual numbers. The difference is that the simplified expression will still be in a format of algebraic terms.

The example below illustrates the rule that if there is a negative sign in front of a set of brackets then the positive and negative signs of the terms within the brackets are reversed if the brackets are removed.

Example 3.10

Simplify the expression

$$16q + 33q - 2q - (15q - 6q)$$

Solution

Removing brackets, the above becomes

$$16q + 33q - 2q - 15q + 6q = 38q$$

QUESTIONS 3.3

1. Simplify the expression $6x - (6 - 24x) + 10$.
2. Simplify the expression

$$4xy + (24x - 13y) - 12 + 3yx - 5y$$

3. A firm produces two goods, x and y, which it sells at prices per unit of £26 and £22 respectively. Good x requires an initial outlay of £400 and then an expenditure of £16 on labour and £4 on raw materials for each unit produced. Good y requires a fixed outlay of £250 plus £14 labour and £3 of raw material for each unit. Write an expression in terms of x and y for the firm's total profit and then simplify it.
4. A worker earns £6 per hour for the first 40 hours a week he works and £9 per hour for any extra hours. Assuming that he works at least 40 hours, write an expression for his gross weekly wage in terms of H, the total hours worked per week, and then simplify it.

3.4 SIMPLIFICATION: MULTIPLICATION

When a set of brackets containing different terms is multiplied by a symbol or a number it may be possible to simplify an expression by multiplying out, i.e. multiplying each term within the brackets by the term outside. In some circumstances, though, it may be preferable to leave brackets in the expression if it makes it clearer to work with.

Example 3.11

$$x(4 + x) = 4x + x^2$$

Example 3.12

$$5(7x^2 - x) - 3(3x^2 + 6x) = 35x^2 - 5x - 9x^2 - 18x$$
$$= 26x^2 - 23x$$

Example 3.13

$$6y(8 + 3x) - 2xy + 12y = 48y + 18xy - 2xy + 12y$$
$$= 60y + 16xy$$

Example 3.14

The basic hourly rate for a weekly paid worker is £8 and any hours above 40 are paid at £12. Tax is paid at a rate of 25% on any earnings above £80 a week. Assuming hours worked per week (H) exceed 40, write an expression for net weekly wage in terms of H and then simplify it.

Solution

$$\text{gross wage} = 40 \times 8 + (H - 40)\,12$$
$$= 320 + 12H - 480$$
$$= 12H - 160$$
$$\text{net wage} = 0.75(\text{gross wage} - 80) + 80$$
$$= 0.75(12H - 160 - 80) + 80$$
$$= 9H - 120 - 60 + 80$$
$$= 9H - 100$$

If you are not sure whether the expression you have derived is correct you can try to check it by substituting numerical values for unknown variables. In the above example, if 50 hours per week were worked, then

$$\text{gross pay} = (40\,\text{hours} @ \pounds 8) + (10\,\text{hours} @ \pounds 12)$$
$$= \pounds 320 + \pounds 120$$
$$= \pounds 440$$

Introduction to algebra

$$\text{tax payable} = (£440 - £80)0.25 = (£360)0.25 = £90$$

Therefore

$$\text{net pay} = £440 - £90 = £350$$

Using the expression derived in Example 3.14, if $H = 50$ then

$$\text{net pay} = 9H - 100 = 9(50) - 100 = 450 - 100 = £350$$

This checks out with the answer above and so we know our expression works.

It is rather more complicated to multiply pairs of brackets together. One method that can be used is rather like the long multiplication that you probably learned at school, but instead of keeping all units, tens, hundreds etc. in the same column it is like algebraic terms that are kept in the same column during the multiplying process so that they can be added together.

Example 3.15

Simplify $(6 + 2x)(4 - 2x)$.

Solution

Writing this as a long multiplication problem:

$$6 + 2x \times$$
$$4 - 2x$$

Multiplying $(6 + 2x)$ by $-2x$ $-12x - 4x^2$

Multiplying $(6 + 2x)$ by 4 $24 + 8x$

Adding together gives the answer $24 - 4x - 4x^2$

One does not have to use the long multiplication format for multiplying out sets of brackets. The basic principle is that each term in one set of brackets must be multiplied by each term in the other set. Like terms can then be collected together to simplify the resulting expression.

Example 3.16

Simplify $(3x + 4y)(5x - 2y)$.

Solution

Multiplying the terms in the second set of brackets by $3x$ gives:

$$15x^2 - 6xy \tag{1}$$

Multiplying the terms in the second set of brackets by 4y gives:

$$20xy - 8y^2 \qquad (2)$$

Therefore, adding (1) and (2) the whole expression is

$$15x^2 - 6xy + 20xy - 8y^2 = 15x^2 + 14xy - 8y^2$$

Example 3.17

Simplify $(x+y)^2$.

Solution

$$(x+y)^2 = (x+y)(x+y) = x^2 + xy + yx + y^2 = x^2 + 2xy + y^2$$

The above answer can be checked by referring to Figure 3.1. The area enclosed in the square with sides of length $x+y$ can be calculated by squaring the lengths of the sides, i.e. $(x+y)^2$. One can also see that this square is made up of the four rectangles A, B, C and D whose areas are x^2, xy, xy and y^2 respectively – in other words, $x^2 + 2xy + y^2$, which is the answer obtained above.

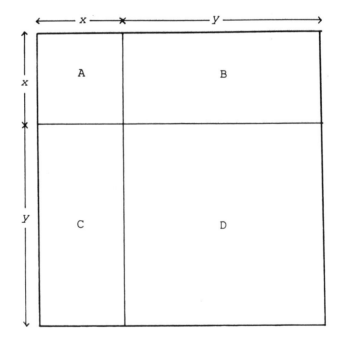

Figure 3.1

Example 3.18

Simplify $(6 - 5x)(10 - 2x + 3y)$.

Solution

Multiplying out gives

$$60 - 12x + 18y - 50x + 10x^2 - 15xy$$
$$= 60 - 62x + 18y + 10x^2 - 15xy$$

Although you need to know how to multiply out pairs of brackets, some expressions may be best left with the brackets still in.

Example 3.19

If a sum of £x is invested at an interest rate of $R\%$ write an expression for the value of the investment at the end of 2 years.

Solution

After 1 year the investment's value is $x(1 + R/100)$.
After 2 years the investment's value is $x(1 + R/100)^2$.

One could multiply out but in this particular case the expression is probably clearer, and also easier to evaluate, if the brackets are left in. In fact, in some cases, 'simplification' may mean the reverse of multiplying out pairs of brackets. The next section explains how some expressions may be reformatted into two expressions in brackets multiplied together. This is called 'factorization'.

QUESTIONS 3.4

Simplify the following expressions:
1. $6x(x - 4)$
2. $(x + 3)^2 - 2x$
3. $(2x + y)(x + 3)$
4. $(6x + 2y)(7x - 8y) + 4y + 2y$
5. $(4x - y + 7)(2y - 3) + (9x - 3y)(5 + 6y)$
6. $(12 - x + 3y + 4z)(10 + x + 2y)$
7. A good costs a basic £180 a unit but if an order is made for more than 10 units this price is reduced by a discount of £2 for every one unit increase in the size of an order (up to a maximum of 60 units purchased), i.e. if the order size is 11, price is £178; if it is 12, price is

£176 etc. Write an expression for the total cost of an order in terms of order size and simplify it. Assume order size is between 10 and 60 units.

8. A holiday excursion costs £8 per person for transport plus £5 per adult and £3 per child for meals. Write an expression for the total cost of an excursion for x adults and y children and simplify it.

9. A firm is building a car park for its employees. Assume that a car park to accommodate x cars must have a length (in metres) of $4x + 10$ and a width of $2x + 10$. If 24 square metres will be specifically allocated for visitors' cars, write an expression for the amount of space available for the cars of the workforce in terms of x if x is the planned capacity of the car park.

10. A firm buys a raw material that costs £220 a tonne for the first 40 tonnes, £180 a tonne for the next 40 tonnes and £150 for any further quantities. Write an expression for the firm's total expenditure on this input in terms of the total amount used (which can be assumed to be greater than 80 tonnes), and simplify.

3.5 SIMPLIFICATION: FACTORIZING

For some purposes (see, for example, Section 3.6 below and Chapter 6 on quadratic equations) it may be helpful if an algebraic expression can be simplified into a format of two sets of brackets multiplied together. For example, the expression $x^2 + 4x + 4$ can be rewritten as $(x + 2)(x + 2)$. This is rather like the arithmetical process of factorizing a number, which means finding all the prime numbers which when multiplied together equal that number, e.g.

$$126 = 2 \times 3 \times 3 \times 7$$

If there is only one unknown variable, x, in an expression then, if it is possible to factorize it into two sets of brackets that do not contain terms in x to a power other than 1, the expression must be in the form

$$ax^2 + bx + c$$

However, not all expressions in this form can be factorized into sets of brackets that only involve integers. In this text the analysis will be confined to situations where a, b and c are positive or negative integers and b may also possibly be zero.

There are no set rules for working out if and how an expression may be factorized. However, if the term in x does not have a number in front of it (i.e. $a = 1$) then the expression can be factorized if there are two numbers which

1 give c when multiplied together, and
2 give b when added together.

Example 3.20

Attempt to factorize the expression $x^2 + 6x + 9$.

Solution

In this example $a = 1$, $b = 6$ and $c = 9$. Since $3 \times 3 = 9$ and $3 + 3 = 6$, it can be factorized. Thus

$$x^2 + 6x + 9 = (x + 3)(x + 3)$$

This can be checked as

$$
\begin{array}{r}
x + 3 \times \\
x + 3 \\
\hline
3x + 9 \\
x^2 + 3x \\
\hline
x^2 + 6x + 9
\end{array}
$$

Example 3.21

Attempt to factorize the expression $x^2 - 2x - 80$.

Solution

Since $-10 \times 8 = -80$ and $-10 + 8 = -2$ then the expression can be factorized and

$$x^2 - 2x - 80 = (x - 10)(x + 8)$$

Check this answer yourself by multiplying out.

Example 3.22

Attempt to factorize the expression $x^2 + 3x + 11$.

Solution

There are no two numbers which when multiplied together give 11 and when added together give 3. Therefore this expression cannot be factorized.

It is sometimes possible to simplify an expression before factorizing if all the terms are divisible by the same number.

Example 3.23

Attempt to factorize the expression $2x^2 - 10x + 12$.

Solution

$$2x^2 - 10x + 12 = 2(x^2 - 5x + 6)$$

The term in brackets can be simplified as

$$x^2 - 5x + 6 = (x - 3)(x - 2)$$

Therefore

$$2x^2 - 10x + 12 = 2(x - 3)(x - 2)$$

When the term in x^2 is multiplied by a number other than 1, then one still has to find two numbers which multiply together to give c. However, one also has to find two numbers for the coefficients of the two terms in x within the two sets of brackets that when multiplied together equal a, and both these sets of numbers should allow the coefficient b to be derived when multiplying out.

Example 3.24

Attempt to factorize the expression $30x^2 + 52x + 14$.

Solution

If we use the results that $6 \times 5 = 30$ and $2 \times 7 = 14$ we can try multiplying

$$
\begin{array}{r}
5x + 7 \times \\
6x + 2 \\
\hline
10x + 14 \\
30x^2 + 42x \\
\hline
30x^2 + 52x + 14
\end{array}
$$

This gives

Thus

$$30x^2 + 52x + 14 = (5x + 7)(6x + 2)$$

Similar rules apply when one attempts to factorize an expression with two unknown variables, x and y. This will usually be in the format

$$ax^2 + bxy + cy^2$$

where a, b and c are specified parameters.

Example 3.25

Attempt to factorize the expression $x^2 - y^2$.

Solution

In this example $a = 1$, $b = 0$ and $c = -1$. The two numbers -1 and 1 give -1 when multiplied together and 0 when added. Thus

$$x^2 - y^2 = (x-y)(x+y)$$

To check this multiply out:

$$(x-y)(x-y) = x^2 - xy + y^2 - yx = x^2 - y^2$$

Example 3.26

Attempt to factorize the expression $3x^2 + 8x + 23$.

Solution

As 23 is a positive prime number, the only pairs of positive integers that could possibly be multiplied together to give 23 are 1 and 23. Thus, whatever permutations of combinations with terms in x that we try, the term in x when brackets are multiplied out will be at least $23x$, e.g. $(3x + 23)(x + 1) = 3x^2 + 26x + 23$, whereas the given expression contains the term $8x$. It is therefore not possible to factorize this expression.

Unfortunately, it is not always so obvious whether or not an expression can be factorized.

Example 3.27

Attempt to factorize the expression $3x^2 + 24 + 16$.

Solution

Although the numbers look promising, if you try various permutations you will find that this expression does not factorize.

There is no easy way of factorizing expressions and it is just a matter of trial and error. Do not despair though! As you will see later on, factorizing may help you to use short-cut methods of solving certain problems. If you spend ages trying to factorize an expression then this will defeat the object of using the short-cut method. If it is not obvious how an expression can be factorized after a few minutes thought and experimentation with some potential possible solutions then it is usually more efficient to forget factorization and use some other method of solving the problem. We shall return to this topic in Chapter 6.

QUESTIONS 3.5

Attempt to factorize the following expressions:
1. $x^2 + 8x + 16$
2. $x^2 - 6xy + 9y^2$
3. $x^2 + 7x + 22$
4. $8x^2 - 10x + 33$
5. Make up your own expression in the format $ax^2 + bx + c$ and attempt to factorize it. Check your answer by multiplying out.

3.6 SIMPLIFICATION: DIVISION

To divide an algebraic expression by a number one divides every term in the expression by the number, cancelling where appropriate.

Example 3.28

$$\frac{15x^2 + 2xy + 90}{3} = 5x^2 + \frac{2}{3}xy + 30$$

To divide by an unknown variable the same rule is used although, of course, where the numerator of a fraction does not contain that variable it cannot be simplified any further.

Example 3.29

$$\frac{2x^2}{x} = 2x$$

Example 3.30

$$\frac{4x^3 - 2x^2 + 10x}{x} = \frac{x(4x^2 - 2x + 10)}{x}$$
$$= 4x^2 - 2x + 10$$

Example 3.31

$$\frac{16x + 120}{x} = 16 + \frac{120}{x}$$

Example 3.32

A firm's total costs are $25x + 2x^2$, where x is output. Write an expression for average cost.

Average cost is total cost divided by output.

Solution

Average cost is total cost divided by output.
Therefore

$$\text{average cost} = \frac{25x + 2x^2}{x} = 25 + 2x$$

If one expression is divided by another expression with more than one term in it then terms can only be cancelled top and bottom if the numerator and denominator have terms in common. Factorizing the numerator may be helpful in this sort of problem.

Example 3.33

$$\frac{x^2 + 2x}{x + 2} = \frac{x(x + 2)}{x + 2} = x$$

Example 3.34

$$\frac{x^2 + 5x + 6}{x + 3} = \frac{(x + 2)(x + 3)}{x + 3} = x + 2$$

Example 3.35

$$\frac{x^2 + 5x + 6}{x^2 + x - 2} = \frac{(x + 2)(x + 3)}{(x + 2)(x - 1)} = \frac{x + 3}{x - 1}$$

QUESTIONS 3.6

1. Simplify

$$\frac{6x^2 + 14x - 40}{2x}$$

2. Simplify

$$\frac{x^2 + 12x + 27}{x + 3}$$

3. Simplify

$$\frac{8xy + 2x^2 + 24x}{2x}$$

4. A firm has to pay fixed costs of £200 and then £16 labour plus £5 raw materials for each unit produced of good X. Write an expression for average cost and simplify.

5. A firm sells 40% of its output at £200 a unit, 30% at £180 and 30% at £150. Write an expression for the average revenue received on each unit sold and then simplify it.
6. You have all come across this sort of party trick: Think of a number. Add 3. Double it. Add 4. Take away the number you first thought of. Take away 3. Take away the number you thought of again. Add 2. Your answer is 9. Show how this answer can be derived by algebraic simplification by letting x equal the number first thought of.
7. Make up your own 'think of a number' trick, writing down the different steps in the form of an algebraic expression that checks out the answer.

3.7 SOLVING SIMPLE EQUATIONS

We have seen that evaluating an expression means calculating its value when one is given specific values for unknown variables in the expression. In this section we explain how it is possible to work backwards to discover the value of an unknown variable when the total value of the expression is given or can be found through simplification.

When an algebraic expression is known to equal a number, or another algebraic expression, we can write an equation, i.e. the two concepts are written on either side of an equality sign. For example,

$$45 = 24 + 3x$$

In this chapter we have already written some equations when simplifying algebraic expressions. However, the ones we have come across so far have usually not been in a format where the value of the unknown variables can be worked out. For example,

$$3x + 14x - 5x = 12x$$

The expressions on either side of the equality sign are equal, but x cannot be calculated from the information given.

Some equations are what are known as 'identities', which means that they must always be true. For example, a firm's total costs (TC) can be split into the two components total fixed costs (TFC) and total variable costs (TVC). It must therefore always be the case that

$$TC = TFC + TVC$$

Given that

$$\text{average total cost} = AC = \frac{TC}{Q}$$

where Q is output and that

$$\text{average fixed cost} = AFC = \frac{TFC}{Q}$$

and

$$\text{average variable cost} = \text{AVC} = \frac{\text{TVC}}{Q}$$

then it must also always be true that

$$\text{AC} = \text{AFC} + \text{AVC}$$

Identities are sometimes written with the three bar equality sign '≡' instead of '=', but usually only when it is necessary to distinguish them from other forms of equations, such as functions.

A function is a relationship between two or more variables such that the value of one variable is determined by the values taken by the other variables in the function. (Functions are explained more fully in Chapter 4.) In economics the parameters of functional relationships are usually estimated from empirical data. For example, statistical analysis may show that a demand function takes the form

$$q = 450 - 3p$$

where p is price and q is quantity demanded. This means that the 450 and the 3 have been estimated from data using regression analysis (explained in your statistics course). Thus the expected quantity demanded can be predicted for any given value of p, e.g. if $p = 60$ then

$$q = 450 - 3(60) = 450 - 180 = 270$$

In this section we shall not distinguish between equations that are identities and those that relate to specific values of functions, since the method of solution is the same for both. We shall also mainly confine the analysis to linear equations with one unknown variable whose value can be deduced from the information given. A linear equation is one where the unknown variable does not take any powers other than 1, e.g. there may be terms in x but not x^2, x^{-1} etc.

Before setting out the rules for solving single linear equations let us look at some simple examples.

Example 3.36

You go into a foreign exchange bureau to buy US dollars for your holiday. You exchange £200 and receive $343. When you get home you discover that you have lost your receipt. How can you find out the exchange rate used for your money if you know that the bureau charges a fixed £4 fee on all transactions?

Solution

The amount of pounds actually exchanged into dollars will be

$$£200 - £4 = £196$$

Let x be the exchange rate of pounds into dollars. Therefore

$$343 = 196x$$

$$\frac{343}{196} = x$$

$$1.75 = x$$

The exchange rate is \$1.75 to the pound.

This illustrates the fundamental principle that one can divide both sides of an equation by the same number.

Example 3.37

If $62 = 34 + 4x$ what is x?

Solution

$28 = 4x$ if 34 is subtracted from both sides.
Thus $7 = x$ when both sides are divided by 4.
In this case the principle that one can subtract the same amount from both sides of an equation is utilized.

The basic principles for solving equations are that all the terms in the unknown variable have to be brought together by themselves on one side of the equation. In order to do this one can add, subtract, multiply or divide both sides of an equation by the same number or algebraic term. One can also perform other arithmetical operations, such as finding the square root of both sides of an equality sign.

Once the equation is in the form

$$ax = b$$

where a and b are numbers, then x can be found by dividing b by a.

Example 3.38

Solve for x if $16x - 4 = 68 + 7x$.

Solution

Subtracting $7x$ from both sides

$$9x - 4 = 68$$

Adding 4 to both sides

$$9x = 72$$

Dividing both sides by 9

$$x = 8$$

Example 3.39

Solve for x if

$$4 = \frac{96}{x}$$

Solution

Multiplying both sides by x

$$4x = 96$$

Dividing both sides by 4

$$x = 24$$

Example 3.40

Solve for x if $6x^2 + 12 = 162$.

Solution

Subtracting 12 from both sides

$$6x^2 = 150$$

Dividing through by 6

$$x^2 = 25$$

Taking square roots

$$x = 5$$

Example 3.41

A firm has to pay fixed costs of £1,500 plus another £60 for each unit produced. How much can it produce for a budget of £4,800?

Solution

$$\text{budget} = \text{total expenditure on production}$$

Therefore if x is output level

$$4,800 = 1,500 + 60x$$

Subtracting 1,500 from both sides

$$3,300 = 60x$$

Dividing by 60

$$55 = x$$

Thus the firm can produce 55 units for a budget of £4,800.

Example 3.42

You sell 500 shares in a company via a stockbroker who charges a flat £20 commission rate on all transactions under £1,000. Your bank account is credited with £692 from the sale of the shares. What price (in pence) were your shares sold at?

Solution

Let price per share be x. Therefore, working in pence,

$$69,200 = 500x - 2,000$$

Adding 2,000 to both sides

$$71,200 = 500x$$

Dividing both sides by 500

$$142.4 = x$$

The share price is 142.4p.

QUESTIONS 3.7

1. Solve for x when $16x = 2x + 56$.
2. Solve for x when

$$14 = \frac{6 + 4x}{5x}$$

3. Solve for x when $45 = 24 + 3x$.
4. Solve for x if $5x^2 + 20 = 1,000$.
5. If $q = 560 - 3p$ solve for p when $q = 314$.
6. You get paid travelling expenses according to the distance you drive in your car plus a weekly sum of £21. You put in a claim for 420 miles travelled and receive an expenses payment of £105. What is the payment rate per mile?

7. In one subject that you are studying, the overall mark for the course is calculated on the basis of a 30:70 weighting between coursework and examination marks. If you have scored 57% for coursework, what examination mark do you need to get to achieve an overall mark of 40%?
8. You sell 900 shares via your broker who charges a flat rate of commission of £20 on all transactions of less than £1,000. Your bank account is credited with £340 from the share sale. What price were your shares sold at?
9. Your net monthly salary is £950. You know that National Insurance and pension contributions take 15% of your gross salary and that income tax is levied at a rate of 25% on gross annual earnings above a £4,000 exemption limit. What is your gross monthly salary?
10. You have 64 square paving stones and wish to lay them to form a square patio in your garden. If each paving stone is 2 feet square, what will the length of a side of your patio be?
11. A firm faces the marginal revenue schedule $MR = 80 - 2q$ and the marginal cost schedule $MC = 15 + 0.5q$ where q is quantity produced. You know that a firm maximizes profit when $MC = MR$. What will the profit-maximizing output be?

3.8 THE SUMMATION SIGN Σ

The summation sign Σ can be used in certain circumstances as a shorthand means of expressing the sum of a number of different terms added together. (Σ is a Greek letter, pronounced 'sigma'.) There are two ways in which it can be used.

The first is when one variable increases its value by 1 in each successive term, as the example below illustrates.

Example 3.43

A firm sells a product at a price of £25. In the first week of business it sells 30 units. Sales then increase at the rate of 30 units per week. If it continues in business for 5 weeks, its total sales revenue will therefore be (in pounds)

$$(30 \times 25 \times 1) + (30 \times 25 \times 2) + (30 \times 25 \times 3) + (30 \times 25 \times 4) + (30 \times 25 \times 5)$$

You can see that the number representing the week is increased by 1 in each successive term. This is rather a cumbersome expression to work with. We can instead write

$$\text{sales revenue} = \sum_{i=1}^{5} (30 \times 25)i$$

57

This means that one is summing all the terms $(30 \times 25)i$ for values of i from 1 to 5.

If the price of the product and the number of weeks of business were not known we could write

$$\text{sales revenue} = \sum_{i=1}^{n} (30pi)$$

where n is the weeks of business and p is the product price.

To evaluate an expression containing a summation sign, one may still have to calculate the value of each term separately and then add up, although in some cases short-cut formulae may be used (see Chapter 7 on series).

Example 3.44

Evaluate

$$\sum_{i=3}^{n} (20 + 3i) \quad \text{for } n = 6$$

Solution

Note that in this example i starts at 3. Thus

$$\sum_{i=3}^{6} (20 + 3i) = (20 + 9) + (20 + 12) + (20 + 15) + (20 + 18)$$
$$= 29 + 32 + 35 + 38$$
$$= 134$$

Some suggested Lotus 1–2–3 templates for calculating summations involving the stream of income from an investment over a number of years are given in Chapter 7.

The second way in which the summation sign can be used requires a set of data where observations are specified in numerical order.

Example 3.45

Assume that a researcher finds a random group of twelve students and observes their weight and height as shown in Table 3.1. If we let H_i represent the height and W_i represent the weight of student i, then the total weight of the first six students can be specified as

Table 3.1

Student no.	1	2	3	4	5	6	7	8	9	10	11	12
Height (cm)	178	175	170	166	168	185	169	189	175	181	177	180
Weight (kg)	72	68	58	52	55	82	55	86	70	71	65	68

$$\sum_{i=1}^{6} W_i$$

In this method the i refers to the number of the observation and the value of i is not incorporated into the actual calculations.

Staying with the same example, the average weight of the first n students could be specified as

$$\frac{1}{n} \sum_{i=1}^{n} W_i$$

When no superscript or subscripts are shown with the Σ sign it usually means that all possible values are summed. For example, a price index is constructed by working out how much a weighted average of prices rises over time. One method of measuring how much, on average, prices rise between year 0 and year 1 is to use the formula

$$\frac{\Sigma p_i^1 x_i}{\Sigma p_i^0 x_i}$$

where p_i^1 is the price of good i in year 1, p_i^0 is the price of good i in year 0 and x_i is the percentage of consumer expenditure on good i. If all goods are in the index then $\Sigma x_i = 100$ by definition.

Example 3.46

Given the figures in Table 3.2 for expenditure proportions and prices calculate the rate of inflation between year 0 and year 1 and compare the price rise of food with the weighted average price rise.

Solution

Note that in this example we are just assuming one price for each category of expenditure. In reality, of course, the prices of several individual goods are included in a price index. It must be stressed that these are *prices* not measures of expenditure on these goods and services.

The weighted average price increase will be

Table 3.2

	Percentage of expenditure (x_i)	Prices, year 0 (p_i^0)	Prices, year 1 (p_i^1)
Durable goods	9	200	216
Food	17	80	98
Alcohol and tobacco	11	70	92
Footwear and clothing	7	120	130
Energy	8	265	270
Other goods	11	62	71
Rent, rates, water	12	94	98
Other services	25	52	60
	100		

Note: All prices are in £.

$$\frac{\Sigma p_i^1 x_i}{\Sigma p_i^0 x_i} = \frac{1,944+1,666+1,012+910+2,160+781+1,176+1,500}{1,800+1,360+770+840+2,120+682+1,128+1,300}$$

$$= \frac{11,149}{10,000} = 1.115$$

This means that, on average, prices in year 1 are 111.5% of prices in year 0, i.e. the inflation rate is 11.5%. The price of food went up from 80 to 98, i.e. by 22.5%, which is greater than the inflation rate. Adjusted for inflation, the real price increase for food is thus

$$100\left(\frac{1.225}{1.115}-1\right)=100(1.099-1)=9.9\%$$

QUESTIONS 3.8

1. Refer to Table 3.1 above and write an expression for the average height of the first n students observed and evaluate for $n = 6$.
2. Evaluate

$$\sum_{i=1}^{5}(4+i)$$

3. Evaluate

$$\sum_{i=2}^{5}2^i$$

(Note that i is an exponent in this question.)
4. A firm sells 6,000 tonnes of its output in its first year of operation. Sales then decrease each year by 10% of the previous year's sales figure. Write an expression for the firm's total sales over n years and evaluate for $n = 3$.

5. Observations of a firm's sales revenue per month are as follows.

Month	1	2	3	4	5	6	7	8	9	10	11	12
Revenue	4.5	4.2	4.6	4.4	5.0	5.3	5.2	4.9	4.7	5.4	5.3	5.8

(£'000)

  (a) Write an expression for average monthly sales revenue for the first *n* months and evaluate for $n = 4$.
  (b) Write an expression for average monthly sales revenue over the preceding 3 months for any given month *n*, assuming that *n* is not less than 4. Evaluate for $n = 10$.

6. Assume that the expenditure and price data given in Example 3.46 above all still hold except that the price of alcohol and tobacco rises to £108 in year 1. Work out the new inflation rate and the new real price increase in the price of food.

3.9 INEQUALITY SIGNS

As well as the equality sign (=), the following four inequality signs are used in algebra:

> which means 'is always greater than'
< which means 'is always less than'
⩾ which means 'is greater than or equal to'
⩽ which means 'is less than or equal to'

The last two are sometimes called 'weak inequality' signs.

Example 3.47

If we let the number of days in any given month be represented by *N*, then whatever month is chosen it must be true that

$$N > 27$$
$$N < 32$$
$$N \geqslant 28$$
$$N \leqslant 31$$

Special care has to be taken when using inequality signs if unknown variables can take negative values. For example, the inequality

$$2x < 3x$$

only holds if

$$x > 0$$

If *x* took a negative value, then the inequality would be reversed. For example, if $x = -5$, then $2x = -10$, $3x = -15$ and so $2x > 3x$.

When considering inequality relationships, it can be useful to work in terms of the absolute value of a variable x. This is written $|x|$ and, as you will recall from Chapter 2, is defined as

$$|x| = x \text{ when } x \geqslant 0$$

$$|x| = -x \text{ when } x < 0$$

i.e. the absolute value of a positive number is the number itself and the absolute value of a negative number is the same number but with the negative sign removed.

If an inequality is specified in terms of the absolute value of an unknown variable, then the inequality will not be reversed if the variable takes on a negative value. For example: $|2x| < |3x|$ for all non-zero values of x.

In economic applications, the unknown variable in an algebraic expression often represents a concept (such as quantity produced or price) that cannot normally take on a negative value. In these cases, the use of inequality signs is therefore usually more straightforward than in cases where negative values are possible.

In order to make clearer the relationship between different variables, it is possible to simplify an inequality relationship by performing the same arithmetical operation on both sides of the inequality sign. However, the rules for doing this differ from those that apply when manipulating both sides of an ordinary equality sign.

One can add any number to or subtract any number from both sides of an inequality sign.

Example 3.48

If

$$x + 6 > y + 2$$

then

$$x + 4 > y$$

One can multiply or divide both sides of an inequality sign by a positive number, but if one multiplies or divides by a negative number the direction of the inequality is reversed. (If one multiplies both sides by zero then they both become zero and the inequality becomes an equality.)

Example 3.49

If $x < y$ then $-x > -y$ (multiplying through by -1).

Example 3.50

If $3x < 18y$ then $-x > -6y$ (dividing both sides by -3).

If both sides of an inequality sign are squared, the same inequality sign only holds if both sides are initially positive values. This is because a negative number squared becomes a positive number.

Example 3.51

If

$$x + 3 < y$$

then

$$(x + 3)^2 < y^2 \text{ if } (x + 3) \geqslant 0 \text{ and } y > 0$$

Note that in this example x could possibly lie between -3 and 0 when $x + 3 \geqslant 0$.

Example 3.52

$$-6 < -4$$

But

$$(-6)^2 > (-4)^2$$

since

$$36 > 16$$

If both sides of an inequality sign are positive and are raised to the same negative power, then the direction of the inequality will be reversed.

Example 3.53

If

$$x > y$$

then

$$x^{-1} < y^{-1}$$

if x and y are positive,
i.e.

$$\frac{1}{x} < \frac{1}{y}$$

Example 3.54

Two leisure park owners A and B have the same weekly running costs of £8,000. The numbers of customers visiting the two parks are x and y respectively. If $x > y$, what can be said about comparative average costs per customer?

Solution

Since

$$x > y$$

then

$$x^{-1} < y^{-1}$$

Thus

$$\frac{£8,000}{x} < \frac{£8,000}{y}$$

and so

average cost for A < average cost for B

QUESTIONS 3.9

1. You are studying a subject which is assessed by coursework and examination with the total mark for the course being calculated on a 30:70 weighting between these two components. Assuming you score 60% in coursework, insert the appropriate inequality sign between your possible overall mark for the course and the percentage figures below.
   (a) 18% ? overall mark   (b) 16% ? overall mark
   (c) 88% ? overall mark   (d) 90% ? overall mark
2. If $x \geqslant 1$, insert the appropriate inequality sign between:
   (a) $(x+2)^2$ and 3   (b) $(x+2)^2$ and 9
   (c) $(x+2)^2$ and $3x$ (d) $(x+2)^2$ and $6x$
3. If Q_1 and Q_2 represent positive production levels of a good and the equality $Q_2 = Z^n Q_1$ always holds where $Z > 1$, what can be said about the relationship between Q_1 and Q_2 if
   (a) $n > 0$ (b) $n = 0$ (c) $n < 0$?
4. If a monopolist can operate price discrimination and charge separate prices P_1 and P_2 in two different markets, it can be proved that for profit maximization the monopolist should choose values for P_1 and P_2 that satisfy the equation

$$P_1\left(1 - \frac{1}{e_1}\right) = P_2\left(1 - \frac{1}{e_2}\right)$$

where e_1 and e_2 are elasticities of demand in the two markets. If $|e_1| > |e_2|$, in which market should price be higher?

4

Graphs and functions

4.1 FUNCTIONS

Suppose that average weekly household expenditure on food (C) depends on average net household weekly income (I) according to the relationship

$$C = 12 + 0.3I$$

For any given value of I, one can evaluate what C will be. For example, if $I = 90$ then

$$C = 12 + 27 = 39$$

Whatever value of I is chosen there will be one unique corresponding value of C. This is an example of a function.

A relationship between the values of two or more variables can be defined as a function when the value of one of the variables is determined by the value of the other variable or variables.

If the precise mathematical form of the relationship is not actually known then a function may be written in what is called a general form. For example, a demand function may be written as

$$Q_d = f(P)$$

This particular general form just tells us that quantity demanded of a good (Q_d) depends on its price (P). The 'f' is not an algebraic symbol in the usual sense and so f(P) means 'is a function of P' and not 'f multiplied by P'. In this case P is what is known as the 'independent variable' because its value is given and is not dependent on the value of Q_d, i.e. it is exogenously determined. On the other hand Q_d is the 'dependent variable'. Its value depends on the value of P.

Functions may have more than one independent variable. For example, the general form production function

$$Q = f(K, L)$$

tells us that output (Q) depends on the values of the two independent variables capital (K) and labour (L).

The specific form of a function tells us exactly how the value of the dependent variable is determined from the values of the independent variable or variables. A specific form for a demand function might be $Q_d = 120 - 2P$. For any given value of P the specific function allows us to calculate the value of Q_d, e.g.

$$\text{when } P = 10, \text{ then } Q_d = 120 - 20 = 100$$
$$\text{when } P = 45, \text{ then } Q_d = 120 - 90 = 30 \text{ etc.}$$

In economic applications of functions it may make sense to restrict the 'domain' of the function, i.e. the range of possible values of the variables. For example, variables that represent price or output may be restricted to positive values. Strictly speaking the 'domain' limits the values of the independent variables and the 'range' governs the possible values of the dependent variable.

For more complex functions with more than one independent variable it may be helpful to draw up a table to show the relationship of different values of the independent variables to the value of the dependent variable.

Table 4.1 shows some possible different values for the specific form production function $Q = 4K^{0.5}L^{0.5}$. It is implicitly assumed that Q, $K^{0.5}$ and $L^{0.5}$ only take positive values.

Table 4.1

K	L	$K^{0.5}$	$L^{0.5}$	Q
1	1	1	1	4
4	1	2	1	8
9	25	3	5	60
7	11	2.64575	3.31662	35.0998

When defining the specific form of a function it is important to make sure that only one unique value of the dependent variable is determined from each given value of the independent variable(s). Consider the equation

$$y = 80 + x^{0.5}$$

This does not define a function because any given value of x corresponds to two possible values for y. For example, if $x = 25$, then $25^{0.5} = 5$ or -5 and so $y = 75$ or 85. However, if we define $y = 80 + x^{0.5}$ for $x^{0.5} \geq 0$, then this does constitute a function. When domains are not specified then one should assume a sensible range for functions representing economic variables. For example, one usually assumes $K^{0.5} > 0$ and $L^{0.5} > 0$ in a production function, as in Table 4.1 above.

QUESTIONS 4.1

1. An economist researching the market for tea assumes that

$$Q_t = f(P_t, \ Y, \ A, \ N, \ P_c)$$

where Q_t is the quantity of tea demanded, P_t is the price of tea, Y is the average household expenditure, A is the tea advertising expenditure, N is population and P_c is the price of coffee.

(a) What does $Q_t = f(P_t, Y, A, N, P_c)$ mean in words?
(b) Identify the dependent and independent variables.
(c) Make up a specific form for this function. (Use your knowledge of economics to deduce whether the coefficients of the different independent variables should be positive or negative.)

2. If a firm faces the total cost function

$$TC = 6 + x^2$$

where x is output, what is TC when x is (a) 14? (b) 1? (c) 0? What restrictions on the domain of this function would it be reasonable to make?

3. A firm's total expenditure E on inputs is determined by the formula

$$E = P_K K + P_L L$$

where K is the amount of input K used, L is the amount of input L used, P_K is the price per unit of K and P_L is the price per unit of L. Is one unique value for E determined by any given set of values for K, L, P_K, and P_L? Does this mean that any one particular value for E must always correspond to the same set of values for K, L, P_K and P_L?

4.2 INVERSE FUNCTIONS

Mathematically it may be possible to work out an inverse function. If we confine the analysis to the case of a function with only one independent variable, x, this means that if y is a function of x, i.e. $y = f(x)$, then x can be written as a function of y, i.e. $x = g(y)$ (the g being used to show that we are talking about a different function).

Example 4.1

If

$$y = 4 + 5x$$

then

$$y - 4 = 5x$$
$$0.2y - 0.8 = x$$

The mathematical condition necessary for a function to have a corresponding inverse function is that the original function must be 'monotonic'. This means that, as the value of the independent variable x is increased, the value of the dependent variable y must either always increase or always decrease. It cannot first increase and then decrease, or vice versa. This will ensure that, as well as there being one unique value of y for any given value of x, there will also be one unique value of x for any given value of y. This point will probably become clearer to you in the following sections on graphs of functions but it can be illustrated here with a simple example.

Example 4.2

Consider the function $y = 9x - x^2$ restricted to the domain $0 \leqslant x \leqslant 9$. Each value of x will determine a unique value of y. However, some values of y will correspond to two values of x, e.g. when $x = 3$ then $y = 27 - 9 = 18$; when $x = 6$ then $y = 54 - 36 = 18$. This is because the function $y = 9x - x^2$ is not monotonic. This can be established by calculating y for a few selected values of x:

x	1	2	3	4	5	6	7
y	8	14	18	20	20	18	14

These figures show that y first increases and then decreases in value as x is increased and so there is no inverse for this function.

Although mathematically it may be possible to derive an inverse function it may not always make sense to derive the inverse of an economic function, or many other functions that are based on empirical data. It is not always a clear-cut case though. If we take the geometric function that the area A of a square is related to the length L of its sides by the function $A = L^2$, then we can also write the inverse function that relates the length of a square's side to its area: $L = A^{0.5}$ (assuming that L can only take non-negative values.) Once one value is known then the other is determined by it. However, suppose that someone investigating expenditure on holidays abroad (H) finds that the level of average annual household income (M) is the main influence and the relationship can be explained by the function

$$H = 0.01M + 100 \quad \text{(for } M \geqslant £10,000)$$

This mathematical equation could be rearranged to give

$$M = 100H - 10,000$$

but to say that H determines M obviously does not make sense. The amount of holidays taken abroad does not determine the level of average household income.

The cause and effect relationship within an economic model is not always obviously in one direction only. Consider the relationship between price and quantity in a demand function. A monopoly may set the product's price and then see how much consumers are willing to buy, i.e. $Q = f(P)$. In a competitive

Graphs and functions

industry firms may first decide how much they are going to produce and then see what price they can get for this output, i.e. $P = f(Q)$. Thus it can be legitimate to derive the inverse of a demand function.

Example 4.3

Given the demand function $Q = 200 - 4P$, write P as a function of Q.

Solution

$$Q = 200 - 4P$$
$$4P + Q = 200$$
$$4P = 200 - Q$$
$$P = 50 - 0.25Q$$

QUESTIONS 4.2

1. To convert temperature from degrees Fahrenheit to degrees Celsius one uses the formula

$$°C = \frac{5}{9}(°F - 32)$$

   What is the inverse of this function?
2. What is the inverse of the demand function

$$Q = 1,200 - 0.5P?$$

3. The total revenue (TR) that a monopoly receives from selling different levels of output (q) is given by the function $TR = 60q - 4q^2$ for $0 \leqslant q \leqslant 15$. Explain why one cannot derive the inverse function $q = f(TR)$.
4. An empirical study suggests that a brewery's weekly sales of beer are determined by the average air temperature given that the price of beer, income, adult population and most other variables are constant in the short run. This functional relationship is estimated as

$$X = 400 + 16T^{0.5}$$

   where X is the number of barrels sold per week and T is the mean average air temperature, in °C. What is the mathematical inverse of this function? Does it make sense to specify such an inverse function in economics?
5. Make up your own examples for:
   (a) a function that has an inverse, and then derive the inverse function;
   (b) a function that does not have an inverse and then explain why this is so.

4.3 GRAPHS OF LINEAR FUNCTIONS

We are all familiar with graphs of the sort illustrated in Figure 4.1. This shows a firm's annual sales figures. To find what its sales were in 1990 you first find 1990 on the horizontal axis, move vertically up to the line marked 'sales' and read off the corresponding figure on the vertical axis, which in this case is £120,000. These graphs are often used as an alternative to tables of data as

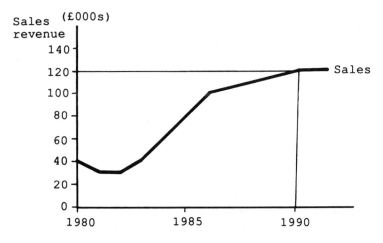

Figure 4.1

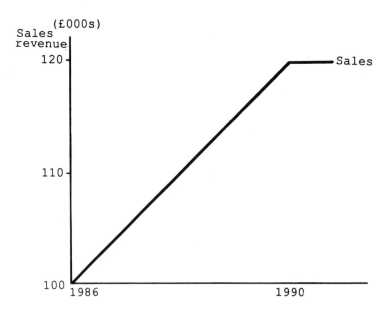

Figure 4.2

they make trends in the numbers easier to identify visually. It is common practice to shift the starting point or to alter the graduations on the axes to emphasize a trend. Figure 4.2 uses information from the graph in Figure 4.1 but it paints a much rosier picture of sales growth in the late 1980s. Note that instead of starting at zero the vertical axis commences at £100,000. These, however, are *not* graphs of functions. Sales are not determined by 'time'.

Mathematical functions are mapped out on what is known as a set of 'cartesian axes'. A two-dimensional set of cartesian axes is shown in Figure 4.3. Variable x is measured by equal increments on the horizontal axis and variable y by equal increments on the vertical axis. Both x and y can be measured in positive or negative directions. Although obviously only a limited range of values can be shown on the page of a book, the cartesian axes theoretically range from $-\infty$ to $+\infty$. If no restrictions are placed on the domain of the independent variable in a function then it could possibly take any positive or negative value. The range of values of the dependent variable will depend on both the domain of the independent variable and the nature of the function.

Any point on the graph will have two 'coordinates', i.e. corresponding values on the x and y axes. For example, to find the coordinates of point A one needs to draw a vertical line down to the x axis and read off the value of 20 and draw a horizontal line across to the y axis and read off the value 17. The coordinates (20, 17) determine point A.

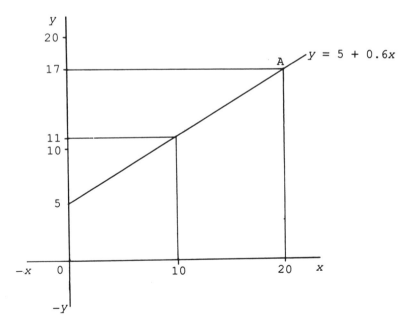

Figure 4.3

71

As only two variables can be measured on the two axes in Figure 4.3, this means that only functions with one independent variable can be illustrated by a graph on a two-dimensional sheet of paper. One axis measures the dependent variable and the other measures the independent variable. (However, in Section 4.9 below a method of illustrating a two-independent-variable function is explained.)

Having set up the cartesian axes in Figure 4.3, let us use it to illustrate the function $y = 5 + 0.6x$. To determine the shape of this function let us work out a few values of y for different values of x:

$$\text{when } x = 0, \quad \text{then } y = 5 + 0.6(0) = 5$$
$$\text{when } x = 10, \text{ then } y = 5 + 0.6(10) = 5 + 6 = 11$$
$$\text{when } x = 20, \text{ then } y = 5 + 0.6(20) = 5 + 12 = 17$$

These points are plotted in Figure 4.3 and it is obvious that they lie along a straight line. The rest of the function can be shown by drawing a straight line through the points that have been plotted.

Any function that takes the format $y = a + bx$ (where a and b can be any positive or negative numbers) will correspond to a straight line when

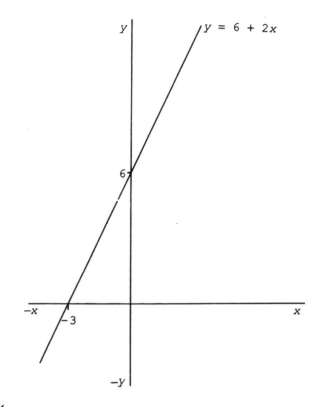

Figure 4.4

represented by a graph. This is because, if $y = a + bx$, then the value of y will change by the same amount, b, for every one unit increment in x.

Usually the easiest way to plot a linear function is to find the points where it cuts the two axes and draw a straight line through them.

Example 4.4

Plot the graph of the function $y = 6 + 2x$.

Solution

When $x = 0$ then $y = 6$ and so this function must cut the y axis at $y = 6$.
    When $y = 0$ then

$$0 = 6 + 2x$$
$$-6 = 2x$$
$$-3 = x$$

and so this function must cut the x axis at $x = -3$. The graph is therefore as shown in Figure 4.4.

However, in economics some variables may only take on positive values. To illustrate a linear function that applies only to positive values of all the variables concerned one may sometimes only be able to use one intercept. In such cases, all one has to do is simply plot another point and draw a line through the two points obtained.

Example 4.5

Plot the graph of the function $C = 200 + 0.6Y$ where C is consumer spending and Y is income.

Solution

When $Y = 0$, then $C = 200$, and so the line cuts the vertical axis at 200.
    However, when $C = 0$, then

$$0 = 200 + 0.6y$$
$$-0.6Y = 200$$
$$Y = -\frac{200}{0.6}$$

As negative values of Y are unacceptable, just choose another pair of values, e.g. when $Y = 500$, then

$$C = 200 + 0.6(500)$$
$$= 200 + 300$$
$$= 500$$

This graph is shown in Figure 4.5.

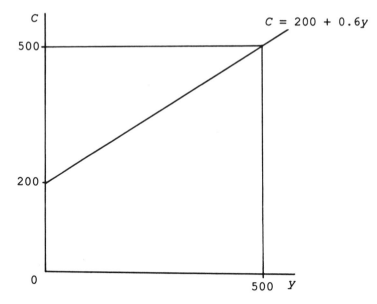

Figure 4.5

In mathematics the usual convention when drawing graphs is to measure the independent variable x along the horizontal axis and the dependent variable y along the vertical axis. However, in economics the usual convention is to measure price P on the vertical axis and quantity Q along the horizontal axis. This sometimes confuses students when a function in economics is specified with Q as the dependent variable, e.g. the demand function

$$Q = 800 - 4P$$

which is illustrated in Figure 4.6. (Before you proceed, check that you understand why the intercepts on the two axes are as shown.)

Theoretically, it does not matter which axis is used to measure which variable. However, one of the main reasons for using graphs is to make certain aspects of mathematical analysis clearer to understand. Therefore, if one always has to keep checking which axis measures which variable this defeats the objective of the exercise. Thus, even though it may upset some mathematical purists, in this text we shall stick to the economist's convention of measuring quantity on the horizontal axis and price on the vertical axis, even if price is the independent variable in a function.

This means that care has to be taken when performing certain operations on functions. If necessary, one can transform monotonic functions to obtain the inverse function (as already explained) if this helps the analysis. For example, the demand function $Q = 800 - 4P$ gives the inverse function

Graphs and functions

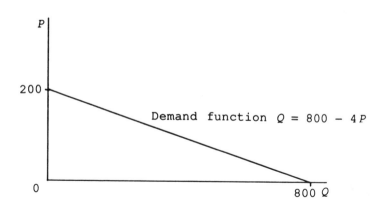

Figure 4.6

$$P = \frac{800 - Q}{4} = 200 - 0.25Q$$

(Check again in Figure 4.6 for the intercepts of the graph of this function.)

QUESTIONS 4.3

Sketch the graphs of the linear functions 1 to 9 below, identifying the relevant intercepts on the axes. Assume that variables represented by letters that suggest they are economic variables (i.e. all variables except x and y) are restricted to non-negative values.

1. $y = 6 + 0.5x$
2. $y = 12x - 40$
3. $P = 60 - 0.2Q$
4. $Q = 750 - 5P$
5. $1200 = 50K + 30L$
   (Note that this is an accounting identity rather than a function although a given value of K will still determine a unique value of L, and vice versa.)
6. $TR = 8Q$
7. $TC = 200 + 5Q$
8. $TFC = 75$
9. Make up your own example of a linear function and then sketch its graph.
10. Which of the following functions do you think realistically represents the supply schedule of a competitive industry? Why?

(a) $P = 0.6Q + 2$ (b) $P = 0.5Q - 10$
(c) $P = 4Q$ (d) $Q = -24 + 0.2P$

Assume $P \geqslant 0$, $Q \geqslant 0$ in all cases.

75

4.4 FITTING LINEAR FUNCTIONS

If you know that two points lie on a straight line then you can draw the rest
of the line. You simply put your ruler on the page, join the two points and then
extend the line in either direction as far as you want to go. For example,
suppose that you know that a firm faces a linear demand schedule and that
400 units of output Q are sold when price is £40 and 500 units are sold when
price is £20. These two price and quantity combinations are marked as points
A and B in Figure 4.7 and then the rest of the demand schedule is drawn in.

One can then use this graph to predict the amounts sold at other prices. For
example, when price is £29.50, one can read off the corresponding quantity,
which is approximately 450. It is more accurate, however, to determine the
algebraic format of a function from the information that is initially given and
then use this function to make predictions of other values. If it is linear then
a demand function must be in the format $P = a - bQ$ where a and b are
parameters that we wish to determine the value of. Therefore, from Figure 4.7
we can see that

$$\text{when } P = 40, \text{ then } Q = 400, \text{ and so } 40 = a - 400b \qquad (1)$$

$$\text{when } P = 20, \text{ then } Q = 500, \text{ and so } 20 = a - 500b \qquad (2)$$

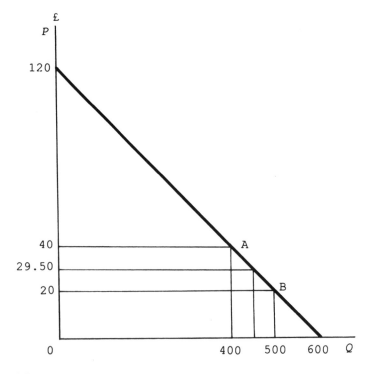

Figure 4.7

Equations (1) and (2) are what is known as simultaneous linear equations. Various methods of solving such sets of simultaneous equations (i.e. finding the values of *a* and *b*) are explained later in Chapter 5. Here we shall just use an intuitively obvious method of deducing the values of *a* and *b* from the graph in Figure 4.7.

Between points B and A we can see that a £20 rise in price causes a 100 unit decrease in quantity demanded. As this is a linear function then we know that further price rises of £20 will also cause quantity demanded to fall by 100 units. At A, quantity is 400 units. Therefore a rise in price of £80 is required to reduce quantity demanded from 400 to zero, i.e. a rise in price of $4 \times £20 = £80$ will reduce quantity demanded by $4 \times 100 = 400$ units. This means that the intercept of this function on the price axis is £40 (the price at A) plus £80, which is £120. This is the value of the parameter *a*.

To find the value of the parameter *b* we need to ask 'what will be the fall in price necessary to cause quantity to increase by one unit?' Given that a £20 price fall causes quantity to rise by 100 units then it must be the case that a price fall of $£20/100 = £0.2$ will cause quantity to rise by one unit. This also means that a price rise of £0.2 will cause quantity demanded to fall by one unit. Therefore, $b = 0.2$. Our function can now be written as

$$P = 120 - 0.2Q$$

We can check that this is correct against the values of *P* and *Q* that were given originally by substituting the values of *Q* into the function.

$$\text{If } Q = 400, \text{ then } P = 120 - 0.2(400) = 120 - 80 = 40$$
$$\text{If } Q = 500, \text{ then } P = 120 - 0.2(500) = 120 - 100 = 20$$

These are the values of *P* originally specified and so we are satisfied that the linear function that passes through points A and B in Figure 4.7 is $P = 120 - 0.2Q$.

The inverse of this function will be $Q = 600 - 5P$. Precise values of *Q* can now be derived for given values of *P*. For example,

$$\text{when } P = £29.50, \text{ then } Q = 600 - 5(29.50) = 452.5$$

This is a more accurate figure than the one read off the graph as approximately 450.

Having learned how to deduce the parameters of a linear downward-sloping demand function, let us now try to fit another linear function, this time one that slopes upward.

Example 4.6

It is assumed that consumption *C* depends on income *Y* and that this relationship takes the form of the linear function $C = a + bY$.

When Y is £600, C is observed to be £660

When Y is £1,000, C is observed to be £900

What are the values of a and b in this function?

Solution

We expect b to be positive, i.e. consumption increases with income, and so our function will slope upwards as shown in Figure 4.8. A decrease in Y of £400, from £1,000 to £600, causes C to fall by £240, from £900 to £660. As this is a linear function then equal changes in Y will cause the same changes in C. If Y is decreased by a further £600 (i.e. to zero) then the corresponding fall in C will be 1.5 times the fall caused by an income fall of £400, since £600 = $1.5 \times$ £400. Therefore the fall in C is $1.5 \times$ £240 = £360, and so the value of C when Y is zero is £660 − £360 = £300. Thus $a = 300$.

A rise in Y of £400 causes C to rise by £240. Therefore a rise in Y of £1 will cause C to rise by £240/400 = £0.6. Thus $b = 0.6$.

The function can therefore be specified as

$$C = 300 + 0.6Y$$

Checking against original values:

When $Y = 600$, then predicted $C = 300 + 0.6(600)$

$$= 300 + 360 = 660. \text{ Correct.}$$

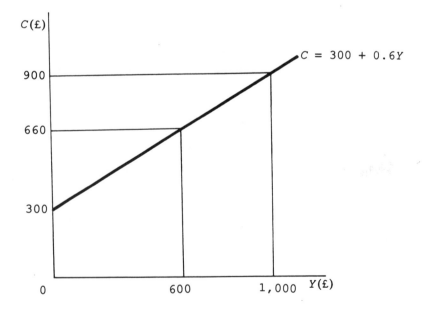

Figure 4.8

Graphs and functions

When $Y = 1,000$, then predicted $C = 300 + 0.6(1,000)$

$$= 300 + 600 = 900. \text{ Correct.}$$

QUESTIONS 4.4

1. A monopoly sells 30 units of output when price is £12 and 40 units when price is £10. If its demand schedule is linear what is the specific form of the actual demand function?

   Use this function to predict quantity sold when price is £8. What domain restrictions would you put on this demand function?
2. Assume that consumption C depends on income Y according to the function $C = a + bY$, where a and b are parameters. If C is £60 when Y is £40 and C is £90 when Y is £80, what are the values of the parameters a and b?
3. On a linear demand schedule quantity sold falls from 90 to 30 when price rises from £40 to £80. How much further will price have to rise for quantity sold to fall to zero?
4. A firm knows that its demand schedule takes the form $P = a - bQ$. If 200 units are sold when price is £9 and 400 units are sold when price is £6, what are the values of the parameters a and b?
5. A firm notices that its total production costs are £3,200 when output is 85 and £4,820 when output is 130. If total cost is assumed to be a linear function of output what expenditure will be necessary to manufacture 175 units?

4.5 SLOPE

British road signs used to give warning of steep hills by specifying their slope in a format such as '1 in 10', meaning that the road rose vertically by 1 foot for every 10 feet travelled in a horizontal direction. Now the continental format is used and so instead of '1 in 10' a road sign will say 10%.

In mathematics the same concept of slope is used but it is expressed as a decimal fraction rather than in percentage terms.

The graph in Figure 4.9 shows the function $y = 2 + 0.1x$. The slope is obviously the same along the whole length of this straight line and it does not matter where the slope is measured. To measure the slope along the stretch AB, draw a horizontal line across from A and drop a vertical line down from B. These intersect at C, forming the triangle ABC with a right angle at C. The horizontal distance AC is 20 and the vertical distance BC is 2, and so if this was a cross-section of a hill you would clearly say that the slope is 1 in 10, or 10%.

The way that the slope of a line is defined in mathematics is

$$\text{slope} = \frac{\text{height}}{\text{base}}$$

79

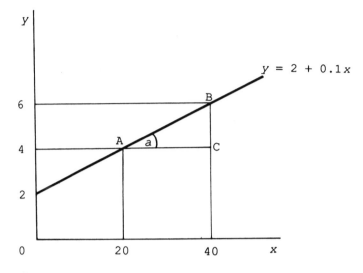

Figure 4.9

where height and base measure the sides of a right-angled triangle drawn as above. (Note that this only applies to lines that slope upwards from left to right.) Thus in this example

$$\text{slope} = \frac{2}{20} = 0.1$$

This is also known as the tangent of the angle *a*.

One can see that the slope of this function (0.1) is the same as the coefficient of *x*. This is a general rule. For any linear function in the format $y = a + bx$, then *b* will always represent its slope.

Example 4.7

Find the slope of the function $y = -2 + 3x$.

Solution

The value of *y* increases by 3 for every 1 unit increase in *x* and so the slope of this linear function is 3.

When a line slopes downwards from left to right it has a negative slope. Thus the *b* in the function $y = a + bx$ will take a negative value.

Consider the function $P = 60 - 0.2Q$ where *P* is price and *Q* is quantity demanded. This is illustrated in Figure 4.10. As *P* and *Q* can be assumed not

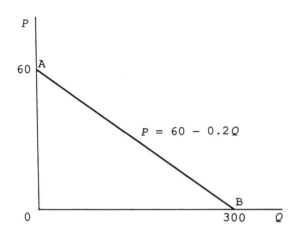

Figure 4.10

to take negative values, the whole function can be drawn by joining the intercepts on the two axes which are found as follows.

When $Q = 0$ then $P = 60$

When $P = 0$ then

$$0 = 60 - 0.2Q$$
$$0.2Q = 60$$
$$Q = \frac{60}{0.2} = 300$$

The slope of a function which slopes down from left to right is found by applying the formula

$$\text{slope} = (-1)\frac{\text{height}}{\text{base}}$$

to the relevant right-angled triangle. Thus, using the triangle 0AB, the slope of the function in Figure 4.10 is

$$(-1)\frac{60}{300} = (-1)0.2 = -0.2$$

This, of course, is the same as the coefficient of Q in the definition of the function.

Remember that in economics the usual convention is to measure P on the vertical axis of a graph. If you are given a function in the format $Q = f(P)$ then you would need to derive the inverse function to read off the slope.

Example 4.8

What is the slope of the demand function $Q = 830 - 2.5P$ when P is measured on the vertical axis of a graph?

Solution

If

$$Q = 830 - 2.5P$$

then

$$2.5P = 830 - Q$$
$$P = 332 - 0.4Q$$

Therefore slope $= -0.4$.

If the coefficient of x in a linear function is zero then the slope is also zero, i.e. the line is horizontal. For example, the function $y = 20$ means that y takes a value of 20 for every value of x. This function could be written

$$y = 20 + 0x$$

to bring home the point but, of course, the second term disappears.

Conversely, a vertical line will have an infinitely large slope. (Note, though, that such a line on a cartesian set of axes would not represent y as a function of x as no unique value of y is determined by a given value of x.)

Now that you understand how the slope of a line is derived we can return to the concept of elasticity mentioned earlier in Chapter 2, when measures of arc elasticity were calculated. Since elasticity of demand can alter along the length of a demand schedule, economists sometimes use the idea of 'point elasticity'. The definition of point elasticity is

$$e = (-1)\frac{P}{Q}\frac{1}{\text{slope}}$$

where P and Q are the price and quantity at the point in question and the slope refers to the slope of the demand schedule at this point. The derivation of this formula and its application to non-linear demand schedules is explained later in Chapter 8. Here we shall just consider its application to linear demand schedules.

Example 4.9

Calculate the point elasticity of demand for the demand schedule

$$P = 60 - 0.2Q$$

where price is (i) zero, (ii) £20, (iii) £40, (iv) £60.

Solution

This is the demand schedule referred to earlier and illustrated in Figure 4.10. The slope of this demand function must be -0.2 at all points as it is a linear function and this is the coefficient of Q.

To find the values of Q corresponding to the given prices we need to derive the inverse function. Given that

$$P = 60 - 0.2Q$$

then

$$0.2Q = 60 - P$$
$$Q = 300 - 5P$$

(i) When P is zero, at point B, then $Q = 300$. The point elasticity will therefore be

$$e = (-1)\frac{P}{Q}\frac{1}{\text{slope}} = (-1)\frac{0}{300}\frac{1}{(-0.2)} = 0$$

(ii) When $P = 20$, then $Q = 200$.

$$e = (-1)\frac{20}{200}\frac{1}{(-0.2)} = \frac{1}{10}\frac{1}{0.2} = \frac{1}{2} = 0.5$$

(iii) When $P = 40$, then $Q = 100$.

$$e = (-1)\frac{40}{100}\frac{1}{(-0.2)} = \frac{2}{5}\frac{1}{0.2} = \frac{2}{1} = 2$$

(iv) When $P = 60$, then $Q = 0$. If $Q = 0$, then $P/Q \to \infty$. Therefore, $e \to \infty$.

QUESTIONS 4.5

1. In Figure 4.11, what are the slopes of the lines 0A, 0B, 0C and EF?
2. A market has a linear demand schedule with a slope of -0.3. When price is £3, quantity sold is 30 units. Where does this demand schedule hit the price and quantity axes?

   What is price if quantity sold is 25 units? How much would be sold at a price of £9?
3. For the demand schedule $P = 60 - 0.2Q$ illustrated in Figure 4.10, calculate point elasticity of demand when price is (a) £24 and (b) £45.
4. Consider the three demand functions

   (a) $P = 8 - 0.75Q$
   (b) $P = 8 - 1.25Q$
   (c) $Q = 12 - 2P$

   Which has the flattest demand schedule, assuming that P is measured on the vertical axis?

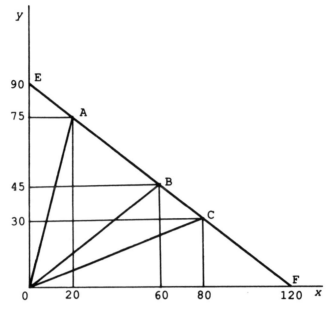

Figure 4.11

In which case is quantity sold the greatest when price is (i) £1 and (ii) £5?

5. For positive values of x which, if any, of the functions below will intersect with the function $y = 1 + 0.5x$?

(a) $y = 2 + 0.4x$ (c) $y = 4 + 0.5x$
(b) $y = 2 + 1.5x$ (d) $y = 4$

6. In macroeconomics the average propensity to consume (APC) and the marginal propensity to consume (MPC) are defined as follows:

APC $= C/Y$ where $C =$ consumption, $Y =$ income
MPC $=$ increase in C from a 1 unit increase in Y

Explain why the slope of the APC function will always be greater than the MPC function if $C = 400 + 0.5Y$.

7. For the demand schedule $P = 24 - 0.125Q$, calculate point elasticity of demand when price is

(a) £5   (b) £10   (c) £15

8. Make up your own examples of linear functions that will

(a) slope upwards and go through the origin;
(b) slope downwards and cut the price axis at a positive value;
(c) be horizontal.

4.6 BUDGET CONSTRAINTS

A frequently used application of the concept of slope in economics is the relationship between prices and the slope of a budget constraint. A budget constraint shows the combinations of two goods that it is possible to buy with a given budget and a given set of prices. The same concept is used in production theory when a firm has a fixed budget to spend on two inputs, and also in other areas of economics such as labour supply analysis.

Assume that a firm has a budget of £3,000 to spend on the two inputs K and L and that input K costs £50 and input L costs £30 a unit. If it spends the whole £3,000 on K then it can buy

$$\frac{3,000}{50} = 60 \text{ units of K}$$

and if it spends all its budget on L it can buy

$$\frac{3,000}{30} = 100 \text{ units of L}$$

These two quantities are marked on the axes of the graph in Figure 4.12. The firm could also split the budget between K and L. Many different combinations are possible, e.g.

30 of K and 50 of L

or

48 of K and 20 of L

If it is assumed that K and L are divisible into fractions of a unit then all the combinations of K and L that can be bought with the given budget of £3,000 can be shown by the line AB which is known as the 'budget constraint' or 'budget line'. The firm could in fact also purchase any of the combinations of K and L within the triangle 0AB but only combinations along the budget constraint AB would entail it spending its entire budget.

Along the budget constraint any pairs of values of K and L must satisfy the equation $50K + 30L = 3,000$ where K is the number of units of K bought and L is the number of units of L bought. All this equation says is that total expenditure on K (price of K × amount bought) plus total expenditure on L (price of L × amount bought) must sum to the total budget available.

We can check that this holds for the combinations of K and L shown in Figure 4.12.

At A	$50 \times 60 + 30 \times 0 = 3,000 + 0 = 3,000$
At B	$50 \times 0 + 30 \times 100 = 0 + 3,000 = 3,000$
At C	$50 \times 30 + 30 \times 50 = 1,500 + 1,500 = 3,000$
At D	$50 \times 48 + 30 \times 20 = 2,400 + 600 = 3,000$

From the graph one can see that this budget constraint has a slope of

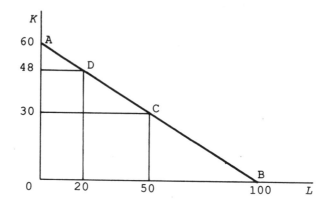

Figure 4.12

$$\frac{-60}{100} = -0.6$$

(Note that budget lines usually slope down from left to right and so have a negative slope.)

It is not necessary to draw a budget constraint in order to determine its slope. This can simply be deduced from the values of the prices of the two goods or inputs concerned.

Consider the general case where the budget is M and the prices of the two goods X and Y are P_X and P_Y respectively. The maximum amount of X that can be bought will be M/P_X which will be the intercept on the horizontal axis. Similarly the maximum amount of Y that can be purchased will be M/P_Y, the intercept on the vertical axis. Thus the slope of the budget constraint will be

$$(-)\frac{M/P_Y}{M/P_X} = (-)\frac{M}{P_Y}\frac{P_X}{M} = (-)\frac{P_X}{P_Y}$$

Thus for any budget constraint the slope will be the negative of the price ratio. Note in particular that it is the price of the good measured on the *horizontal* axis that is at the top in this formula.

From this result we can also see that

1 if the price ratio changes the slope of the budget line changes;
2 if the budget alters, the slope of the budget line does not alter.

Example 4.10

A consumer has an income of £160 to spend on the two goods X and Y whose prices are £20 and £5 each, respectively.

(i) What is the slope of the budget constraint?
(ii) What happens to this slope if P_Y rises to £10?
(iii) What happens if income then falls to £100?

Solution

(i) slope $= -\dfrac{P_X}{P_Y} = -\dfrac{20}{5} = -4$

This can be checked by considering the intercepts on the X and Y axes shown in Figure 4.13 by points B and A. If the total budget is spent on X then $160/20 = 8$ units are bought. If the total budget is spent on Y then $160/5 = 32$ units are bought. Therefore,

$$\text{slope} = (-)\,\frac{\text{intercept on } Y \text{ axis}}{\text{intercept on } X \text{ axis}} = (-)\,\frac{32}{8} = -4$$

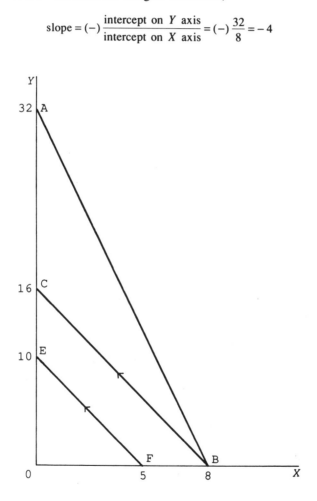

Figure 4.13

87

(ii) The new slope of the budget constraint (shown by BC in Figure 4.13) becomes

$$-\frac{P_X}{P_Y} = -\frac{20}{10} = -2$$

(iii) The price ratio remains unchanged. There is a parallel shift inwards of the budget constraint to EF. The new intercepts are

$$\frac{M}{P_Y} = \frac{100}{10} = 10 \text{ on the } Y \text{ axis}$$

and

$$\frac{M}{P_X} = \frac{100}{20} = 5 \text{ on the } X \text{ axis}$$

The slope is thus $-10/5 = -2$, as before.

Example 4.11

A consumer can buy the two goods A and B at prices per unit of £6 and £4 respectively, and initially has an income of £120.

(i) Show that a 25% rise in all prices will have a lesser effect on the consumer's purchasing possibilities than would a 25% reduction in money income with prices unchanged.
(ii) What is the opportunity cost of buying an extra unit of A? (Assume units of A and B are divisible.)

Solution

(i) The original intercepts on the A and B axes are 20 and 30 respectively. If the price of A rises by 25% to £7.50 and the price of B rises by 25% to £5, then the new intercepts become 16 and 24 respectively. If income is reduced by 25% to 90 then the new intercept on the A axis is $90/6 = 15$ and the new intercept on the B axis is $90/4 = 22.5$. Thus the 25% fall in income shifts the budget constraints towards the origin slightly more than does the 25% rise in prices, i.e. it reduces the consumer's purchasing possibilities by a greater amount.

   Note that the slope of the budget constraint remains constant throughout at $-6/4 = -1.5$.
(ii) The opportunity cost of something is the next best alternative that one has to forgo in order to obtain it. In this context, the opportunity cost of an extra unit of A will be the amount of B the consumer has to forgo.

Graphs and functions

One unit of A costs £6 and one unit of B costs £4. Therefore, the opportunity cost of A in terms of B is 1.5, which is the negative of the slope of the budget line.

QUESTIONS 4.6

1. A consumer can buy good A at £3 a unit and good B at £2 a unit and has a budget of £60. What is the slope of the budget constraint if quantity of A is measured on the horizontal axis?
   What happens to this slope if

   (a) the price of A falls to £2?
   (b) with A at its original price the price of B rises to £3?
   (c) both prices double?
   (d) the budget is cut by 25%?

2. A firm has a budget of £800 per week to spend on the two inputs K and L. One week it is observed to buy 120 units of L and 25 of K. Another week it is observed to buy 80 units of L and 50 of K. Find out what the intercepts of its budget line on the K and L axes will be and use this information to deduce the prices of K and L, which are assumed to be unchanged from one week to the next.

3. A firm can buy the two inputs K and L at £60 and £40 per unit respectively, and has a budget of £480. Explain why it would not be able to purchase 6 units of K plus 4 units of L and then calculate what price reduction in L would make this input combination a feasible purchase.

4. If a firm buys the two inputs X and Y, what would the slope of its budget constraint be if the price of Y was £10 and

   (a) the price of X was £100   (b) the price of X was £10
   (c) the price of X was £1   (d) the price of X was 25p
   (e) X was free?

5. If a consumer's income doubles and the prices of the two goods that she spends her entire income on also double, what happens to her budget constraint?

6. An hourly paid worker can chose the number of hours per day worked, up to a maximum of 12, and gets paid £10 an hour. Leisure hours are assumed to be any hours not worked out of this 12. On a graph with leisure hours on the horizontal axis and total pay on the vertical axis draw in the budget constraint showing the feasible combinations of leisure and pay that this worker might choose from. Show that the slope of this budget constraint equals − 1 multiplied by the hourly wage rate.

89

7. A firm has a limited budget to spend on inputs K and L. Make up your own values for the budget and the prices of K and L and then say what the slope of the budget constraint and its intersection points on the *K* and *L* axes will be.

4.7 NON-LINEAR FUNCTIONS

If the function $y = f(x)$ has a term with x to the power of anything other than 1, then it is called a 'power function' and will be non-linear. (Remember that $x^1 = x$.) For example,

$$y = x^2 \text{ is a non-linear function}$$

$$y = 6 + x^{0.5} \text{ is a non-linear function}$$

but

$$y = 5 + 0.2x \text{ is linear}$$

Non-linear power functions can take a variety of shapes and here we shall only consider a few possibilities that will be useful at a later stage when looking at functions of economic variables.

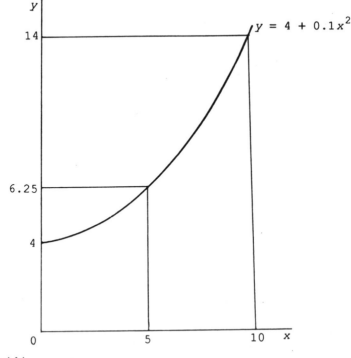

Figure 4.14

Graphs and functions

If the function $y = f(x)$ has one term in x with x to the power of something greater than 1, then it will rise at an increasing rate. This is obvious from Table 4.2. The graphs of the functions $y = x^2$ and $y = x^3$ will curve upwards since y increases at a faster rate than x. The table shows that the greater the power of x then the more quickly y rises.

Table 4.2

x	0	1	2	3	4	5	6
$y = x^2$	0	1	4	9	16	25	36
$y = x^3$	0	1	8	27	64	125	216

These functions all go through the origin, as y is zero when x is zero. Although the intercept may vary if there is a constant term in a function, and the rate of change of y may be modified if the term in x has a coefficient other than 1, the general shape of an upward-sloping curve will still be retained. For example, Figure 4.14 illustrates the function

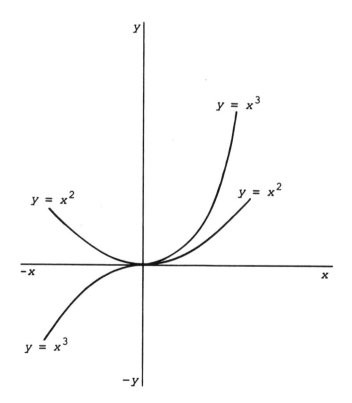

Figure 4.15

91

$$y = 4 + 0.1x^2$$

In economics the quantities one is working with are frequently constrained to positive values, e.g. price and quantity. However, if variables are allowed to take negative values then the functions $y = x^2$ and $y = x^3$ will take the shapes shown in Figure 4.15. Note that, when $x < 0$, $x^2 > 0$ but $x^3 < 0$.

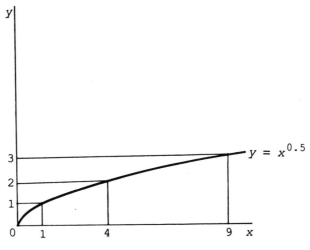

Figure 4.16

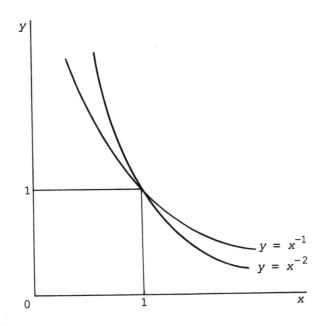

Figure 4.17

If the power of x in a function lies between 0 and 1, then the value of the function increases as x gets larger, but its rate of increase gets smaller and smaller. The values below and Figure 4.16 illustrate this for the function $y = x^{0.5}$ where $x^{0.5} > 0$.

x	0	1	2	3	4	5	6	7	8	9
$y = x^{0.5}$	0	1	1.414	1.732	2	2.236	2.449	2.646	2.828	3

If the power of x in a function is negative then the graph of the function will slope downwards and take the shape of a curve convex to the origin. The examples in Table 4.3 are illustrated in Figure 4.17 for positive values of x. Note that when x is less than 1 the value of y in these functions gets larger as x approaches zero. A firm's average fixed cost (AFC) schedule typically takes this shape.

Table 4.3

x	0	0.1	1	2	3	4	5
$y = x^{-1}$	∞	10	1	0.5	0.33	0.25	0.2
$y = x^{-2}$	∞	100	1	0.25	0.11	0.0625	0.04

Example 4.12

A firm has to pay a fixed annual cost of £90,000 for leasing its premises. Derive its average fixed cost function.

Solution

$$\text{AFC} = \frac{\text{total fixed cost}}{Q} = \frac{90,000}{Q} = 90,000\,Q^{-1}$$

Although all values of y will be multiplied by 90,000, this will not alter the general shape of the function which will be similar to the graph of $y = x^{-1}$ illustrated in Figure 4.17 above.

QUESTIONS 4.7

1. Sketch the approximate shapes of the following functions for positive values of x and y.

    (a) $y = -8 + 0.2x^3$ (b) $y = 250 - 0.01x^2$
    (c) $y = x^{-1.5}$ (d) $y = x^{-0.5}$
    (e) $y = 20 - 0.2x^{-1}$

2. Sketch the approximate shapes of the following functions when x and y are allowed to take both positive and negative values.

(a) $y = 4 + 0.1x^2$ (b) $y = 0.01x^3$
(c) $y = 10 - x^{-1}$

3. A firm faces the non-linear demand schedule

$$p = 570 - 0.4q^2$$

Will this get flatter or steeper as q rises?
4. A firm has to pay fixed costs of £65,000 before it starts production. What will its average fixed cost function look like? What will AFC be when output is 250?
5. Make up your own example of a non-linear function and sketch its approximate shape.

4.8 COMPOSITE FUNCTIONS

When a function has more than one term then one can build up the shape of the overall function from its different components. We have already done this when showing how a constant term determines the starting point of a function on the vertical axis of a graph. Now some more complex functions are considered.

Example 4.13

A firm faces the average fixed cost function

$$AFC = 200x^{-1}$$

where x is output, and the average variable cost (AVC) function

$$AVC = 0.2x^2$$

What shape will its average total cost (AC) schedule take?

Solution

The graphs of AFC and AVC are illustrated in Figure 4.18. By definition,

$$AC = AFC + AVC$$

Therefore

$$AC = 200x^{-1} + 0.2x^2$$

For any given value of x, this means that the position of the AC schedule can be found by vertically summing the corresponding values on the AFC and AVC schedules.

As x gets larger then the value of AFC gets closer and closer to zero. Therefore, the AC schedule gets closer and closer to the AVC schedule as

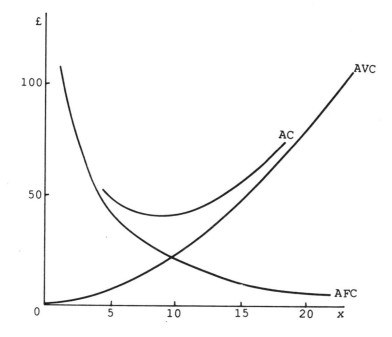

Figure 4.18

the value of AFC diminishes and the AC schedule will take the U-shape shown.

A composite function that takes the form

$$y = a_0 + a_1 x + a_2 x^2 + \ldots a_n x^n$$

where $a_0, a_1, \ldots, a_n$ are constants and n is a non-negative integer is what is called a 'polynomial'. The 'degree' of the polynomial is the highest power value of x. For example, the total cost (TC) function

$$TC = 4 + 6x - 0.2x^2 + 0.1x^3$$

is a polynomial of the third degree. (See if you can sketch the shape of this function. In Chapter 8 the conditions necessary for a TC function to increase continuously will be explained.) Note, however, that not all polynomials are composite functions. If $n = 0$ then the horizontal line $y = a_0$ would still be a special case of the family of polynomial functions.

To see how the graph of a composite function is constructed when one term is subtracted, an example of a total revenue function is worked through below.

Example 4.14

If a demand schedule is represented by the function $P = 80 - 0.2Q$, what shape will the corresponding total revenue function take?

Solution

Total revenue (TR) is simply the total amount of money raised by selling a good and so

$$TR = PQ$$

If we substitute the demand function $P = 80 - 0.2Q$ for P in the TR function then

$$TR = (80 - 0.2Q)Q = 80Q - 0.2Q^2$$

Now that we have derived the function for TR, its shape can be built up as shown in Figure 4.19. The component $80Q$ is clearly a straight line from the origin. The component $0.2Q^2$ is a curve rising at an increasing rate. One can easily see that, for low values of Q, $80Q > 0.2Q^2$. However, as Q becomes larger, the value of Q^2, and hence $0.2Q^2$, rapidly increases and eventually exceeds $80Q$.

Given that TR is the difference between $80Q$ and $0.2Q^2$, its value is the vertical distance between these two functions. This gets larger as Q increases to 200 and then decreases. It is zero when Q is 400 and then becomes negative. Thus we get the inverted U-shape shown.

We can check that this shape makes sense by referring to the demand schedule $P = 80 - 0.2Q$ illustrated at the top of Figure 4.19. When Q is zero, nothing is sold and so TR must be zero. To sell 400, price must fall to zero and so again TR will be zero. Between these two output levels, TR will rise and then fall.

We have seen how the slopes of non-linear composite functions can change along their length, but how can the slope of non-linear functions be measured from a graph? In Chapter 8 a mathematical method for finding the precise value of the slope of a function at any point using calculus is explained. Here we shall just consider an approximate geometrical method assuming that the graph of the function has already been drawn.

Example 4.15

The graph of the composite function

$$y = 40x - 2x^2$$

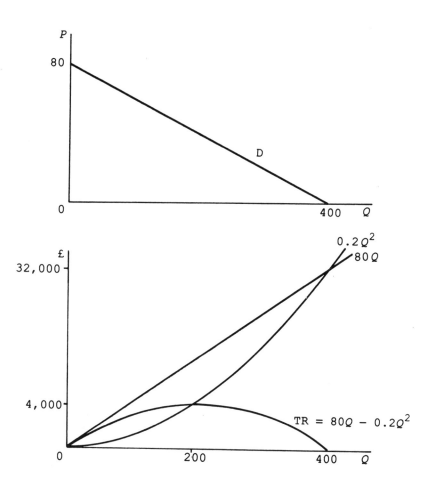

Figure 4.19

is illustrated in Figure 4.20. Find its slope at point A where

$$x = 5$$

and

$$y = 40(5) - 2(5)^2 = 200 - 50 = 150$$

Solution

Draw a straight line that just touches the curve at point A. This line is known as a 'tangent' and is shown by TT' in Figure 4.20. The slope of the line is the same as the slope of the function at A.

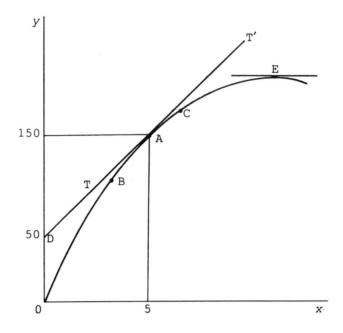

Figure 4.20

To understand why this is so, first consider point B which is slightly to the left of A. The function is steeper than at A and also has a greater slope than the tangent at A. On the other hand, at point C (slightly to the right of A) the function is flatter than at A and has a slope less than that of the tangent TT′. If the slope of the tangent TT′ is less than the slope of the function for points slightly to the left of A and greater than the slope of the function for points slightly to the right of A, then it will be equal to the slope of the function at point A itself.

To determine the actual value of the slope of the tangent TT′ and hence the value of the slope of the function at A, it is necessary to find two sets of coordinates for the line TT′. We already know that it goes through A where $x = 5$ and $y = 150$. If TT′ is extended leftwards, it cuts the y axis at D where $y = 50$ and x is obviously zero.

Using the method explained in Section 4.5 above, the equation $y = 50 + 20x$ can then be fitted to TT′. The slope of this linear function is 20 and so the slope of the function $y = 40 - 2x^2$ will also be 20 at point A.

The slope of the function at other points can be determined in the same way by drawing other tangents. For example, the slope at the highest point of the function E will be the same as the slope of the tangent at E. This tangent is a horizontal line. A horizontal line always has a slope of zero and so at its maximum point the slope of this function will also be zero.

Sometimes you may encounter functions with similar terms that are summed together. The summed function can then usually be simplified so that it does not remain a composite function.

Example 4.16

A firm's product requires two processes. Process A involves a fixed cost of £650 plus £15 for each unit produced and process B involves a fixed cost of £220 plus £45 for each unit. What is the composite total cost function?

Solution

For process A

$$TC_A = 650 + 15Q$$

For process B

$$TC_B = 220 + 45Q$$

The overall total cost is

$$TC = TC_A + TC_B$$
$$= (650 + 15Q) + (220 + 45Q)$$

The summed function is thus

$$TC = 870 + 60Q$$

If the graphs of functions are to be summed 'horizontally' then special care has to be taken, as explained in Section 4.10.

QUESTIONS 4.8

1. Sketch the approximate shape of the following composite functions for positive values of all independent variables.

   (a) $TR = 85q - 2.5q^2$
   (b) $TC = 12 + 9q - 3.6q^2 + 0.8q^3$
   (c) $\pi = -12 + 76q + 1.1q^2 - 0.8q^3$
   (d) $y = 6x - 20x^{-1}$
   (e) $y = 4x^3 - 125x^2$
   (f) $y = 12 + 2x^{-2}$

2. Make up your own example of a composite function and sketch its approximate shape.
3. A firm is able to sell all its output at a fixed price of £20 per unit. If its average cost of production is given by the function

$$AC = 400x^{-1} + 0.2x^2$$

where x is output, derive a function for profit (π) in terms of x. What approximate shape will this profit function take?

4. A small group of companies operate in an industry where all firms face the average cost function $AC = 40 + 1{,}250q^{-1}$ where q is output per week. This function refers only to production costs. They then decide to launch an advertising campaign, not just to try to increase sales but also to try to raise the total average cost of low output levels and deter potential smaller scale rival firms from competing in the same market. The cost of the advertising campaign is £2,000 per week per firm and any competitor would have to spend the same sum on advertising if it wished to compete in this market.

   (a) Derive a function for the new total average cost function including advertising, and sketch its approximate shape.
   (b) Explain why this advertising campaign will deter competition if the original companies sell a 100 units a week at a price of £100 each and new competitors cannot produce more than 25 units a week.

4.9 FUNCTIONS WITH TWO INDEPENDENT VARIABLES

The point has already been made that on a two-dimensional sheet of paper you cannot sketch a function with more than one independent variable as this would require more than two axes (one for the dependent variable and one each for the independent variables). However, in economics we often need to analyse functions with two or more independent variables, e.g. production functions. When there are more than two independent variables then a function cannot really be visually represented and one can only use algebra to analyse it, but when there are only two independent variables a 'contour line' graphing method can be used.

Consider the production function

$$Q = f(K, L)$$

Assume that the way in which Q depends on K and L is represented by the height above the two-dimensional surface on which K and L are measured. To show this production 'height' economics borrows the idea of contour lines from geography. On a map, contour lines join points of equal height. In production theory a line that joins combinations of inputs K and L that will give the same production level (when used efficiently) is known as an 'isoquant'. An 'isoquant map' is shown in Figure 4.21. It is usually drawn with isoquants showing equal increments in output level. This enables one to get an idea of how quickly output responds to changes in the inputs. If isoquants are spaced far apart then output increases relatively slowly, and if they are spaced closely together then output increases relatively quickly.

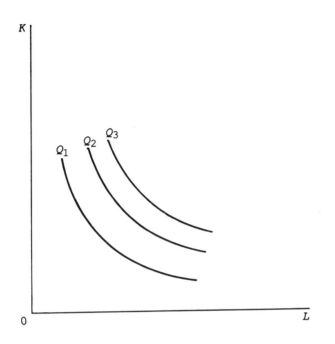

Figure 4.21

One can plot the position of an isoquant map from a production function although this is a rather tedious long-winded business. As we shall see later, it is not usually necessary to draw in all the isoquants in order to tackle some of the resource allocation problems that this concept can be used to illustrate. Examples of some of the different combinations of K and L that would produce an output of 320 with the production function

$$Q = 20K^{0.5}L^{0.5}$$

are shown in Table 4.4. In this particular case one would get a symmetrical curve known as a 'rectangular hyperbola' for the isoquant $Q = 320$.

Table 4.4

K	L	$K^{0.5}$	$L^{0.5}$	Q
64	4	8	2	320
16	16	4	4	320
4	64	2	8	320
256	1	16	1	320
1	256	1	16	320

A quicker way of finding out the shape of an isoquant is to transform it into a function with only two variables.

Example 4.17

For the production function $Q = 20K^{0.5}L^{0.5}$ derive a two-variable function in the form $K = f(L)$ for the isoquant $Q = 100$.

Solution

$$20K^{0.5}L^{0.5} = Q = 100$$

Thus

$$K^{0.5}L^{0.5} = 5$$

$$K^{0.5} = \frac{5}{L^{0.5}}$$

Squaring both sides,

$$K = \frac{25}{L} = 25L^{-1}$$

From Section 4.7 above we know that this form of function will give a curve convex to the origin since K gets closer to zero as L increases in value.

Example 4.18

For the production function $Q = 4.5K^{0.4}L^{0.7}$ derive a function in the form $K = f(L)$ for the isoquant representing an output of 54.

Solution

$$Q = 54 = 4.5K^{0.4}L^{0.7}$$
$$12 = K^{0.4}L^{0.7}$$
$$12L^{-0.7} = K^{0.4}$$

Taking both sides to the power 2.5

$$12^{2.5}L^{-1.75} = K$$
$$K = 498.83L^{-1.75}$$

The production functions given in this section are examples of what are known as 'Cobb–Douglas' production functions. The general format of a Cobb–Douglas production function with two inputs K and L is

$$Q = AK^{\alpha}L^{\beta}$$

where A, α and β are parameters. Many years ago, the two economists Cobb and Douglas found this form of function to be a good match to the statistical

Graphs and functions

evidence on input and output levels that they studied. Although economists have since developed more sophisticated forms of production functions, this basic Cobb–Douglas production function is a good starting point for students to examine the relationship between a firm's output level and the inputs required, and hence costs. Here we shall examine how the parameters of a Cobb–Douglas production relate to the degree of returns to scale present.

Cobb–Douglas production functions fall into the mathematical category of *homogeneous* functions. In general terms, a function is said to be homogeneous of degree n if, when all inputs are multiplied by any given positive constant λ, the value of y increases by the proportion λ^n. Thus if

$$y = f(x_1, x_2, \ldots, x_n)$$

then

$$y\lambda^n = f(\lambda x_1, \lambda x_2, \ldots, \lambda x_n)$$

(Note: λ is the Greek letter lambda.)

An example of a function that is homogeneous of degree 1 is the production function $Q = 20K^{0.5}L^{0.5}$.

Assume that initially the input levels are K_1 and L_1, giving production level $Q_1 = 20K_1^{0.5}L_1^{0.5}$. If input levels are doubled (i.e. $\lambda = 2$) then the new input levels are $K_2 = 2K_1$ and $L_2 = 2L_1$, giving the new output level

$$Q_2 = 20K_2^{0.5}L_2^{0.5} \tag{1}$$

This can be compared with the original production by substituting $2K_1$ for K_2 and $2L_1$ for L_2. Thus

$$Q_2 = 20(2K_1)^{0.5}(2L_1)^{0.5} = 20(2^{0.5}K_1^{0.5}2^{0.5}L_1^{0.5}) = 2(20K_1^{0.5}L_1^{0.5}) = 2Q_1$$

Thus, when inputs are doubled, output doubles, and so this production function exhibits *constant returns to scale*.

The degree of homogeneity of a Cobb–Douglas production function can easily be determined by adding up the indices of the input variables. This can be demonstrated for the two-input function $Q = AK^\alpha L^\beta$.

If we let initial input levels be K_1 and L_1, then

$$Q_1 = AK_1^\alpha L_1^\beta$$

If all inputs are multiplied by the constant λ then new input levels will be $K_2 = \lambda K_1$ and $L_2 = \lambda L_1$. The new output level will then be

$$Q_2 = AK_2^\alpha L_2^\beta = A(\lambda K_1)^\alpha(\lambda L_1)^\beta = \lambda^{\alpha+\beta}AK_1^\alpha L_1^\beta = \lambda^{\alpha+\beta}Q_1$$

Given that λ, α and β are all assumed to be positive numbers there are three possible categories of returns to scale.

103

1 If $\alpha + \beta = 1$ then $\lambda^{\alpha+\beta} = \lambda$ and so $Q_2 = \lambda Q_1$, i.e. constant returns to scale.
2 If $\alpha + \beta > 1$ then $\lambda^{\alpha+\beta} > \lambda$ and so $Q_2 > \lambda Q_1$, i.e. increasing returns to scale.
3 If $\alpha + \beta < 1$ then $\lambda^{\alpha+\beta} < \lambda$ and so $Q_2 < \lambda Q_1$, i.e. decreasing returns to scale.

Example 4.19

What type of returns to scale does the production function $Q = 45K^{0.4}L^{0.4}$ exhibit?

Solution

Indices sum to $0.4 + 0.4 = 0.8$. Thus the degree of homogeneity is less than 1 and so there are decreasing returns to scale.

In your economics course you should learn how the optimum input combination for a firm can be discovered using budget constraints, production functions and isoquant maps. We shall return to these concepts in Chapters 8 and 11, when mathematical solutions to optimization problems using calculus are explained.

QUESTIONS 4.9

For the production functions below

(a) derive a function for the isoquant representing the specified output level in the form $K = f(L)$,
(b) find the level of K required to achieve the given output level if $L = 100$, and
(c) say what type of returns to scale are present.

1. $Q = 9K^{0.5}L^{0.5}$, $Q = 36$
2. $Q = 0.3K^{0.4}L^{0.6}$, $Q = 24$
3. $Q = 25K^{0.6}L^{0.6}$, $Q = 800$
4. $Q = 42K^{0.6}L^{0.75}$, $Q = 5,250$
5. $Q = 0.4K^{0.3}L^{0.5}$, $Q = 65$
6. $Q = 2.83K^{0.35}L^{0.62}$, $Q = 52$

(Assume fractions of a unit of K can be used.)

4.10 SUMMING FUNCTIONS HORIZONTALLY

In economics, there are several occasions when theory requires one to sum certain functions 'horizontally'. Students are most likely to encounter this concept when studying the theory of third-degree price discrimination and the theory of multiplant monopoly and/or cartels. By 'horizontally' summing a function we mean summing it along the horizontal axis. This idea is best explained with an example.

Example 4.20

A price-discriminating monopolist sells in two separate markets at prices P_1 and P_2, measured in pounds. The relevant demand and marginal revenue schedules are (for positive values of Q)

$$P_1 = 12 - 0.15Q_1 \qquad P_2 = 9 - 0.075Q_2$$
$$MR_1 = 12 - 0.3Q_1 \qquad MR_2 = 9 - 0.15Q_2$$

It is assumed that output is allocated between the two markets according to the revenue-maximizing criterion that $MR_1 = MR_2$. Derive a formula for the aggregate marginal revenue schedule which is the horizontal sum of MR_1 and MR_2.

(Note: In Chapter 5, we shall return to this example to find out how this summed MR schedule can help determine P_1 and P_2 when marginal cost is known.)

Solution

The two schedules MR_1 and MR_2 are illustrated in Figure 4.22. What we are required to do is find a formula for the schedule MR. This tells us what aggregate output will correspond to a given level of marginal revenue and vice versa, assuming that output is adjusted so that the marginal revenue from the last unit sold in each market is the same.

As you can see in Figure 4.22, the summed MR schedule is in fact kinked at point K. This is because the MR schedule sums the horizontal distances of MR_1 and MR_2 from the price axis. Given that MR_2 starts from a price of £9, then above £9 the only distance being summed is the distance between MR_1 and the price axis. Thus between £12 and £9 MR is the same as MR_1, i.e.

$$MR = 12 - 0.3Q$$

where Q is aggregate output. If MR = £9 then

$$9 = 12 - 0.3Q$$
$$0.3Q = 3$$
$$Q = 10$$

Thus the coordinates of the kink K are £9 and 10 units of output.

The proper summation occurs below £9. We are given the schedules

$$MR_1 = 12 - 0.3Q_1$$

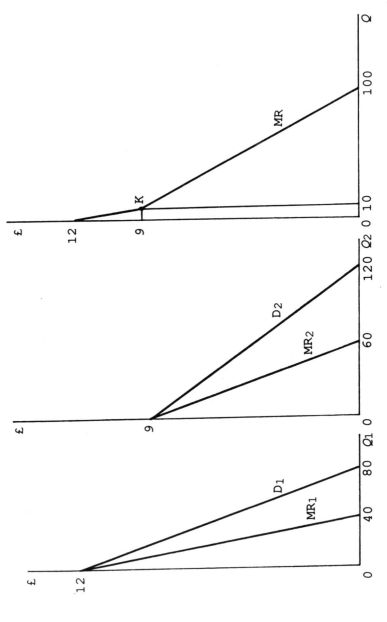

Figure 4.22

and

$$MR_2 = 9 - 0.15Q_2$$

but if we simply added MR_1 and MR_2 we would be summing vertically instead of horizontally. To be summed horizontally, these marginal revenue functions first have to be transposed to obtain their inverse functions as follows:

$$MR_1 = 12 - 0.3Q_1 \qquad MR_2 = 9 - 0.15Q_2$$

$$0.3Q_1 = 12 - MR_1 \qquad 0.15Q_2 = 9 - MR_2$$

$$Q_1 = 40 - 3\tfrac{1}{3}MR_1 \tag{1}$$

$$Q_2 = 60 - 6\tfrac{2}{3}MR_2 \tag{2}$$

Given that the theory of price discrimination assumes that a firm will adjust the amount sold in each market until $MR_1 = MR_2 = MR$, then

$$
\begin{aligned}
Q &= Q_1 + Q_2 \\
&= (40 - 3\tfrac{1}{3}MR) + (60 - 6\tfrac{2}{3}MR) \qquad \text{substituting (1) and (2)} \\
&= 100 - 10MR
\end{aligned}
$$

$$10MR = 100 - Q$$

$$MR = 10 - 0.1Q$$

This function will apply above an aggregate output of 10.

From the above example it can be seen that the basic procedure for summing functions horizontally is as follows:

1 transform the functions so that quantity is the dependent variable;
2 sum the functions representing quantities;
3 transform the function back so that quantity is the independent variable again;
4 note the quantity range that this summed function applies to, given the intersection points of the functions to be summed on the price axis.

This procedure can also be applied to multiplant monopoly examples where it is necessary to find the horizontally summed marginal cost schedule.

Example 4.21

A monopoly operates two plants whose marginal cost schedules are

$$MC_1 = 2 + 0.2Q_1$$

$$MC_2 = 6 + 0.04Q_2$$

Find the function which describes the horizontal summation of these two functions.

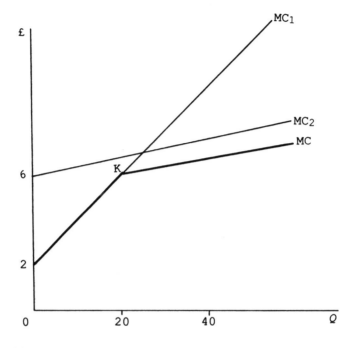

Figure 4.23

(As with the previous example, we shall return to the use of the summed function in determining profit-maximizing price and output levels in Chapter 5.)

Solution

The relevant schedules are illustrated in Figure 4.23. The horizontal sum of MC_1 and MC_2 will be the function MC which is kinked at K. Below £6 only MC_1 is relevant. Therefore, MC is the same as MC_1 from £2 to £6. The corresponding output range can be found by substituting £6 for MC_1. Thus

$$MC_1 = 6 = 2 + 0.2Q_1$$
$$4 = 0.2Q_1$$
$$20 = Q_1$$

Thus $MC = 2 + 0.2Q$ between $Q = 0$ and $Q = 20$.

Above this output we need to derive the proper sum of the two functions. Given

$$MC_1 = 2 + 0.2Q_1 \qquad MC_2 = 6 + 0.04Q_2$$

then

108

$$MC_1 - 2 = 0.2Q_1 \qquad MC_2 - 6 = 0.04Q_2$$

$$5MC_1 - 10 = Q_1 \tag{1}$$

$$25MC_2 - 150 = Q_2 \tag{2}$$

Summing the functions (1) and (2) gives

$$Q = Q_1 + Q_2 = (5MC_1 - 10) + (25MC_2 - 150) \tag{3}$$

A profit-maximizing monopoly will adjust output between two plants until $MC_1 = MC_2 = MC$. Substituting MC into (3) gives

$$Q = 5MC - 10 + 25MC - 150$$

$$Q = 30MC - 160$$

$$160 + Q = 30MC$$

$$5\tfrac{1}{3} + \tfrac{1}{30}Q = MC$$

This summed MC function applies above an output level of 20.

In the examples above the summation of only two linear functions was considered. The method can easily be adapted to situations when three or more linear functions are to be summed. However, the inverses of some non-linear functions are not in forms that can easily be summed and so this method is best confined to applications involving linear functions.

QUESTIONS 4.10

Sum the following sets of marginal revenue and marginal cost schedules horizontally to derive functions in the form $MR = f(Q)$ or $MC = f(Q)$ and define the output ranges over which the summed function applies.

1. $MR_1 = 30 - 0.01Q_1$
   $MR_2 = 40 - 0.02Q_2$
2. $MR_1 = 80 - 0.4Q_1$
   $MR_2 = 71 - 0.5Q_2$
3. $MR_1 = 48.75 - 0.125Q_1$
   $MR_2 = 75 - 0.3Q_2$
   $MR_3 = 120 - 0.15Q_3$
4. $MC_1 = 20 + 0.25Q_1$
   $MC_2 = 34 + 0.1Q_2$
5. $MC_1 = 60 + 0.2Q_1$
   $MC_2 = 48 + 0.4Q_2$
6. $MC_1 = 3 + 0.02Q_1$
   $MC_2 = 1.75 + 0.25Q_2$
   $MC_3 = 4 + 0.2Q_3$

5

Linear equations

5.1 SIMULTANEOUS LINEAR EQUATION SYSTEMS

The way to solve single linear equations with one unknown was explained in Chapter 3. We now turn to sets of equations with more than one unknown.

A simultaneous linear equation system exists when

1 there is more than one functional relationship between a set of specified variables, and
2 all the functional relationships are in linear form.

The solution to a set of simultaneous equations involves finding values for all the unknown variables. Where only two variables and equations are involved, a simultaneous equation system can be related to familiar graphical solutions. For example, assume that in a competitive market the demand schedule is

$$p = 420 - 0.2q \tag{1}$$

and the supply schedule is

$$p = 60 + 0.4q \tag{2}$$

If this market is in equilibrium then the equilibrium price and quantity will be where the demand and supply schedules intersect. As this will correspond to a point which is on both the demand schedule and the supply schedule then the equilibrium values of p and q will be such that both equations (1) and (2) hold. In other words, when the market is in equilibrium (1) and (2) above form a set of simultaneous linear equations.

Simultaneous linear equations systems often involve more than two unknown variables in which case no graphical illustration of the problem will be possible. It is also possible that a set of simultaneous equations may contain non-linear functions, but these are left until the next chapter.

5.2 SOLVING SIMULTANEOUS LINEAR EQUATIONS

The basic idea involved in all the different methods of algebraically solving simultaneous linear equation systems is to manipulate the equations until there is a single linear equation with one unknown. This can then be solved using the methods explained in Chapter 3. The value of the variable that has been found can be substituted back into the other equations to solve for the other unknown values.

It is important to realize that not all sets of simultaneous linear equations have solutions. The general rule is that the number of unknowns must be equal to the number of equations for there to be a unique solution. However, even if this condition is met, one may still come across systems that cannot be solved, e.g. functions which are geometrically parallel and therefore never intersect (see Example 5.2 below).

We shall first consider four different methods of solving a 2×2 set of simultaneous linear equations, i.e. one in which there are two unknowns and two equations, and then look at how some of these methods can be employed to solve simultaneous linear equation systems with more than two unknowns.

5.3 GRAPHICAL SOLUTION

The graphical solution method can be used when there are only two unknown variables. It will not always give 100% accuracy but it can be useful for checking that algebraic solutions are not widely inaccurate owing to analytical or computational errors.

Example 5.1

Solve for p and q in the set of simultaneous equations given in Section 5.1 above:

$$p = 420 - 0.2q \qquad (1)$$
$$p = 60 + 0.4q \qquad (2)$$

Solution

These two functional relationships are plotted in Figure 5.1. Both hold at the intersection point X which corresponds to the values

$$p = 300 \text{ and } q = 600$$

read off the graph.

A graph can also illustrate why some simultaneous linear equation systems cannot be solved.

Example 5.2

Solve for y and x if

111

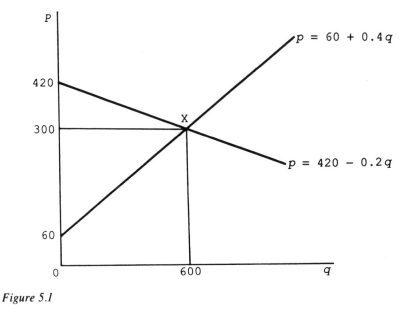

Figure 5.1

$$y = 2 + 2x$$

and

$$y = 5 + 2x$$

Solution

These two functions are plotted in Figure 5.2. They are obviously parallel lines which never intersect. This problem therefore does not have a solution.

QUESTIONS 5.1

Solve the following (if a solution exists) using graph paper.

1. In a competitive market, the demand and supply schedules are respectively

$$p = 9 - 0.075q$$

and

$$p = 2 + 0.1q$$

Find the equilibrium values of p and q.

2. Find x and y when

$$x = 80 - 0.8y$$

112

and

$$y = 10 + 0.1x$$

3. Find x and y when

$$y = -2 + 0.5x$$

and

$$x = 2y - 9$$

5.4 EQUATING TO SAME VARIABLE

The method of equating to the same variable involves rearranging both the equations so that the same unknown variable appears by itself on one side of the equality sign. This variable can then be eliminated by setting the other two sides of the equality sign in the two equations equal to each other. One then has an equation in one unknown which can easily be solved.

Example 5.3

Solve the set of simultaneous equations in Example 5.1 above by the equating method.

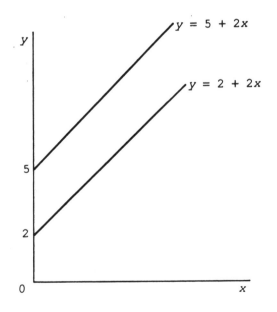

Figure 5.2

Solution

In this example no preliminary rearranging of the equations is necessary because a single term in p appears on the left-hand side of both. As

$$p = 420 - 0.2q \qquad (1)$$

and

$$p = 60 + 0.4q \qquad (2)$$

then it must be true that

$$420 - 0.2q = 60 + 0.4q$$

Therefore

$$360 = 0.6q$$

$$600 = q$$

The value of p can be found by substituting this value for q back into either of the two original equations:

(1) $p = 420 - 0.2q = 420 - 0.2(600) = 420 - 120 = 300$

or (2) $p = 60 + 0.4q = 60 + 0.4(600) = 60 + 240 = 300$.

Example 5.4

Assume that a firm can sell as many units of its product as it can manufacture in a month at £18 each. It has to pay out £240 fixed costs in addition to a marginal cost of £14 for each unit produced. How much does it need to produce to break even?

Solution

According to these assumptions, this firm faces the total revenue function $TR = 18q$ where q is output and the total cost function $TC = 240 + 14q$. These functions are plotted in Figure 5.3, which is an example of what is known as a break-even chart. This is a rough guide to the profit that can be expected for any given production level. (Note that in reality at some point the TR schedule will start to flatten out when the firm has to reduce price to sell more, and TC will get steeper when diminishing marginal productivity causes marginal cost to rise. If this did not happen, then the firm could make infinite profits by indefinitely expanding output.)

The break-even point is clearly at B, where the TR and TC schedules intersect. Since

$$TC = 240 + 14q$$

and

$$TR = 18q$$

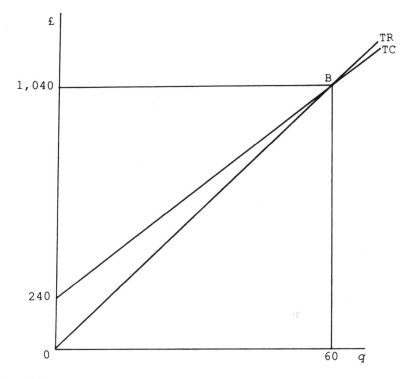

Figure 5.3

and the break-even point is where TR = TC, then

$$18q = 240 + 14q$$
$$4q = 240$$
$$q = 60$$

Therefore the output required to break even is 60 units.

What happens if you try to use this algebraic method when no solution exists, as in Example 5.2 above?

Example 5.5

Solve for y and x if

$$y = 2 + 2x$$

and

$$y = 5 + 2x$$

115

Solution

Eliminating y from the system we get

$$2 + 2x = 5 + 2x$$

Subtracting $2x$ from both sides gives $2 = 5$. This is clearly impossible, and hence no solution can be found.

QUESTIONS 5.2

1. A competitive market has the demand schedule $p = 610 - 3q$ and the supply schedule $p = 20 + 2q$. Calculate equilibrium price and quantity.
2. A competitive market has the demand schedule $p = 610 - 3q$ and the supply schedule $p = 50 + 4q$ where p is measured in pounds.

   (a) Find the equilibrium values of p and q.
   (b) What will happen to these values if the government imposes a tax of £14 per unit on q?

3. Make up your own linear functions for a supply schedule and a demand schedule and then

   (a) plot them on graph paper and read off the values of price and quantity where they intersect, and
   (b) algebraically solve your set of linear simultaneous equations and compare your answer with the values you got for (a).

4. A firm manufactures produce x and can sell any amount at a price of £25 a unit. The firm has to pay fixed costs of £200 plus a marginal cost of £20 for each unit produced.

   (a) How much of x must be produced to make a profit?
   (b) If price is cut to £24 what happens to the break-even output?

5. If $y = 16 + 22x$ and $y = -2.5 + 30.8x$, solve for x and y.

5.5 SUBSTITUTION

The substitution method involves rearranging one equation so that one of the unknown variables appears by itself on one side. The other side of the equation can then be substituted into the second equation, thus eliminating one unknown.

Example 5.6

Solve the linear simultaneous equation system

$$20x + 6y = 500 \tag{1}$$

$$10x - 2y = 200 \tag{2}$$

Linear equations

Solution

Equation (2) can be rearranged to give

$$10x - 200 = 2y$$
$$5x - 100 = y \tag{3}$$

If we substitute (3) into equation (1) we get

$$20x + 6(5x - 100) = 500$$
$$20x + 30x - 600 = 500$$
$$50x = 1,100$$
$$x = 22$$

To find y substitute this value of x into (1) or (2).
Thus, in (1)

$$20(22) + 6y = 500$$
$$440 + 6y = 500$$
$$6y = 60$$
$$y = 10$$

Example 5.7

Find the equilibrium level of national income in the simple macroeconomic model

$$Y = C + I \tag{1}$$
$$C = 0.5Y \tag{2}$$
$$I = 200 \tag{3}$$

Solution

Substituting the functions (2) and (3) into (1) we get

$$Y = 0.5Y + 200$$

Therefore

$$0.5Y = 200$$
$$Y = 400$$

QUESTIONS 5.3

1. A consumer has a budget of £240 and spends it all on the two goods A and B whose prices are initially £5 and £10 per unit respectively. The

price of A then rises to £6 and the price of B falls to £8. What combination of A and B that uses up all the budget is it possible to purchase at both sets of prices?

2. In a simple Keynesian macroeconomic model it is assumed that

$$Y = C + I$$ the accounting identity

$$C = 20 + 0.6Y$$ the consumption function

$$I = 60$$ exogenously determined

Find the equilibrium value of y.

3. Solve for x and y when

$$600 = 3x + 0.5y$$
$$52 = 1.5y - 0.2x$$

5.6 ROW OPERATIONS

Row operations entail multiplying or dividing all the terms in one equation by whatever number is necessary to get the coefficient of one of the unknowns equal to the coefficient of that same unknown in another equation. Then, by subtraction of one equation from the other, this unknown can be eliminated.

Alternatively, if two rows have the same absolute value for the coefficient of an unknown but one coefficient is positive and the other is negative, then this unknown can be eliminated by adding the two rows.

Example 5.8

Given the equations below, solve for x and y.

$$10x + 3y = 250 \qquad (1)$$
$$5x + y = 100 \qquad (2)$$

Solution

Multiplying (2) by 3 $15x + 3y = 300$

Subtracting (1) $10x + 3y = 250$

gives $5x + 0 = 50$

        $x = 10$

Substituting this value of x back into (1),

$$10(10) + 3y = 250$$
$$100 + 3y = 250$$

Linear equations

$$3y = 150$$
$$y = 50$$

Example 5.9

A firm makes two goods A and B which require two inputs K and L as shown below:

1 unit of A requires 6 units of K plus 3 units of L
1 unit of B requires 4 units of K plus 5 units of L

The firm has 420 units of K and 300 units of L at its disposal. How much of A and B should it produce if it wishes to exhaust its supplies of K and L totally?

Solution

The total requirements of K are 6 for every unit of A and 4 for each unit of B, which can be written as

$$K = 6A + 4B$$

Similarly

$$L = 3A + 5B$$

As we know that $K = 420$ and $L = 300$ because all resources are used up, then

$$420 = 6A + 4B \qquad (1)$$

and

$$300 = 3A + 5B \qquad (2)$$

Multiplying (2) by 2 $\qquad 600 = 6A + 10B$

Subtracting (1) $\qquad\quad \underline{420 = 6A + 4B}$

gives $\qquad\qquad\quad\; 180 = 0 + 6B$

$$30 = B$$

Substituting into (1)

$$420 = 6A + 4(30)$$
$$420 = 6A + 120$$
$$300 = 6A$$
$$50 = A$$

Thus the firm should produce 50 units of A and 30 units of B.

(Note that the method of setting up this problem will be used again when we get to linear programming in the Appendix to this chapter.)

QUESTIONS 5.4

1. Solve for x and y if

$$420 = 4x + 5y$$

and

$$600 = 2x + 9y$$

2. A firm produces the two goods A and B using inputs K and L.

   Each unit of A requires 2 units of K plus 6 units of L
   Each unit of B requires 3 units of K plus 4 units of L

   The amounts of K and L available to the firm are 120 and 180 respectively. What output levels of A and B will use up all the available K and L?

3. Solve for x and y when

$$160 = 8x - 2y$$

and

$$295 = 11x + y$$

5.7 MORE THAN TWO UNKNOWNS

With more than two unknowns it is usually best to use the row operations method. The basic idea is to use one pair of equations to eliminate one unknown and then utilize another equation to do the same, repeating the process until a single equation in one unknown is obtained. The exact operations necessary will depend on the format of the particular problem. There are several ways in which row operations can be used to solve most problems and you will only learn which is the quickest method for any specific problem through practice.

Example 5.10

Solve for x, y and z, given that

$$x + 12y + 3z = 120 \tag{1}$$
$$2x + y + 2z = 80 \tag{2}$$
$$4x + 3y + 6z = 219 \tag{3}$$

Solution

Multiplying (2) by 2	$4x + 2y + 4z = 160$	(4)
Subtracting (4) from (3)	$y + 2z = 59$	(5)

120

We have now eliminated x from equations (2) and (3) and so the next step is to eliminate x from equation (1) by row operations with one of the other two equations. In this example the easiest way is

Multiplying (1) by 2	$2x + 24y + 6z = 240$	
Substracting (2)	$2x + \quad y + 2z = \quad 80$	
	$23y + 4z = 160$	(6)

We now have a 2×2 set of simultaneous equations (5) and (6) to solve. Writing these out again, we use row operations to solve for y and z.

$$y + 2z = 59 \tag{5}$$
$$23y + 4z = 160 \tag{6}$$

Multiplying (5) by 2	$2y + 4z = 118$
Subtracting (6)	$23y + 4z = 160$
gives	$-21y \qquad = -42$

Substituting into (5),
$$y = 2$$
$$2 + 2z = 59$$
$$2z = 57$$
$$z = 28.5$$

Substituting these values for y and z into (1):
$$x + 12(2) + 3(28.5) = 120$$
$$x + 24 + 85.5 = 120$$
$$x = 120 - 109.5$$
$$x = 10.5$$

Therefore, solutions are $x = 10.5$, $y = 2$, $z = 28.5$.

Example 5.11

Solve for x, y and z in the following set of simultaneous equations:

$$14.5x + 3y + 45z = 340 \tag{1}$$
$$25x - 6y - 32z = 82 \tag{2}$$
$$9x + 2y - 3z = 16 \tag{3}$$

Solution

Multiplying (1) by 2	$29x + 6y + 90z = 680$	
Adding (2)	$25x - 6y - 32z = \quad 82$	
gives	$54x \qquad + 58z = 762$	(4)

Multiplying (3) by 3	$27x + 6y - 9z = 48$	
Adding (2)	$25x - 6y - 32z = 82$	
gives	$52x \quad -41z = 130$	(5)

Multiplying (5) by 27	$1{,}404x - 1{,}107z = 3{,}510$
Multiplying (4) by 26	$1{,}404x + 1{,}508z = 19{,}812$
Subtracting gives	$-2{,}615z = -16{,}302$
	$z = 6.2340344$

(Note that although final answers are more neatly specified to one or two decimal places, more accuracy will be maintained if the full value of z above is entered when substituting to calculate remaining values of unknown variables.)

Substituting the above value of z into (5) gives

$$52x - 41(6.2340344) = 130$$
$$52x = 130 + 255.59541$$
$$x = 7.4152964$$

Substituting for both x and z in (1) gives

$$14.5(7.4152964) + 3y + 45(6.2340344) = 340$$
$$3y = -48.053346$$
$$y = -16.017782$$

Thus, solutions to 2 decimal places are

$$x = 7.42 \qquad y = -16.02 \qquad z = 6.23$$

The above examples show how the solution to a 3×3 set of simultaneous equations can be solved by row operations. The same method can be used for larger sets but obviously more stages will be required to eliminate the unknown variables one by one until a single equation with one unknown, is arrived at.

It must be stressed that it is only practical to use the methods of solution for linear equation systems explained here where there are a relatively small number of equations and unknowns. For large systems of equations with more than a handful of unknowns it is more appropriate to use other methods based on matrix algebra. Matrix algebra and its application to various economic problems involving large numbers of variables may be studied by those of you who go on to take further courses in mathematical economics after mastering the basic methods covered in this text.

QUESTIONS 5.5

1. Solve for x, y and z when

$$2x + 4y + 2z = 144 \qquad (1)$$
$$4x + y + 0.5z = 120 \qquad (2)$$
$$x + 3y + 4z = 144 \qquad (3)$$

2. Solve for x, y and z when

$$12x + 15y + 5z = 158 \qquad (1)$$
$$4x + 3y + 4z = 58 \qquad (2)$$
$$5x + 20y + 2z = 148 \qquad (3)$$

3. Solve for A, B and C when

$$32A + 14B + 82C = 644 \qquad (1)$$
$$11.5A + 8B + 52C = 349 \qquad (2)$$
$$18A + 26.2B - 62C = 560.4 \qquad (3)$$

4. Find the values of x, y and z when

$$4.5x + 7y + 3z = 158.5$$
$$6x + 18.2y + 12z = 390.8$$
$$3x + 8y + 7z = 209$$

5. Solve for A, B, C and D when

$$4A + 6B + 25C + 17D = 843$$
$$3A + 14B + 60C + 21D = 1,286.5$$
$$10A + 3B + 4C + 28D = 1,206$$
$$6A + 2B + 12C + 51D = 1,096$$

5.8 WHICH METHOD?

There is no hard and fast rule regarding which of the different methods for solving simultaneous equations should be used in different circumstances. The row operations method can be used for most problems but sometimes it will be quicker to use one of the other methods, particularly in 2×2 systems. It may also be quicker to change methods midway. For example, one may find that in a 3×3 problem it may be quicker to revert to the substitution method after one of the unknowns has been eliminated by row operations. Only by practising solving problems will you learn how to spot the quickest methods of solving them.

Not all economic problems are immediately recognizable as linear simultaneous equation systems and one first has to apply economic analysis to set up a problem. Try solving Questions 5.6 below when you have covered the relevant topics in your economics course.

Example 5.12

A firm uses the three inputs K, L and R to manufacture its final product. The prices per unit of these inputs are £20, £4 and £2 respectively. If the other two inputs are held fixed then the marginal product functions are

$$MP_K = 200 - 5K$$
$$MP_L = 60 - 2L$$
$$MP_R = 80 - R$$

What combination of inputs should the firm use to maximize output if it has a fixed budget of £390?

Solution

The basic rule for optimal input determination is that the last £1 spent on each input should add the same amount to output, i.e.

$$\frac{MP_K}{P_K} = \frac{MP_L}{P_L} = \frac{MP_R}{P_R}$$

Therefore

$$\frac{200 - 5K}{20} = \frac{60 - 2L}{4} = \frac{80 - R}{2}$$

Multiplying out two of the three pairwise combinations of equations to get K and R in terms of L we get

$$4(200 - 5K) = 20(60 - 2L) \qquad\qquad 2(60 - 2L) = 4(80 - R)$$
$$800 - 20K = 1{,}200 - 40L \qquad\qquad 120 - 4L = 320 - 4R$$
$$40L - 400 = 20K \qquad\qquad 4R = 4L + 200$$
$$2L - 20 = K \qquad (1) \qquad\qquad R = L + 50 \qquad (2)$$

The third pairwise combination will not add any new information. Instead we use the budget constraint

$$20K + 4L + 2R = 390 \qquad (3)$$

124

Linear equations

Substituting (1) and (2) into (3),

$$20(2L - 20) + 4L + 2(L + 50) = 390$$
$$40L - 400 + 4L + 2L + 100 = 390$$
$$46L = 690$$
$$L = 15$$

Substituting this value for L in (1)

$$K = 2(15) - 20 = 10$$

and in (2)

$$R = 15 + 50 = 65$$

Therefore the optimal input combination is

$$K = 10 \qquad L = 15 \qquad R = 65$$

Example 5.13

In a closed economy where the usual assumptions of the basic Keynesian macroeconomic model apply,

$$C = £60m + 0.7Y_t$$
$$Y = C + I + G$$
$$Y_t = 0.6Y$$

where C is consumption, Y is national income, Y_t is disposable income, I is investment and G is government expenditure. If the values of I and G are exogenously determined as £90 million and £140 million respectively, what is the equilibrium level of national income?

Solution

Once the given values of I and G are substituted, we have a 3×3 set of simultaneous equations with three unknowns:

$$C = 60 + 0.7Y_t \tag{1}$$
$$Y = C + 230 \tag{2}$$
$$Y_t = 0.6Y \tag{3}$$

This sort of problem is best solved by substitution. Substituting (3) into (1) gives

$$C = 60 + 0.7(0.6Y)$$
$$C = 60 + 0.42Y \tag{4}$$

Substituting (4) into (2) gives

$$Y = (60 + 0.42Y) + 230$$
$$0.58Y = 290$$
$$Y = 500$$

Therefore the equilibrium value of the national income is £500 million.

Example 5.14

In a competitive market where the supply price (in £) is

$$p = 3 + 0.25q$$

and demand price (in £) is

$$p = 15 - 0.75q$$

the government imposes a per-unit tax of £4. How much of a price rise will this tax mean to consumers? What will be the tax revenue raised?

Solution

The original equilibrium price and quantity can be found by equating demand and supply price. Hence

$$15 - 0.75q = 3 + 0.25q$$
$$12 = q$$

Substituting this value of q into the supply schedule

$$p = 3 + 0.25(12) = 3 + 3 = 6$$

When a per-unit tax is imposed, the supply schedule shifts upwards by the amount of the tax. In this case the tax is £4 and so each quantity would be offered for sale by suppliers at the old price plus £4. Thus the new supply schedule becomes

$$p = 3 + 0.25q + 4 = 7 + 0.25q$$

Again equating demand and supply price

$$15 - 0.75q = 7 + 0.25q$$
$$8 = q$$

Substituting this value of q into the demand schedule

$$p = 15 - 0.75(8) = 15 - 6 = 9$$

Linear equations

Therefore, consumers see a price rise of £3 from £6 to £9.

$$\text{tax revenue} = \text{quantity sold} \times \text{tax per unit}$$
$$= 8 \times 4 = £32$$

QUESTIONS 5.6

1. A firm faces the demand schedule

$$p = 400 - 0.25q$$

the marginal revenue schedule

$$MR = 400 - 0.5q$$

and the marginal cost schedule

$$MC = 0.3q$$

What price will maximize profit?

2. A firm buys the three inputs K, L and R at prices per unit of £10, £5 and £3 respectively, out of a fixed budget of £273.40. The marginal product functions of these three inputs are

$$MP_K = 150 - 4K$$
$$MP_L = 72 - 2L$$
$$MP_R = 34 - R$$

What input combination will maximize output given this budget?

3. In a competitive market, the supply and demand schedules are

$$p = 4 + 0.25q$$

and

$$p = 16 - 0.5q$$

What would happen to the price paid by consumers and the quantity sold if

(a) a per-unit tax of £3 was imposed, and
(b) a proportional sales tax of 20% was imposed?

4. In a Keynesian macroeconomic model of an economy with no foreign trade it is assumed that

$$Y = C + I + G$$
$$C = 0.75Y_t$$
$$Y_t = (1 - t)Y$$

127

where the usual notation applies and the following are exogenously fixed: $I = £600$ m, $G = £900$ m, $t = 0.2$ is the tax rate. Find the equilibrium value of Y and say whether or not the government's budget is balanced at this value.

5. In an economy which engages in foreign trade, it is assumed that

$$Y = C + I + G + X - M$$
$$C = 0.9Y_t$$
$$M = 0.15Y_t$$
$$Y_t = (1 - t)Y$$

The usual notation applies and the following values are given:

$I = £200$ m $G = £270$ m $X = £180$ m $t = 0.2$

What is the equilibrium value of Y? What is the balance of payments surplus/deficit at this value?

6. In a factor market for labour, a monopsonistic buyer faces the marginal revenue product schedule

$$MRP_L = 244 - 2L$$

the supply of labour schedule

$$w = 20 + 0.4L$$

and the marginal cost of labour schedule

$$MC_L = 20 + 0.8L$$

How much labour should it employ, and at what wage, if MRP_L must equal MC_L in order to maximize profit?

5.9 PRICE DISCRIMINATION AND MULTIPLANT MONOPOLY

In Section 4.10 we examined how linear functions could be summed 'horizontally'. We shall now use this method to help tackle some problems involving price discrimination and multiplant firm/cartel pricing. It is assumed that you will cover these topics in your economics course where the main principles of the models will be explained. Only methods of calculating prices and output are explained here.

In third-degree price discrimination, firms charge different prices in separate markets. To maximize profits the theory of price discrimination says that firms should

1 split total sales between the different markets so that the marginal revenue from the last unit sold in each market is the same, and

Linear equations

2 decide on the total sales level by finding the output level where the aggregate marginal revenue function (derived by horizontally summing the marginal revenue schedules from each individual market) intersects the firm's marginal cost function.

It is usually assumed that the firm practising price discrimination is a monopoly. In the examples in this section we shall assume that the firm faces linear demand schedules in each of the separate markets and make use of the rule that the marginal revenue schedule corresponding to a linear demand schedule will have the same intercept on the price axis but twice the slope, i.e. if the demand schedule is $p = a + bq$ where a and b are constants, then the marginal revenue schedule is

$$MR = a + 2bq$$

(A proof of this rule in given in Section 8.3.)

The method is best explained with some examples.

Example 5.15

A monopoly can sell in two separate markets at different prices and faces the marginal cost schedule

$$MC = 1.75 + 0.05q$$

The two demand schedules are

$$p_1 = 12 - 0.15q_1$$

and

$$p_2 = 9 - 0.075q_2$$

What price should it charge and how much should it sell in each market to maximize profit? (All prices are measured in pounds.)

Solution

It helps to draw a sketch diagram when tackling this type of problem so that you can relate the different quantities to the economic model. This example is illustrated in Figure 5.4. (Note that the demand schedules in this example are the same as those in Example 4.20 in the last chapter when the marginal revenue summation process was explained in more detail. You can refer back if you do not follow the steps below.)

First, the relevant MR schedules and their inverse functions are derived from the demand schedules. Given

$$p_1 = 12 - 0.15q_1 \qquad p_2 = 9 - 0.075q_2$$

then

129

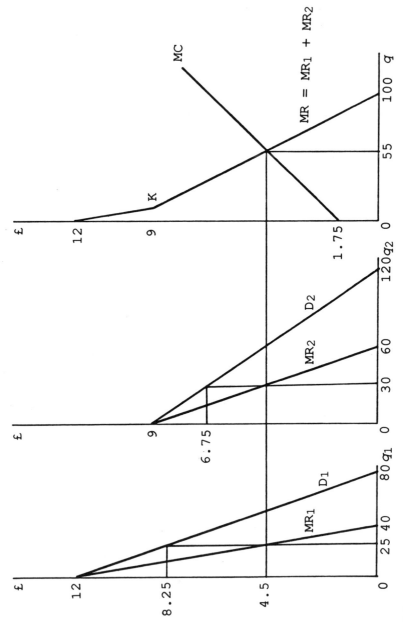

Figure 5.4

Linear equations

$$MR_1 = 12 - 0.3q_1 \qquad MR_2 = 9 - 0.15q_2$$

and so

$$q_1 = 40 - \frac{MR_1}{0.3} \tag{1}$$

$$q_2 = 60 - \frac{MR_2}{0.15} \tag{2}$$

For profit maximization

$$MR_1 = MR_2 = MR \tag{3}$$

and by definition

$$q = q_1 + q_2 \tag{4}$$

Therefore, substituting (1), (2) and (3) into (4)

$$q = \left(40 - \frac{MR}{0.3}\right) + \left(60 - \frac{MR}{0.15}\right)$$

$$= \frac{12 - MR + 18 - 2MR}{0.3}$$

$$= \frac{30 - 3MR}{0.3} = 100 - 10MR$$

and so

$$MR = 10 - 0.1q \tag{5}$$

This function does not apply above £9 as only MR_1 applies above this price. Your sketch diagram will give you an idea of whether or not MC will cut MR in the section below the kink K. Figure 5.4 shows that this is clearly the case.

The aggregate profit-maximizing output is found where

$$MR = MC$$

Thus using (5) above and the MC function given in the question

$$10 - 0.1q = 1.75 + 0.05q$$
$$8.25 = 0.15q$$
$$55 = q$$

Therefore

$$MR = 10 - 0.1(55) = 10 - 5.5 = 4.5$$

and so

$$MR_1 = 4.5 \qquad MR_2 = 4.5$$

131

To determine the prices and output levels in each market we now just substitute these MR values into the inverse marginal revenue functions (1) and (2) derived above. Thus

$$q_1 = 40 - \frac{MR_1}{0.3} = 40 - \frac{4.5}{0.3} = 40 - 15 = 25$$

$$q_2 = 60 - \frac{MR_2}{0.15} = 60 - \frac{4.5}{0.15} = 60 - 30 = 30$$

You can check these output figures to ensure that

$$q_1 + q_2 = q$$

Relating these calculations to Figure 5.4, what we have done is found the intersection point of MR and MC to determine the profit-maximizing levels of q and MR. Then a horizontal line is drawn across to see where this level of marginal revenue cuts MR_1 and MR_2. This enables us to read off q_1 and q_2 and the corresponding prices p_1 and p_2. These prices can be determined by simply substituting the above values of q_1 and q_2 into the demand schedules specified in the question. Thus

$$p_1 = 12 - 0.15q_1 = 12 - 0.15(25) = 12 - 3.75 = £8.25$$
$$p_2 = 9 - 0.075q_2 = 9 - 0.075(30) = 9 - 2.25 = £6.75$$

Finally, refer back to your sketch diagram to ensure that the relative magnitudes of your answer correspond to those read off the graph. In this type of problem it is easy to get mixed up in the various stages of the calculation. From Figure 5.4, we can see that p_1 should be greater than p_2, which checks out with the above answers.

Not all price discrimination models involve the horizontal summation of demand schedules.

In first-degree (perfect) price discrimination each individual unit is sold at a different price. Because the prices of other units do not have to be reduced for a firm to increase sales, the marginal revenue from each unit is the price it sells for. Therefore the marginal revenue schedule is the same as the demand schedule, instead of lying below it.

In second-degree price discrimination a firm breaks the market up into a series of price bands. In a two-part pricing scheme this might mean that the first few units are sold at a previously determined price and then a price is chosen for the remaining units that will maximize profits, given the first price and the marginal cost schedule.

The example below explains how the relevant prices and quantities can be calculated under these different forms of price discrimination.

Example 5.16

A monopoly faces the demand schedule

$$p = 16 - 0.064q$$

and the marginal cost schedule

$$MC = 2.2 + 0.019q$$

It has already decided that the first 60 units will be sold at a price of £12.16. Given this constraint, what price for the remaining units will maximize profits? How will total output compare with output when

(i) the firm can only set a single price?
(ii) perfect price discrimination takes place?

Solution

The demand schedule is illustrated in Figure 5.5. We can check that the price of £12.16 for the first 60 units corresponds to point A on the demand schedule since

$$p = 16 - 0.064q = 16 - 0.064(60) = 16 - 3.84 = £12.16$$

If the firm wishes to sell more output it will not have to reduce the price of these first 60 units. It therefore effectively faces the marginal revenue schedule MR'. This is constructed by assuming that the zero on the quantity axis is moved 60 units to the right to point B. MR' is then drawn in the usual

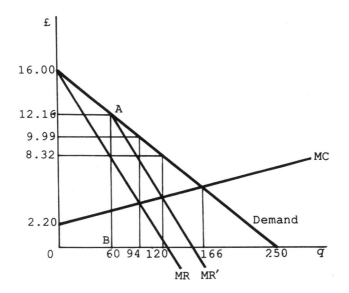

Figure 5.5

way with the same 'intercept' on the price axis (effectively point A) but twice the slope of the demand schedule. The firm should then employ the usual rule for profit maximization, which is to produce the output level at which marginal revenue equals marginal cost, where MR' is the relevant marginal revenue schedule.

To derive a function for MR', define $q' = q - 60$, i.e. q' measures output from point B on the quantity axis. The demand schedule over this output range has the same slope as the original demand schedule (-0.064) but the new 'intercept' value of £12.16. It is therefore described by the function

$$p = 12.16 - 0.064q'$$

and therefore

$$MR' = 12.16 - 0.128q' \qquad (1)$$

using the rule that a marginal revenue function has twice the slope of a linear demand schedule.

Substituting the original definition of output for q' into (1)

$$MR' = 12.16 - 0.128(q - 60)$$
$$= 12.16 - 0.128q + 7.68$$
$$= 19.84 - 0.128q \qquad (2)$$

Profit maximization requires

$$MR' = MC$$

Therefore, equating (2) and the given MC function

$$19.84 - 0.128q = 2.2 + 0.019q$$
$$17.64 = 0.147q$$
$$120 = q$$

This is total output. The amount sold at the second (lower) price will be

$$q' = q - 60 = 120 - 60 = 60$$

The price for these units will be

$$p = 12.16 - 0.064q' = 12.16 - 0.064(60) = 12.16 - 3.84 = £8.32$$

The total output level of 120 units under this second-degree price discrimination policy can be compared with

(a) single-price profit maximization: given the demand schedule

$$p = 16 - 0.064q$$

then

$$MR = 16 - 0.128q$$

Linear equations

Single-price profit maximization occurs when

$$MR = MC$$
$$16 - 0.128q = 2.2 + 0.019q$$
$$13.8 = 0.147q$$
$$93.877 = q \qquad \text{(94 on graph)}$$

This is lower than the output of 120 units produced under a two- part pricing scheme. This is what one would expect given that price discrimination allows a firm to sell extra units of output without reducing the price of all previously sold units and hence shifts the relevant marginal revenue schedule to the right. Price will be higher at £9.99.

(b) Perfect price discrimination: If all units are sold at different prices then marginal revenue is the same as the demand schedule, i.e.

$$MR = 16 - 0.064q$$

and profit-maximizing output is determined where

$$MR = MC$$
$$16 - 0.064q = 2.2 + 0.019q$$
$$13.8 = 0.083q$$
$$166.265 = q \qquad \text{(166 on graph)}$$

This is greater than the two-part pricing discrimination output, which is what is expected. The greater the number of different segments a market can be broken up into the higher will be the profits that can be extracted and the output that can be sold.

Note that in the above example, and in all the others in this section, we cannot be sure that the firm actually makes a profit and should continue to operate in the long run because total costs are not specified. We shall assume that an overall profit is made in each of the examples and questions in this section. Later, in Chapter 12, a method of finding TVC and TR from MC and MR schedules using the mathematical concept of integration will be explained.

The theory of multiplant monopoly is analogous to the model of third-degree price discrimination explained above except that it is marginal cost schedules that are summed rather than marginal revenue schedules. The basic principles of the multiplant model are as follows.

1 The firm should adjust production so that the marginal cost of the last unit produced in each plant is equal to the marginal cost of the last unit produced by the other plant(s).

2 Total output is determined where the aggregate marginal cost schedule (derived by horizontally summing the marginal cost schedules in each individual plant) intersects the firm's marginal revenue schedule.

Again, the firm is usually assumed to be a monopoly so that the demand and marginal schedules can be clearly defined. The same model can be used to determine price and output levels for the different (single-plant) firms in a cartel where perfect collusion takes place. This is a less likely scenario, however, as perfect collusion within cartels is beset with many problems, as you will know from your economics course.

Example 5.17

A firm operates two plants whose marginal cost schedules are

$$MC_1 = 2 + 0.2q_1 \qquad MC_2 = 6 + 0.04q_2$$

It is a monopoly seller in a market where the demand schedule is

$$p = 66 - 0.1q$$

where q is aggregate output. How much should the firm produce in each plant, and at what price should total output be sold, if the firm wishes to maximize profits? (All costs and prices are measured in pounds.)

Solution

(You will note that the marginal cost schedules to be summed are the same as those in Example 4.21 which was illustrated in Figure 4.23.)

We need to derive the horizontally summed marginal cost schedule MC, find where it intersects MR, and then see which output levels this marginal cost value corresponds to in each plant. Price is read off the demand schedule at the aggregate output level.

Given the demand schedule

$$p = 66 - 0.1q$$

we know that

$$MR = 66 - 0.2q \qquad (1)$$

To be able to set $MC = MR$ and solve for q we first need to derive the function

$$MC = f(q)$$

First derive the inverse functions of the individual plant marginal cost schedules. Given

$$MC_1 = 2 + 0.2q_1 \qquad MC_2 = 6 + 0.04q_2$$

136

then

$$MC_1 - 2 = 0.2q_1 \qquad MC_2 - 6 = 0.04q_2$$

and so

$$5MC_1 - 10 = q_1 \qquad 25MC_2 - 150 = q_2$$

Given that $q = q_1 + q_2$ by definition and $MC = MC_1 = MC_2$ for profit maximization then

$$q = (5MC - 10) + (25MC - 150) = 30MC - 160$$

$$q + 160 = 30MC$$

$$\frac{q + 160}{30} = MC \qquad\qquad (2)$$

Setting $MC = MR$ we now get

$$\frac{q + 160}{30} = 66 - 0.2q$$

from (1) and (2) and hence

$$q + 160 = 1{,}980 - 6q$$

$$7q = 1{,}820$$

$$q = 260$$

Substituting this aggregate output level into (2) gives

$$MC = \frac{q + 160}{30} = \frac{260 + 160}{30} = \frac{420}{30} = 14$$

Therefore

$$MC_1 = MC_2 = MC = 14$$

and so

$$q_1 = 5(14) - 10 = 70 - 10 = 60$$

and

$$q_2 = 25(14) - 150 = 350 - 150 = 200$$

We can easily check that these output levels for the individual plants correspond to the aggregate output calculated above since

$$q_1 + q_2 = 60 + 200 = 260 = q$$

To find the price at which this aggregate output is sold, simply substitute this value of q into the demand schedule given in the question. Therefore

$$p = 66 - 0.1q = 66 - 0.1(260) = 66 - 26 = £40$$

The basic principles explained above can also be applied to more complex problems. The following example considers a case where there are more than two plants.

Example 5.18

A firm operates four plants whose marginal cost schedules are

$$MC_1 = 20 + q_1 \qquad MC_3 = 40 + q_3$$
$$MC_2 = 40 + 0.5q_2 \qquad MC_4 = 60 + 0.5q_4$$

and it is a monopoly seller in a market where

$$p = 580 - 0.3q$$

How much should it produce in each plant and at what price should its output be sold if it wishes to maximize profit?

Solution

First find the inverses of the marginal cost functions.

$$MC_1 = 20 + q_1 \qquad MC_2 = 40 + 0.5q_2$$
$$q_1 = MC_1 - 20 \qquad q_2 = 2MC_2 - 80$$
$$MC_3 = 40 + q_3 \qquad MC_4 = 60 + 0.5q_4$$
$$q_3 = MC_3 - 40 \qquad q_4 = 2MC_4 - 120$$

Given that

$$q = q_1 + q_2 + q_3 + q_4$$

and

$$MC = MC_1 = MC_2 = MC_3 = MC_4$$

for profit maximization, then

$$q = (MC - 20) + (2MC - 80) + (MC - 40) + (2MC - 120)$$
$$q = 6MC - 260$$
$$\frac{q + 260}{6} = MC \qquad (1)$$

Since $p = 580 - 0.3q$, then

$$MR = 580 - 0.6q \qquad (2)$$

To maximize profits $MC = MR$ and so equating (1) and (2)

$$\frac{q + 260}{6} = 580 - 0.6q$$

$$q + 260 = 3,480 - 3.6q$$
$$4.6q = 3,220$$
$$q = 700$$
$$MC = \frac{q + 260}{6} = \frac{700 + 260}{6} = \frac{960}{6} = 160$$

Substituting this value of MC into the individual inverse marginal cost functions to find plant output levels gives

$$q_1 = MC_1 - 20 = 160 - 20 = 140$$
$$q_2 = 2MC_2 - 80 = 320 - 80 = 240$$
$$q_3 = MC_3 - 40 = 160 - 40 = 120$$
$$q_4 = 2MC_4 - 120 = 320 - 120 = 200$$

These total to 700, which checks out with the answer for q above.

The price to sell at is found by substituting the total output of 700 units into the demand schedule given in the question. Thus

$$p = 580 - 0.3q = 580 - 0.3(700) = 580 - 210 = £370$$

Note that we did not draw a sketch diagram for the above example to check whether or not the MR schedule cuts the aggregated MC schedule at a level where output by all four plants is positive, i.e. where the value of MC is above the intercept on the vertical axis for each individual MC schedule. However, as all four output levels were calculated as positive numbers we know that this must be the case. In this type of question, if the usual mathematical method throws up a negative quantity for output by one or more plants (or a negative sales figure in a price discrimination model), then this means that output in this plant (or plants) should be zero. The question should then be reworked with the marginal cost schedule for any such plants excluded from the aggregated MC schedule.

Finally, we shall work through an example where both price discrimination and multiplant pricing apply.

Example 5.19

A multiplant monopoly operates two plants whose marginal cost schedules are

$$MC_1 = 42.5 + 0.5q_1 \qquad MC_2 = 130 + 2q_2$$

It also sells its product in two separable markets whose demand schedules are

$$p_A = 360 - q_A \qquad p_B = 280 - 0.4q_B$$

(Note that the subscripts A and B are used to distinguish quantities sold in the two markets from the quantities q_1 and q_2 produced in the two plants.)

Calculate how much it should produce in each plant, how much it should sell in each market, and how much it should charge in each market.

Solution

First derive the aggregate MC function by the usual method. Given

$$MC_1 = 42.5 + 0.5q_1 \qquad MC_2 = 130 + 2q_2$$

then

$$q_1 = 2MC_1 - 85 \qquad q_2 = 0.5MC_2 - 65$$

To maximize profits, output is adjusted so that

$$MC_1 = MC_2 = MC$$

Therefore

$$q = q_1 + q_2 = (2MC - 85) + (0.5MC - 65)$$
$$q = 2.5MC - 150$$
$$60 + 0.4q = MC \tag{1}$$

Next, derive the aggregate MR function. Given

$$p_A = 360 - q_A \qquad p_B = 280 - 0.4q_B$$

then

$$MR_A = 360 - 2q_A \qquad MR_B = 280 - 0.8q_B$$
$$q_A = 180 - 0.5MR_A \qquad q_B = 350 - 1.25MR_B$$

To maximize profits, sales are adjusted so that

$$MR_A = MR_B = MR$$

Therefore

$$q = q_A + q_B = (180 - 0.5MR) + (350 - 1.25MR)$$
$$q = 530 - 1.75MR$$
$$MR = \frac{530 - q}{1.75} \tag{2}$$

To maximize profits MC = MR. Therefore, equating (1) and (2)

$$60 + 0.4q = \frac{530 - q}{1.75}$$
$$105 + 0.7q = 530 - q$$
$$1.7q = 425$$
$$q = 250$$

Thus

$$MC = 60 + 0.4q = 60 + 0.4(250) = 60 + 100 = 160$$

and also

$$MR = MC = 160$$

To find production levels in the two plants, substitute this value of MC into the inverse MC functions above. Thus

$$q_1 = 2MC - 85 = 2(160) - 85 = 320 - 85 = 235$$
$$q_2 = 0.5MC - 65 = 0.5(160) - 65 = 80 - 65 = 15$$

To find sales levels in each market, substitute this value of MR into the inverse MR functions above. Thus

$$q_A = 180 - 0.5MR = 180 - 0.5(160) = 180 - 80 = 100$$
$$q_B = 350 - 1.25MR = 350 - 1.25(160) = 350 - 200 = 150$$

A quick check shows that the two production levels and the two sales levels both add to 250, which is what is expected.

The prices charged in the two markets A and B are found by substituting the above values of q_A and q_B into the demand schedules specified in the question. Thus

$$p_A = 360 - q_A = 360 - 100 = £260$$
$$p_B = 280 - 0.4q_B = 280 - 0.4(150) = 280 - 60 = £220$$

QUESTIONS 5.7

1. A price-discriminating monopoly sells in two markets whose demand schedules are

$$p_1 = 16 - 0.1q_1$$

and

$$p_2 = 12 - 0.05q_2$$

and it faces the marginal cost schedule

$$MC = 1.2 + 0.02q$$

where $q = q_1 + q_2$. How much should it sell in each market, and at what prices, in order to maximize profits?

2. A monopoly operates two plants whose marginal cost schedules are

$$MC_1 = 2 + 0.1q_1 \qquad MC_2 = 4 + 0.08q_2$$

and sells in a market where the demand schedule is

$$p = 58 - 0.05q$$

How much should it produce in each plant and at what price should its product be sold?

3. A multiplant monopoly sells in a market where the demand schedule is

$$p = 253.4 - 0.025q$$

and produces in two plants whose marginal cost schedules are

$$MC_1 = 20 + 0.0625q_1 \qquad MC_2 = 50 + 0.1q_2$$

How should it split output between the two plants in order to maximize profit? What price should it sell at?

4. A monopoly can price-discriminate between the two markets $p_1 = 10 - 0.1q_1$ and $p_2 = 6 - 0.04q_2$. If its marginal cost schedule is

$$MC = 1.1 + 0.01q$$

how much should it sell in each market to maximize profit, and at what prices?

5. A price-discriminating monopoly sells in two markets whose demand schedules are

$$p_1 = 12.5 - 0.0625q_1 \qquad p_2 = 7.2 - 0.002q_2$$

and faces the horizontal marginal cost schedule

$$MC = 5$$

What price and output should it choose for each market?

6. A monopoly can operate a two-part pricing scheme in the market where the demand schedule $p = 180 - 0.6q$ holds. It faces the horizontal marginal cost schedule $MC = 42$. If the first 100 units are sold at a price of £120 each, what price should be charged for the remaining units in order to maximize profit?

7. A monopoly operates second-degree price discrimination in a market where $p = 12 - 0.06q$ and has the marginal cost schedule $MC = 3 + 0.04q$. What price should it sell the remaining units for if it has already been decided to sell the first 50 units for a price of £9?

8. A firm operates two plants whose marginal cost schedules are

$$MC_1 = 22.5 + 0.25q_1 \qquad MC_2 = 15 + 0.25q_2$$

It is also a monopoly which can price-discriminate between two markets whose demand schedules are

$$p_A = 600 - 0.125q_A \qquad p_B = 850 - 0.1q_B$$

If it wishes to maximize profits, how much should it produce in each plant, how much should it sell in each market, and what prices should it sell at?

9. A multiplant monopoly produces using two plants with the marginal cost schedules

$$MC_1 = 8 + 0.2q_1 \qquad MC_2 = 10 + 0.05q_2$$

It can also price-discriminate between three markets whose demand schedules are

$$p_A = 150 - 0.1875q_A$$
$$p_B = 120 - 0.15q_B$$
$$p_C = 80 - 0.1q_C$$

In order to maximize profits, how much should it produce in each plant, how much should it sell in each market, and what prices should it sell at?

10. A price-discriminating monopoly sells in two markets whose demand schedules are

$$q_1 = 120 - 6p_1 \qquad q_2 = 110 - 8p_2$$

If its marginal cost function is

$$MC = 2.26 + 0.02q$$

calculate the profit-maximizing price and sales levels for each market.

11. A monopoly has the demand schedule

$$p = 210 - 0.2q$$

and the marginal cost schedule

$$MC = 20 + 0.8q$$

(a) If it can practise first-degree price discrimination how much should it sell?

(b) If it can practise second-degree price discrimination and it has already made the decision to sell the first 100 units at a price of £190, what price should it charge for the rest of the units it sells?

12. A monopoly operates three plants with marginal cost schedules

$$MC_1 = 1 + 0.1q_1 \qquad MC_2 = 3 + 0.05q_2 \qquad MC_3 = 5 + 0.05q_3$$

How much should it make in each plant to maximize profit if its market demand schedule is

$$p = 28 - 0.02q$$

and what price will the total output be sold at?

APPENDIX: Linear Programming

Although basically an extension of the linear algebra covered in the main body of the chapter, the technique of linear programming involves special features which distinguish it from other linear algebra applications. When all the relevant functions are linear, it enables one to calculate the profit-maximizing output mix of a multiproduct firm subject to restrictions on input availability, or to calculate the input mix that will minimize costs subject to minimum quality standards being met. This makes it an extremely useful tool for managerial decision-making.

The techniques used are explained in the following sections. It should be noted, though, that from a pure economic theory viewpoint linear programming cannot make any general predictions about price or output for a large number of firms. Its usefulness lies in the realm of managerial (or business) economics where economic techniques can help an individual firm to make efficient decisions.

Constrained maximization

Some firms produce a variety of different products using basically the same inputs. A typical resource allocation problem that such a firm may come across is how to decide on the product mix which will maximize profits when it has limited amounts of the various inputs required for the different products that it makes. This is what is known as a 'constrained optimization' problem.

In this type of problem the firm's objective is to maximize profit and so profit is what is known as the 'objective function'. The firm is trying to optimize this function subject to the constraint of limited input availability.

When both the objective function and the constraints can be expressed in a linear form then the technique of linear programming can be used to try to find a solution. (Constrained optimization of non-linear functions is explained in Chapter 11.) We shall restrict the analysis here to objective functions which have only two variables, e.g. when only two goods contribute to a firm's profit. This enables us to use graphical analysis to help find a solution, as explained in the example below.

Example 5A.1

A firm manufactures two goods A and B using three inputs K, L and R as shown below:

   A needs 3 units of K, 4 units of L plus 2 units of R per unit
   B needs 5 units of K, 3 units of L and none of R per unit

The firm has at its disposal 150 units of K, 120 units of L and 40 units of R. The net profit contributed by each unit sold is £4 for A and £1 for B. What combination of A and B should the firm manufacture to maximize profits given these constraints on input availability?

144

Solution

From the per-unit profit figures in the question we can see that total profit is

$$\pi = 4A + B$$

where A and B represent the quantities of goods A and B that are produced. This is the linear objective function which the firm wishes to maximize. The total amount of input K required will be 3 for each unit of A plus 5 for each unit of B and we know that only 150 units of K are available. The constraint on this input is thus

$$3A + 5B \leqslant 150 \tag{1}$$

Similarly, for L

$$4A + 3B \leqslant 120 \tag{2}$$

and for R

$$2A \leqslant 40 \tag{3}$$

As the firm cannot produce negative quantities of the two goods, we can also add the two non-negativity constraints on the solutions for the optimum values of A and B, i.e.

$$A \geqslant 0 \tag{4}$$

$$B \geqslant 0 \tag{5}$$

Now turn to the graph in Figure 5A.1 which measures A and B on its axes. The first step in the graphical solution of a linear programming problem is to mark out what is know as the 'feasible area'. This will contain all the values of A and B that satisfy all the above constraints (1)–(5). This is done by eliminating the areas which could not possibly contain the solution.

We can easily see that the non-negativity constraints (4) and (5) mean that the solution must lie on, or above, the A axis and on, or to the right of, the B axis.

To mark out the other constraints we consider in turn what would happen if the firm entirely used up its quota of each of the inputs K, L and R.

If all the available K was used up, the constraint (1) would become the function

$$3A + 5B = 150 \tag{6}$$

with the equality sign replacing the $\leqslant$ sign. This linear constraint can easily be marked out by joining its intercepts on the two axes. When $A = 0$ then $B = 30$ and when $B = 0$ then $A = 50$. Thus the constraint will be the straight line marked (K). This is rather like a budget constraint. If all the available K is used then the firm's production mix will correspond to a point some-

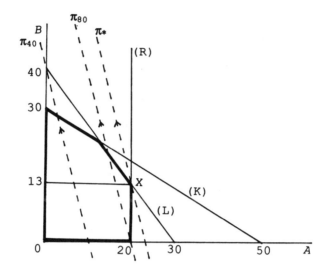

Figure 5A.1

where on the constraint line (K). It is also possible to use less than the total amount available, in which case the firm would produce a combination of *A* and *B* below this constraint. Points above this constraint are not feasible as they correspond to more than 150 units of K.

In a similar fashion we can deduce that all points above the constraint line (L) are not feasible given that when all the available L is used up then

$$4A + 3B = 120 \qquad (7)$$

The constraint on R is shown by the vertical line (R) since when all available R is used

$$2A = 40 \qquad (8)$$

Points to the right of this line will not be feasible.

Having marked out the individual constraints, we can now delineate the area which contains combinations of *A* and *B* which satisfy all five constraints. This is shown by the heavier black lines in Figure 5A.1.

The firm's objective function is $\pi = 4A + B$. We cannot draw in this function without having a value for π. But how can we do this if we do not yet know what the maximum profit is? To overcome this problem, first make up a figure for profit, preferably a number which when divided by the two per-unit profit figures (£4 and £1) will give numbers within the range shown on the graph. If we suppose profit is £40, say, then we can draw in the broken line π_{40} corresponding to the function

$$40 = 4A + B$$

146

If we had chosen a figure for profit of more than £40 then we would have obtained a parallel line but further away from the origin; e.g. the line π_{80} corresponds to the function $80 = 4A + B$. If the firm is seeking to maximize profit then it needs to find the furthest profit line from the origin that passes through the feasible area. All profit lines will have the same slope and so, using π_{40} as a guideline, we can see that the highest feasible profit line is π_*, which just touches the edge of the feasible area at X. The optimum values of A and B can then simply be read off the graph, giving $A = 20$ and $B = 13$ (approximately).

A more accurate answer may be obtained algebraically, once the graph has been used to determine which is the optimum point, since the solution to a linear programming problem will nearly always be at the intersection of two or more constraints. (Exceptionally the objective function may be parallel to a constraint – see Example 5A.3.)

The graph in Figure 5A.1 tells us that the solution to this problem is where the constraints (L) and (R) intersect. Thus we have the simultaneous equations

$$4A + 3B = 120 \tag{7}$$

$$2A = 40 \tag{8}$$

which can easily be solved to find the optimum values of A and B. From (8)

$$A = 20$$

Substituting in (7)

$$4(20) + 3B = 120$$
$$3B = 40$$
$$B = 13.33 \qquad \text{(to 2 dp)}$$

Thus maximum profit is

$$\pi = 4A + B = 4(20) + 13.33 = 80 + 13.33 = £93.33$$

The optimum position was on the constraints for L and R, but it was below the constraint for K. Thus, as the K constraint did not 'bite' there must be some spare capacity or 'slack' for K, which can also be calculated. If the firm produces 20 of A and 13.33 of B, then its usage of K is

$$3A + 5B = 3(20) + 5(13.33) = 60 + 66.67 = 126.67$$

The amount of K available is 150 units; therefore slack is

$$150 - 126.67 = 23.33 \text{ units of K}$$

Now that the different steps involved in solving a linear programming problem have been explained let us work through another problem.

Example 5A.2

A firm produces two goods A and B, which each contribute a net profit of £1 per unit sold, using the two inputs K and L. The input requirements are

3 units of K plus 2 units of L for each unit of A
2 units of K plus 3 units of L for each unit of B

If the firm has 600 units of K and 600 units of L at its disposal, how much of A and B should it produce to maximize profit?

Solution

Using the same method as in the previous example we can see that the constraints are, for input K,

$$3A + 2B \leqslant 600 \qquad\qquad (1)$$

and, for input L,

$$2A + 3B \leqslant 600 \qquad\qquad (2)$$
$$A \geqslant 0 \qquad\qquad \text{(non-negativity)}$$
$$B \geqslant 0 \qquad\qquad \text{(non-negativity)}$$

The feasible area is therefore as marked out by the heavy black lines in Figure 5A.2.

As profit is £1 per unit for both A and B, the objective function is

$$\pi = A + B$$

If we suppose profit is £200, then

$$200 = A + B$$

This function corresponds to the line π_{200} which can be used as a guideline for the slope of the objective function. The line parallel to π_{200} that is furthest away from the origin but still within the feasible area will represent the maximum profit. This is the line π_* through point M. The optimum values of A and B can thus be read off the graph as 120 of each.

Alternatively, once we know that the optimum combination of A and B is at the intersection of the constraints (K) and (L), the values of A and B can be found from the simultaneous equations

$$3A + 2B = 600 \qquad\qquad (1)$$
$$2A + 3B = 600 \qquad\qquad (2)$$

148

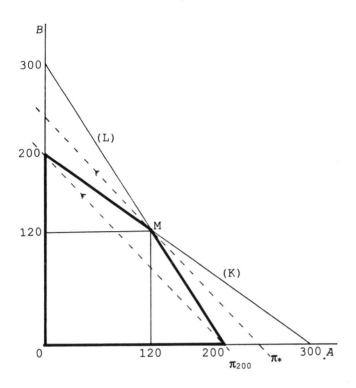

Figure 5A.2

From (1)

$$2B = 600 - 3A$$
$$B = 300 - 1.5A \qquad (3)$$

Substituting (3) into (2)

$$2A + 3(300 - 1.5A) = 600$$
$$2A + 900 - 4.5A = 600$$
$$300 = 2.5A$$
$$120 = A$$

Substituting this value of A into (3)

$$B = 300 - 1.5(120) = 120$$

As both A and B equal 120 then

$$\pi_* = 120 + 120 = £240$$

149

The optimum combination at M is where both constraints (K) and (L) bite. There is therefore no slack for either K or L.

It is possible that the objective function will have the same slope as one of the constraints. In this case there will not be one optimum combination of the inputs as all points along the section of this constraint that forms part of the boundary of the feasible area will correspond to the same value of the objective function.

Example 5A.3

A firm produces two goods x and y which require inputs of raw material (R), labour (L) and manufactured components (K) in the following quantities:

1 unit of x requires 12 kg of R, 10 hours of L and 15 of K
1 unit of y requires 21 kg of R, 10 hours of L and 6 of K

Both x and y add £200 per unit sold to the firm's profits. The firm can use up to 252 kg of R, 150 hours of L and 180 of K. What production mix of x and y will maximize profits?

Solution

The constraints can be written as

$$12x + 21y \leq 252 \qquad \text{(R)}$$
$$10x + 10y \leq 150 \qquad \text{(L)}$$
$$15x + 6y \leq 180 \qquad \text{(K)}$$
$$x \geq 0, \quad y \geq 0$$

These are shown in Figure 5A.3 where the feasible area is marked out by the shape ABCD0. The objective function is

$$\pi = 200x + 200y$$

To find the slope of this objective function, assume profit is £2,000. This could be achieved by producing 10 of x and none of y, or 10 of y and no x, and is therefore shown by the broken line π_{2000}. You will see that this line is parallel to the constraint (L). Therefore if we slide out the objective function π to find the maximum value of profit within the feasible area we can see that it coincides with the boundary of the feasible area all along the stretch BC.

What this means is that both points B and C, and anywhere along the portion of the constraint line (L) between these points, will give the same (maximum) profit figure.

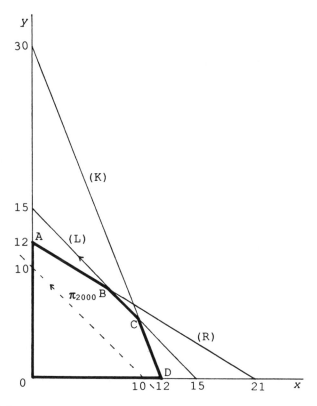

Figure 5A.3

At B the constraints (R) and (L) intersect. Therefore

$$12x + 21y = 252 \qquad (1)$$
$$10x + 10y = 150 \qquad (2)$$

From (2)

$$x = 15 - y \qquad (3)$$

Substituting (3) in (1)

$$12(15 - y) + 21y = 252$$
$$180 - 12y + 21y = 252$$
$$9y = 72$$
$$y = 8$$

Substituting this value of y into (3)

$$x = 15 - 8 = 7$$

Thus profit at B is

$$\pi = 200x + 200y = 200(7) + 200(8) = £1{,}400 + £1{,}600 = £3{,}000$$

At C the constraints (L) and (K) intersect. Therefore

$$10x + 10y = 150 \qquad\qquad (2)$$
$$15x + 6y = 180 \qquad\qquad (4)$$

Using (3) again to substitute for x in (4),

$$15(15 - y) + 6y = 180$$
$$225 - 15y + 6y = 180$$
$$45 = 9y$$
$$5 = y$$

Substituting this value of y into (3)

$$x = 15 - 5 = 10$$

Thus, profit at C is

$$\pi = 200x + 200y = 200(10) + 200(5) = 2{,}000 + 1{,}000 = £3{,}000$$

which is the same as the profit achieved at B, as expected, illustrating how a linear programming problem may not have a unique solution when the objective function has the same slope as one of the constraints that bounds the feasible area.

It is important to note that the solution to a linear programming problem may be on one of the axes, where a non-negativity constraint operates. Some students who do not fully understand linear programming sometimes manage to draw in the constraints correctly but then incorrectly assume that the solution must lie where the constraints they have drawn intersect. It is, of course, necessary to draw in the objective function to find the solution. The example below illustrates such a case.

Example 5A.4

A company uses inputs K and L to manufacture goods A and B. It has available 200 units of K and 180 units of L and the input requirements are

10 units of K plus 30 units of L for each unit of A
25 units of K plus 15 units of L for each unit of B

If the per-unit profit is £80 for A and £30 for B, what combination of A and B should it produce to maximize profit and how much of K and L will be used in doing this?

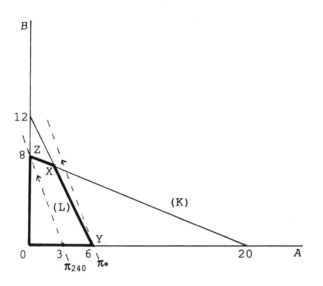

Figure 5A.4

Solution

The resource constraints are, for K,

$$10A + 25B \leq 200 \tag{1}$$

and, for L,

$$30A + 15B \leq 180 \tag{2}$$

The corresponding feasible area 0ZXY is marked out in Figure 5A.4.

To find the slope of the objective function assume total profit is £240. This could be obtained by selling 8 of B or 3 of A, and so the broken line π_{240} in Figure 5A.4 illustrates the combination of A and B that would yield this level of profit. The output combination that produces maximum profit is obtained when a line parallel to π_{240} is drawn as far out from the origin as possible but still within the feasible area. This will be the line π_* through point Y.

At Y, no B is produced and 6 units of A are produced.

Therefore, profit is $6 \times £80 = £480$.

In this example only the constraint (L) bites and so there will be slack in the (K) constraint. The total requirement of K to produce 6 units of A will be 60. There are 200 units of K available and so 140 remain unused.

QUESTIONS 5A.1

1. A firm manufactures products A and B using the two inputs X and Y in the following quantities:

1 tonne of A requires 80 units of X plus 148 units of Y
1 tonne of B requires 200 units of X plus 120 units of Y

The profit per unit of A is £20, and the per-unit profit of B is £30. If the firm has at its disposal 1,600 units of X and 1,800 units of Y, what combination of A and B should it manufacture in order to maximize profit? (Fractions of a tonne may be produced.)

Should the firm change its production mix if per-unit profits alter to (a) £25 each for both A and B or (b) £30 for A and £20 for B?

2. A firm produces the goods A and B using the four inputs W, X, Y and Z in the following quantities:

   1 unit of A requires 9 units of W, 30 of X, 20 of Y and 20 of Z
   1 unit of B requires 13 units of W, 55 of X, 28 of Y and 20 of Z

   The firm has available 468 units of W, 1,980 units of X, 1,120 units of Y and 800 units of Z. What production mix will maximize its total profit if each unit of A adds £60 to profit and each unit of B adds £75?

3. A firm sells two versions of a device for cutting and drilling. Version A is sold direct to the public in DIY stores, yielding a profit per unit of £50, and version B is sold to other firms for industrial use, yielding a per-unit profit of £20. Each day the firm is able to use 400 hours of labour, 750 kg of raw material and 240 metres of packaging material. These inputs are required to produce A and B in the following quantities: one version A device requires 20 hours of labour, 50 kg of raw material and 20 metres of packaging, whilst one of version B only requires 20 hours of labour plus 30 kg of raw material. How many of each version should be produced each day in order to maximize profit?

4. A firm uses three inputs X, Y and Z to manufacture two goods A and B. The requirements per tonne are as follows.

   A: 5 loads of X, 4 containers of Y and 6 hours of Z
   B: 5 loads of X, 6 containers of Y and 2 hours of Z

   Each tonne of A brings in £400 profit and each tonne of B brings in £300. What combination of A and B should the firm produce to maximize profit if it has at its disposal 150 loads of X, 240 containers of Y and 150 hours of Z?

5. A firm makes the two food products A and B and the contribution to profit is £2 per unit of A and £3 per unit of B. There are three stages in the production process: cleaning, mixing and tinning. The number of hours of each process required for each product and the total number of hours available for each process are given in Table 5A.1. Given these constraints what combination of A and B should the firm produce to maximize profit?

6. Make up your own values for the per-unit profit of A and B in the above question and then say what the optimum production combination is.

Table 5A.1

	Hours of		
	Cleaning	*Mixing*	*Tinning*
1 unit of A requires	3	6	2
1 unit of B requires	6	2	1.5
Total hours available	210	120	60

7. A firm manufactures two compounds A and B using two raw materials R and Q, in addition to labour and a mixing additive.

   1 tonne of A requires 1 container of R, 3 sacks of Q, 4 hours labour and 2 tins of mixing additive

   1 tonne of B requires 2 containers of R, 5 sacks of Q and 3 hours labour, but no mixing additive

   Both A and B add £200 per tonne to the firm's profits and it has at its disposal 60 containers of R, 150 sacks of Q, 120 hours of labour and 50 tins of mixing additive.

   What combination of A and B should it produce to maximize profits, assuming that fractions of a tonne can be manufactured? What will these profits be? What surplus amounts of the inputs will there be?

8. A firm uses the four processes cutting, drilling, finishing and assembly to manufacture its two products A and B.

   1 unit of A requires 5 hours cutting, 18 hours drilling, 9 hours finishing and 10 hours assembly

   1 unit of B requires 15 hours cutting, 7 hours drilling, 15 hours finishing and 10 hours assembly

   Products A and B sell for respectively £900 and £2,000 each.

   How can this firm maximize its sales revenue if the capacity of its factory is limited to 390 hours cutting, 630 hours drilling, 450 hours finishing and 400 hours assembly?

9. If a firm is faced with the constraints described in Questions 5.4, number 2 (page 120), what combination of A and B will maximize profit if A contributes £30 per unit to profit and B contributes £10?

10. Show that more than one solution exists if one is trying to maximize the objective function

$$\pi = 4A + 4B$$

subject to the constraints

$$20A + 20B \leqslant 60$$

$$20A + 80B \leqslant 120$$

155

11. A firm has £120,000 to invest. It can buy shares in company X which cost £2 each and give an expected annual return of 6%, or shares in company Y which cost £4 each and give an expected annual return of 8%. It is advised not to put more than 60% of its total investments into any one type of share. What investment portfolio will maximize the expected return? (You may answer this question with or without a diagram.)

12. Make up your own linear programming problem involving the constrained maximization of an objective function with two variables and at least two constraints, and solve it.

Constrained minimization

Another problem a firm might be faced with is how to minimize the cost of producing a good subject to constraints regarding its quality. If the objective function and the constraints can be expressed in linear form then the method used for constrained minimization is analogous to that used in the maximization problems described in the previous section. However, in constrained minimization problems the feasible area is usually above the constraint lines and one needs to find the objective function line that is nearest to the origin within the feasible area, as the following examples show.

Example 5A.5

A firm manufactures a product containing three ingredients X, Y and Z. Each unit produced must contain at least 100 mg of X, 30 mg of Y and 75 mg of Z. The product is made by mixing the inputs A and B which come in containers costing respectively £3 and £6 each.
These contain X, Y and Z in the following quantities:

1 container of A contains 50 mg of X, 10 mg of Y and 15 mg of Z
1 container of B contains 20 mg of X, 10 mg of Y and 50 mg of Z

What mix of A and B will minimize the cost per unit of the product subject to the above quality constraints? (Assume all other costs can be ignored.)

Solution

Total usage of X will be 50 mg for each container of A plus 20 mg for each container of B, i.e. $50A + 20B$. Total usage must be at least 100 mg. This quality constraint can thus be written as

$$50A + 20B \geqslant 100 \qquad (X)$$

Similarly, the quality constraints on Y and Z can be written as

156

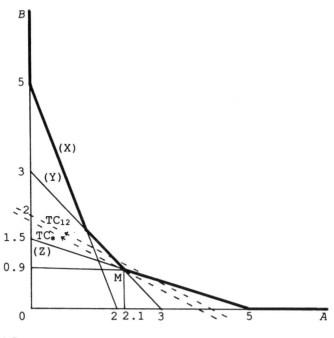

Figure 5A.5

$$10A + 10B \geqslant 30 \qquad (Y)$$
$$15A + 50B \geqslant 75 \qquad (Z)$$

(Note that the constraints have the $\geqslant$ sign instead of $\leqslant$.) As negative amounts of the inputs A and B are assumed not to be feasible, then there are also the two non-negativity constraints

$$A \geqslant 0 \qquad B \geqslant 0$$

If the quality constraint for X is only just met then

$$50A + 20B = 100 \qquad (X)$$

The line representing this function is drawn as (X) in Figure 5A.5. Any combination of A and B above this line will more than satisfy the quality constraint for X. Any combination of A and B below this line will not satisfy this constraint and will therefore not be feasible.

In a similar fashion the constraints for Y and Z are shown by the lines representing the functions

$$10A + 10B = 30 \qquad (Y)$$

and

$$15A + 50B = 75 \qquad (Z)$$

157

Taking all the constraints into account, the feasible area is marked out by the heavy black lines in Figure 5A.5, or at least its lower bounds are. As these are minimum constraints then theoretically there are no upper limits to the amounts of A and B that could be used to make a unit of the final product.

The objective function is total cost (TC) which the firm is seeking to minimize. Given the prices of A and B of £3 and £6 respectively, then

$$TC = 3A + 6B$$

To obtain a guideline for the slope of the TC function assume any value for TC that is easily divisiole by £3 and £6, i.e. the prices of the two inputs A and B. For example, if TC is assumed to be £12 then the line TC_{12} representing the function

$$12 = 3A + 6B$$

can be drawn, which has a slope of -0.5. One now needs to ask the question 'can a line with this slope be drawn closer to the origin (thus representing a smaller value for TC) but still going through the feasible area?' In this case the answer is 'yes'. The line TC_* through M represents the lowest cost method of combining A and B that still satisfies the three quality constraints.

The optimum amounts of A and B can be read off the graph as approximately 2.1 and 0.9 respectively. More accurate answers can be obtained algebraically. The optimum combination M is where the quality constraints for Y and Z intersect. These correspond to the linear equations

$$10A + 10B = 30 \tag{1}$$
$$15A + 50B = 75 \tag{2}$$

Dividing (2) by 5 we get $\qquad 3A + 10B = 15$

Subtracting (1) $\qquad \underline{10A + 10B = 30}$

$$-7A = -15$$
$$A = \frac{15}{7} = 2\frac{1}{7}$$

Substituting into (1)

$$10\left(\frac{15}{7}\right) + 10B = \frac{150}{7} + 10B = 30 \tag{3}$$

Multiplying (3) by 7

$$150 + 70B = 210$$
$$70B = 60$$
$$B = \frac{6}{7}$$

158

Thus the firm should use $2\frac{1}{7}$ containers of A plus $\frac{6}{7}$ of a container of B for every unit of the final product it makes. As long as large quantities of the product are made, the firm does not have to worry about unused fractions of containers. It just needs to use containers A and B in the ratio $2\frac{1}{7} : \frac{6}{7}$ which is the same as the ratio 2.5 : 1.

The constraint on X does not bite and so there is some slack. In a minimization problem slack means overabundance. The total amount of X contained in a unit of the final product will be

$$50A + 20B = 50\left(\tfrac{15}{7}\right) + 20\left(\tfrac{6}{7}\right) = \frac{750 + 120}{7} = \frac{970}{7} = 138.57\,\text{mg}$$

This exceeds the minimum requirement of 100 mg of X by 38.57 mg.

Example 5A.6

A firm produces a product that has minimum input requirements for the four ingredients W, X, Y and Z. These cannot be manufactured individually and can only be supplied as part of the composite inputs A and B.

1 litre of A includes 20 g of W, 5 g of X, 5 g of Y and 20 g of Z
1 litre of B includes 90 g of W, 7 g of X and 4 g of Y but no Z

One drum of the final product must contain at least 7,200 g of W, 1,400 g of X, 1,000 g of Y and 1,200 g of Z. (The volume of the drum is fixed and not related to the volume of inputs A and B as evaporation occurs during the production process.) If a litre of A costs £9 and a litre of B costs £16 how many litres of A and B should the firm use to minimize the cost of a drum of the final product? Assume that all other costs can be ignored.

Solution

The minimum input requirements can be written as

$$20A + 90B \geqslant 7,200 \qquad \text{(W)}$$
$$5A + 7B \geqslant 1,400 \qquad \text{(X)}$$
$$5A + 4B \geqslant 1,000 \qquad \text{(Y)}$$
$$20A \geqslant 1,200 \qquad \text{(Z)}$$

There are also the non-negativity conditions $A \geqslant 0$, $B \geqslant 0$. These constraints are shown in Figure 5A.6.

If only the minimum 7,200 g of W is included in the final product then $20A + 90B = 7,200$. If no B was used then one would need $7,200/20 = 360$ litres of A to satisfy this constraint. If no A was used then $7,200/90 = 80$ litres of B would be needed. Thus the values where the linear constraint (W)

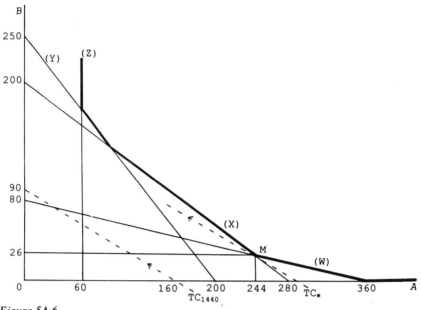

Figure 5A.6

hits the A and B axes are 360 and 80 respectively. Combinations of A and B below this line do not satisfy the minimum amount of W requirement. The other constraints, for X, Y and Z, are constructed in a similar fashion and the feasible area is marked out by the heavy black lines in Figure 5A.6.

To find the value of the objective function that gives the lowest total cost but still allows a combination of A and B within the feasible area one has to find a guideline for the slope of the objective function. Assume that the total cost (TC) of A and B is £1,440, giving the budget constraint

$$1,440 = 9A + 16B$$

This particular budget constraint shown by the broken line TC_{1440}, does not go through the feasible area and so total cost must be greater than £1,440. An increased budget will mean a budget line further from the origin but still with the same slope as TC_{1440}. The budget line with this slope that is closest to the origin and that also passes through the feasible area is TC_*.

The minimum TC is therefore achieved by using the combination of A and B corresponding to point M. Approximate values for these quantities of A and B, read off the graph, are 244 litres of A and 26 litres of B.

More accurate answers can be obtained algebraically as we know that M is at the intersection of the constraints for W and X. This means that the minimum requirements for W and X are only just met and so

$$20A + 90B = 7,200 \text{ for W} \tag{1}$$
$$5A + 7B = 1,400 \text{ for X} \tag{2}$$

Multiplying (2) by 4 gives
$$20A + 28B = 5,600$$

Subtracting (1)
$$\underline{20A + 90B = 7,200}$$
$$-62B = -1,600$$

$$B = \frac{1,600}{62} = 25.8 \qquad \text{(to 1 dp)}$$

Substituting this value for B in (1) gives

$$20A + 90(25.8) = 7,200$$
$$20A + 2,322 = 7,200$$
$$A = \frac{4,878}{20} = 243.9$$

Therefore the firm should use 243.9 litres of A and 25.8 litres of B for each drum of the final product.

The total input cost will be

$$243.9 \times £9 + 25.8 \times £16 = £2,195.10 + £412.80 = £2,607.90$$

QUESTIONS 5A.2

1. Find the minimum value of the function $C = 40A + 20B$ subject to the constraints $A \geqslant 0$, $B \geqslant 0$,

$$10A + 40B \geqslant 40 \qquad \text{(x)}$$
$$30A + 20B \geqslant 60 \qquad \text{(y)}$$
$$10A \geqslant 10 \qquad \text{(z)}$$

Will there be slack in any of the constraints at the optimum combination of A and B? If so, what is the spare capacity?

2. A firm manufactures a product that, per litre, must contain at least 18 g of chemical X and 10 g of chemical Y. The rest of the product is water whose costs can be ignored. The two inputs A and B contain X and Y in the following quantities:

1 unit of A contains 6 g of X and 5 g of Y
1 unit of B contains 9 g of X and 2 g of Y

The per-unit costs of A and B are £2 and £6 respectively. What combination of A and B will give the cheapest way of producing a litre of the final product?

161

3. A firm mixes the two inputs Q and R to make a vitamin supplement in liquid form. The inputs Q and R contain the four vitamins A, B, C and D in the following amounts:

> 6 mg of A, 50 mg of B, 35 mg of C and 12 mg of D per unit of Q
> 30 mg of A, 25 mg of B, 30 mg of C and 20 mg of D per unit of R

The inputs Q and R cost respectively 5 p and 12 p per unit. Each centilitre of the final product must contain at least 60 mg of A, 100 mg of B, 105 mg of C and 60 mg of D. What is the cheapest way of making the final product? Which vitamins will exceed the minimum amount per centilitre using this method?

4. A delivery firm has two types of van, A and B, and carries three types of load, X, Y and Z. Each van is capable of carrying a mixed load, but only in certain proportions, given the special size and weight of the different loads. When fully loaded,

> type A can carry 20 of X, 15 of Y and 15 of Z
> type B can carry 10 of X, 60 of Y and 15 of Z

A typical daily delivery schedule requires the firm to carry 200 loads of X, 450 loads of Y and 225 loads of Z. Each van is only loaded for deliveries once a day. The smaller van, A, costs £50 a day to run and the larger van, B, costs £100 a day. How many of each type of van should the firm use to minimize total running costs? Will there be space in the vans for any more of any of the loads X, Y or Z should more orders be placed?

5. A firm uses two inputs R and T which cost £40 each per tonne. They both contain the chemical compounds G and H in the following quantities:

> 1 tonne of R contains 6 kg of G and 3 kg of H
> 1 tonne of T contains 15 kg of G and 4 kg of H

The final product must contain at least 180 kg of G and 60 kg of H per batch. How many tonnes of R and T should the firm use to minimize the cost of a batch of the final product? Will the amount of G or H it contains exceed the minimum requirement?

6. An aircraft manufacturer fitting out the interior of a plane can use two fitments A and B, which contain components X, Y and Z in the following quantities:

> 1 unit of A contains 3 units of X, 4 units of Y plus 2 units of Z
> 1 unit of B contains 6 units of X, 5 units of Y plus 8 units of Z

The aircraft design is such that there must be at least 540 units of X, 600 units of Y and 480 units of Z in total in the plane. If each unit of A weighs 4 kg and each unit of B weighs 6 kg what combination of A and B will minimize the total weight of these fitments in the plane?

Linear equations

7. Construct your own linear programming problem involving the minimization of an objective function and then solve it.

Mixed constraints

Some linear programming problems may contain both 'less than or equal to' and 'greater than or equal to' constraints. It is also possible to have equality constraints, i.e. where one variable must equal a specified quantity.

Example 5A.7

Minimize the objective function $C = 12A + 8B$ subject to the constraints

$$10A + 40B \geqslant 40 \qquad (1)$$
$$12A + 16B \leqslant 48 \qquad (2)$$
$$A = 1.5 \qquad (3)$$

Solution

The constraints are marked out in Figure 5A.7. Constraint (1) means that the feasible area must be above the line

$$10A + 40B = 40$$

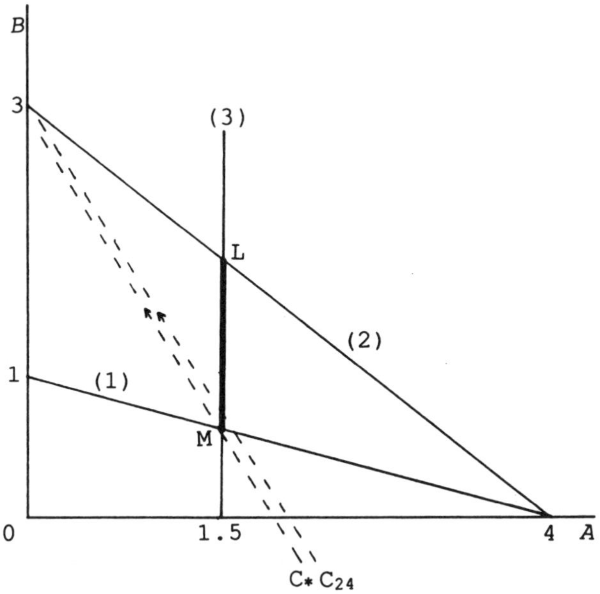

Figure 5A.7

163

Constraint (2) means that the feasible area must be below the line

$$12A + 16 = 48$$

Constraint (3) means that the feasible area must be along the vertical line through $A = 1.5$. The only section of the graph that satisfies all three of these constraints is the heavy black section LM of the vertical line through $A = 1.5$.

If C is assumed to be 24 then the line C_{24} representing the objective function

$$24 = 12A + 8B$$

can be drawn in and has a slope of -1.5. To minimize C, one needs to find the closest line to the origin that has this slope and also passes through the feasible area. This will be the line C_* through M. The optimum value of A is therefore obviously 1.5.

The optimum value of B occurs where the two lines

$$A = 1.5$$

and

$$10A + 40B = 40$$

intersect. Thus

$$10(1.5) + 40B = 40$$
$$15 + 40B = 40$$
$$40B = 25$$
$$B = 0.625$$

QUESTIONS 5A.3

1. A firm makes two goods A and B using the three inputs X, Y and Z in the following quantities:

    20 units of X, 8 units of Y and 20 units of Z per unit of A
    20 units of X, 20 units of Y and 14 units of Z per unit of B

    The per-unit profit of A is £1,500, and for B the figure is £1,000. Input availability is restricted to 60 units of X, 40 units of Y and 70 units of Z. The firm has already committed itself to a contract to supply one customer with 1 unit of B. What combination of A and B should it produce to maximize total profit?
2. A company produces two industrial compounds X and Y that are mixed in a final product. They both contain one common input, R. The amount of R in one tonne of X is 8 litres and the amount of R in one tonne of Y

is 12 litres. A load of the final product must contain at least 240 litres of R to ensure that its quality level is met. No R is lost in the production process of combining X and Y.

The total cost of a tonne of X is £30 and the total cost of a tonne of Y is £15. If the firm has already signed a contract to buy 7.5 tonnes of X per week, what mix of X and Y should the firm use to minimize the cost of a load of the final product?

3. A firm manufactures two goods A and B which require the two inputs K and L in the following amounts:

   1 unit of A requires 6 units of K and 4 of L
   1 unit of B requires 8 units of K and 10 of L

   The firm has at its disposal 96 units of K and 100 of L. The per- unit profit of A is £600 and for B the figure is £300. The firm is under contract to produce a minimum of 6 units of B. How many units of A should it make to maximize profit?

4. Construct and solve your own linear programming problem that has two variables in the objective function and three constraints of at least two different types.

More than two variables

When the objective function in a linear programming problem contains more than two variables then it cannot be solved by graphical analysis. An advanced mathematical technique known as the simplex method can be used for these problems. This is based on the principle that the optimum value of the objective function will usually be at the intersection of two or more constraints.

It is an iterative method that can be very time consuming to use manually and for most practical purposes it is best to use a computer program package to do the necessary calculations. If you have access to a linear programming computer package then you may try to use it now that you understand the basic principles of linear programming. The way that data are entered will depend on the computer package you use and you will need to consult the relevant handbook.

6

Quadratic equations

6.1 SOLVING QUADRATIC EQUATIONS

A quadratic equation is one that can be written in the form

$$ax^2 + bx + c = 0$$

where x is an unknown variable with $a \neq 0$. For example,

$$6x^2 + 2.5x + 7 = 0$$

Because a quadratic equation involves an unknown variable to the power of 2 in one of its terms, it is not possible to rearrange to get a single term in x, as one would when solving a linear equation.

There are three possible methods one might try to use to solve for the unknown in a quadratic equation: (i) graphical, (ii) by factorization and (iii) by using the quadratic 'formula'. In the next three sections we shall see how each can be used to tackle the following question.

A firm faces the linear demand function

$$p = 85 - 2q \tag{1}$$

At what output will total revenue be 200?

It is not immediately obvious that this question involves a quadratic equation. We first need to use economic analysis to set up the mathematical problem to be solved. By definition we know that

$$\text{TR} = pq \tag{2}$$

So, substituting the function for p from (1) into (2), we get

$$\text{TR} = (85 - 2q)q = 85q - 2q^2$$

This is a quadratic function that cannot be 'solved' as it stands. It just tells us the value of TR for any given output. What the question asks is 'at what value of q will this function be equal to 200'? The mathematical problem is therefore to solve the quadratic equation

$$200 = 85q - 2q^2 \tag{3}$$

All three solution methods require all like terms to be brought together on one side of the equality sign, leaving a zero on the other side. It is also necessary to put the terms in the order given in the above definition of a quadratic equation, i.e.

$$\text{unknown squared } (q^2), \text{ unknown } (q), \text{ constant}$$

Thus (3) above can be rewritten as

$$2q^2 - 85q + 200 = 0$$

It is this quadratic equation that each of the three methods will be used to solve.

6.2 GRAPHICAL SOLUTION

The graphical method is not very practical. Drawing a quadratic function is a long winded and not very accurate process that involves separately plotting each individual value of the variable within the range that is being considered. The graphical method can be useful, however, not so much for finding an approximate value for the solution, but for explaining why certain quadratic equations do not have a solution whilst others have two solutions. Only a rough sketch diagram is necessary for this purpose.

Example 6.1

Show graphically that the quadratic equation

$$2q^2 - 85q + 200 = 0$$

can be solved.

Solution

We first need to define the function

$$y = 2q^2 - 85q + 200$$

If the graph of this function cuts the q axis then $y = 0$ and we have a solution. Next, we calculate a few values of the function to get an approximate idea of its shape.

When $q = 0$, then $y = 200$
When $q = 1$, then $y = 2 - 85 + 200 = 117$

and so the graph falls.

When $q = 3$, then $y = 18 - 255 + 200 = -37$

and so it cuts q axis.

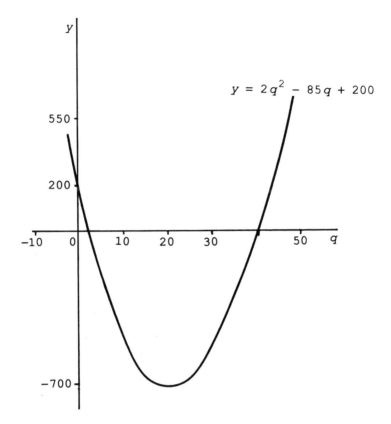

$$y = 2q^2 - 85q + 200$$

Figure 6.1

When $q = 50$ then $y = 5,000 - 4,250 + 200 = 550$

and so it rises again.

These values indicate that the graph is a U-shape that cuts the q axis twice, and this is shown in Figure 6.1. There are therefore two values of q for which y is zero, and therefore there are two solutions to the question. The precise values of these solutions, 2.5 and 40, can be found by the other two methods explained in the following sections or by computation of y for different values of q.

If we slightly change the problem in Example 6.1 we can see why there may not always be a solution to a quadratic equation.

Example 6.2

Given the total revenue function

168

Quadratic equations

$$TR = 85q - 2q^2$$

when will total revenue be 1,500?

Solution

The quadratic equation to be solved is

$$1,500 = 85q - 2q^2$$

which can be rewritten as

$$2q^2 - 85q + 1,500 = 0$$

Calculating a few values of the function $y = 2q^2 - 85q + 1,500$, we can see that it falls and then rises again but never cuts the q axis, as Figure 6.2 shows.

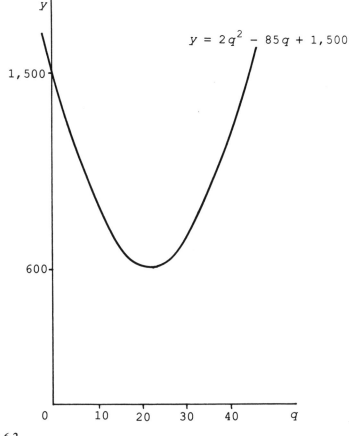

Figure 6.2

$$\text{When } q = 0, \text{ then } y = 1{,}500$$
$$\text{When } q = 10, \text{ then } y = 850$$
$$\text{When } q = 20, \text{ then } y = 600$$
$$\text{When } q = 25, \text{ then } y = 625$$

There are therefore no solutions to this quadratic equation, i.e. there is no output at which total revenue will be 1,500.

Although one would never try to plot the whole graph of a quadratic function manually, one may of course get a computer plot. The accuracy of the answer you obtain will depend on the graphics package that you use.

A worksheet for calculating the different values of y in Example 6.1 using the Lotus 1–2–3 spreadsheet program can be constructed as follows.

Enter the labels below and right justify (RJ) where indicated.

QUADRATIC	in cell	A1
SOLUTION	in cell	C1
$y = 2q \char`\^ 2$	in cell	D1	(RJ)
$-85q + 200$	in cell	E1
q	in cell	A3	(RJ)
y	in cell	B3	(RJ)

Enter 0 in cell A4 and then the formula $+A4 + 0.5$ in cell A5 and copy down the q column until $q = 44.5$, to get $q = 0$ to 44.5 in increments of 0.5. To calculate the corresponding value of y, enter the formula $+2*A4\char`\^2 - 85*A4 + 200$ in cell B4. Copy this down the column. You should now get a

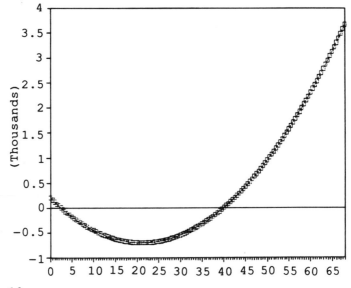

Figure 6.3

170

series of values as shown in the spreadsheet in Table 6.1. (Note that this has been slightly amended from the format specified above in order to get three columns of values to fit on one page.) You can now just read off the q values of 2.5 and 40 which correspond to a y value of zero.

Table 6.1

QUADRATIC		SOLUTION	$y=2q^2 -85q+200$		
q	y	q	y	q	y
0	200	15	−625	30	−550
0.5	158	15.5	−637	30.5	−532
1	117	16	−648	31	−513
1.5	77	16.5	−658	31.5	−493
2	38	17	−667	32	−472
2.5	0	17.5	−675	32.5	−450
3	−37	18	−682	33	−427
3.5	−73	18.5	−688	33.5	−403
4	−108	19	−693	34	−378
4.5	−142	19.5	−697	34.5	−352
5	−175	20	−700	35	−325
5.5	−207	20.5	−702	35.5	−297
6	−238	21	−703	36	−268
6.5	−268	21.5	−703	36.5	−238
7	−297	22	−702	37	−207
7.5	−325	22.5	−700	37.5	−175
8	−352	23	−697	38	−142
8.5	−378	23.5	−693	38.5	−108
9	−403	24	−688	39	−73
9.5	−427	24.5	−682	39.5	−37
10	−450	25	−675	40	0
10.5	−472	25.5	−667	40.5	38
11	−493	26	−658	41	77
11.5	−513	26.5	−648	41.5	117
12	−532	27	−637	42	158
12.5	−550	27.5	−625	42.5	200
13	−567	28	−612	43	243
13.5	−583	28.5	−598	43.5	287
14	−598	29	−583	44	332
14.5	−612	29.5	−567	44.5	378

You may also use the Lotus 1–2–3 spreadsheet you have created to get a plot of the function $y = 2q^2 - 85q + 200$. Assuming that you have q and y in single columns, then you just use the Lotus 1–2–3 graph command (/G) to obtain a plot with q measured on the X axis and y as variable A on the vertical axis. (It is assumed that you already know how to use this command, or your tutor will explain its operation.) The only special instruction you need to use is the Skip command in the Graphics Options menu so that only every fifth value on the X axis is printed, in order to make the numbering readable. You should obtain a plot similar to that shown in Figure 6.3. This clearly shows how this function

171

cuts the horizontal axis twice, although it is difficult to read off the precise values where this occurs. If you wish to get your own graph printed out you will need to save it as a .PIC file and then go out of the main Lotus 1–2–3 program and into PrintGraph.

This spreadsheet can easily be amended to calculate values and plot graphs of other quadratic functions. One only needs to change the formula in cell B4. For example, to calculate values for the function

$$y = 2q^2 - 85q + 1,500$$

from Example 6.2, the formula in cell B4 should be changed to $+2*A4^2-85*A4+1500$. A computer plot of this function will produce the shape shown in Figure 6.2 above, showing that this function will not cut the horizontal axis, and confirming again that there is no solution to the quadratic equation

$$0 = 2q^2 - 85q + 1,500$$

6.3 FACTORIZATION

In Chapter 3 it was explained how an expression could be factorized, i.e. how it could be broken down into the terms which when multiplied together give this expression. For example,

$$a^2 - 2ab + b^2 = (a - b)(a - b)$$

If a quadratic function which has been rearranged to equal zero can be factorized in this way then one or the other of the two factors must equal zero. (Remember that if $A \times B = 0$ then either A or B must be zero.)

Example 6.3

Solve by factorization the quadratic equation

$$2q^2 - 85q + 200 = 0$$

Solution

This expression can be factorized as

$$(2q - 5)(q - 40) = 2q^2 - 85q + 200$$

Therefore

$$(2q - 5)(q - 40) = 0$$

This means that

$$2q - 5 = 0 \qquad \text{or} \qquad q - 40 = 0$$

so the solutions are

$$q = 2.5 \qquad \text{or} \qquad q = 40$$

It may be the case that mathematically a quadratic equation has one or more solutions with a negative value. These will not apply in an economic problem if it does not make sense to have negative values for the variable in question. One cannot have a negative output, for example.

Example 6.4

Solve by factorization the quadratic equation

$$2x^2 - 6x - 20 = 0$$

Solution

By factorization

$$(2x - 10)(x + 2) = 0$$

Therefore

$$2x - 10 = 0 \qquad \text{or} \qquad x + 2 = 0$$
$$x = 5 \qquad \text{or} \qquad x = -2$$

If x represented output, then $x = 5$ would be the only answer we would use.

If a quadratic equation cannot be factorized then the formula method in Section 6.5 below must be used. The formula method can also be used, however, when an equation can be factorized. Therefore, if you cannot quickly see a way of factorizing then you should use the formula method. Factorization is only useful as a short-cut way of solving certain quadratic equations. It defeats the object of the exercise if you spend half an hour trying to find a way of factorizing an expression.

It should also go without saying that quadratic equations for which no solutions exist cannot be factorized. For example, it is not possible to factorize the equation

$$2q^2 - 85q + 1,500 = 0$$

which we have already shown to have no solution.

QUESTIONS 6.1

1. Solve for x in the equation $x^2 - 5x + 6 = 0$.
2. A firm's demand schedule is given by the function

$$p = 70 - q$$

At what output will total revenue be £600?

3. A firm faces the average cost function

$$AC = 40x^{-1} + 10x$$

where x is output. When will average cost be 40?

4. Is there a positive solution for x when

$$0 = 12x^2 + 90x - 48?$$

5. A firm faces the total cost schedule

$$TC = 6 - 2q + 2q^2$$

when $q > 2$. At what output level will TC = £150?

6.4 THE QUADRATIC FORMULA

Any quadratic equation expressed in the form

$$ax^2 + bx + c = 0$$

where a, b and c are given parameters and for which a solution exists can be solved for x by using the quadratic formula

$$x = \frac{-b \pm \sqrt{b^2 - 4ac}}{2a}$$

(The sign $\pm$ means + or −.) There is no need for you to understand how the formula is derived. You just need to know that it works.

Example 6.5

Use the quadratic formula to solve the quadratic equation

$$2q^2 - 85q + 200 = 0$$

Solution

In the quadratic formula applied to this example $a = 2$, $b = -85$ and $c = 200$ (and, of course, $x = q$). Note that the minus signs for any negative coefficients must be included. One also needs to take special care to remember to use the rules for arithmetic operations using negative numbers.

Substituting these values for a, b and c into the formula we get

$$q = \frac{-(-85) \pm \sqrt{(-85)^2 - 4 \times 2 \times 200}}{2 \times 2}$$

$$= \frac{85 \pm \sqrt{7{,}225 - 1{,}600}}{4}$$

$$= \frac{85 \pm \sqrt{5{,}625}}{4}$$

$$= \frac{85 \pm 75}{4} = \frac{160}{4} \quad \text{or} \quad \frac{10}{4} = 40 \quad \text{or} \quad 2.5$$

These are, of course, the same as the solutions found by factorization in Example 6.3 above.

What happens if you try to use the quadratic formula when no solution exists? We can find out by applying the formula to the quadratic equation in Example 6.2 above, where a sketch graph showed that there was no solution.

Example 6.6

Use the quadratic formula to try to solve

$$2q^2 - 85q + 1{,}500 = 0$$

Solution

In this example $a = 2$, $b = -85$ and $c = 1{,}500$. Therefore

$$q = \frac{-(-85) \pm \sqrt{(-85)^2 - 4 \times 2 \times 1{,}500}}{2 \times 2}$$

$$= \frac{85 \pm \sqrt{7{,}225 - 12{,}000}}{4}$$

$$= \frac{85 \pm \sqrt{-4{,}775}}{4}$$

We are now stuck! It is impossible to find the square root of a negative number. In other words, no solution exists.

It will always be the case that the quadratic formula will require the square root of a negative number if no solution exists.

QUESTIONS 6.2

1. Solve for x if $0 = x^2 + 2.5x - 125$.
2. A firm faces the demand schedule $q = 400 - 2p - p^2$. What price does it need to charge to sell 100 units?

175

3. If a firm's demand function is $p = 100 - q$, what quantities need to be sold to bring in a total revenue of
   (a) £100   (b) £1,000   (c) £10,000?
   (Give answers to 2 decimal places, where they exist.)
4. Make up your own quadratic equation and then find whether a solution exists.

6.5 QUADRATIC SIMULTANEOUS EQUATIONS

It is possible that one or more equations in a simultaneous equation system may be non-linear. If the non-linear equations are in quadratic form then one method of solution is to use the techniques learned in Chapter 5 to eliminate all but one unknown and to reduce the problem to a single quadratic equation. If this can be solved then the other unknowns can be found by substitution.

Example 6.7

In a competitive market demand is given by

$$p = 200q^{-1}$$

and the supply function is

$$p = 30 + 2q$$

Find the equilibrium values of p and q.

Solution

In equilibrium demand price equals supply price. Therefore

$$200q^{-1} = 30 + 2q$$

Multiplying through by q,

$$200 = 30q + 2q^2$$
$$0 = 2q^2 + 30q - 200$$
$$0 = (2q - 10)(q + 20)$$

Therefore $2q - 10 = 0$ or $q + 20 = 0$

$$q = 5 \quad \text{or} \quad q = -20$$

We can ignore the second answer as negative quantities cannot exist. Thus $q = 5$. Substituting this value into the supply function

$$p = 30 + 2 \times 5 = 40$$

Quadratic equations

You should now be able to link the different mathematical techniques you have learned so far to tackle more complex problems. If you have covered the theory of perfect competition in your economics course, then you should be able to follow the analysis in the example below.

Example 6.8

An industry in made up of 100 firms, all with the cost schedules

$$AC = 40q^{-1} + 0.4q^2$$
$$TC = 40 + 0.4q^3$$
$$MC = 1.2q^2$$

They sell in a market where the demand schedule is

$$p = 70 - 0.08Q$$

where Q is industry output (and q is an individual firm's output).

(i) What will be the short-run price, industry output and profit for each firm?
(ii) What will happen to price, industry output and the number of firms in the long run? (Assume new entrants have the same cost structure.)

Solution

(i) The industry supply schedule is the horizontal sum of the individual firms' marginal cost schedules. Given

$$MC = 1.2q^2$$
$$\frac{MC}{1.2} = q^2$$
$$\left(\frac{MC}{1.2}\right)^{0.5} = q$$

There are 100 firms, and so industry supply is

$$Q = 100q = 100\left(\frac{MC}{1.2}\right)^{0.5} \qquad (1)$$

As MC corresponds to the price at which any given quantity will be supplied, (1) can be rewritten as

$$Q = 100\left(\frac{p}{1.2}\right)^{0.5}$$

177

Therefore

$$0.01Q = \left(\frac{p}{1.2}\right)^{0.5}$$

$$(0.01Q)^2 = \frac{p}{1.2}$$

$$0.0001Q^2 = \frac{p}{1.2}$$

$$0.00012Q^2 = p \qquad (2)$$

The function (2) is the industry supply schedule.
  The demand schedule given in the question is

$$p = 70 - 0.08Q \qquad (3)$$

Therefore, in equilibrium, demand price equals supply price and

$$70 - 0.08Q = 0.00012Q^2$$

$$0 = 0.00012Q^2 + 0.08Q - 70$$

Using the quadratic formula

$$Q = \frac{-0.08 \pm \sqrt{0.0064 + 0.0336}}{0.00024}$$

$$= \frac{-0.08 \pm \sqrt{0.04}}{0.00024}$$

$$= \frac{-0.08 + 0.2}{0.00024} \quad \text{or} \quad \frac{-0.08 - 0.2}{0.00024}$$

$$= \frac{0.12}{0.00024} \quad \text{or} \quad \frac{-0.28}{0.00024}$$

$$= 500$$

ignoring the negative answer. Substituting this value of Q into the demand schedule (3) gives

$$p = 70 - 0.08(500) = 70 - 40 = £30$$

Each firm produces the same amount. Therefore,

$$q = \frac{Q}{100} = \frac{500}{100} = 5$$

Each firm's profit will be

Quadratic equations

$$\text{TR} - \text{TC} = pq - (40 + 0.4q^3)$$
$$= 30(5) - (40 + 50)$$
$$= 150 - 90 = £60$$

(ii) If existing firms are making a profit then in the long run new entrants will be attracted and price will be driven down until each firm is only just breaking even. This will be when price equals the lowest value on the firm's U-shaped average cost schedule.

From cost theory you should know that MC always cuts AC at its minimum point. Therefore, if

$$\text{MC} = \text{AC}$$
$$1.2q^2 = 40q^{-1} + 0.4q^2$$
$$0.8q^2 = 40q^{-1}$$
$$q^3 = 50$$
$$q = 3.684 \qquad \text{(to 3 dp)}$$

When $q = 3.684$, then

$$\text{AC} = 40q^{-1} + 0.4q^2$$
$$= 40(3.684)^{-1} + 0.4(3.684)^2$$
$$= 16.2865$$

Therefore

$$p = £16.29 \text{ (to the nearest penny)}$$

The old supply schedule does not now apply because of the increased number of firms in the industry. Therefore, substituting this price into the demand schedule (3) to get total output gives

$$16.29 = 70 - 0.08Q$$
$$0.08Q = 53.71$$
$$Q = 671.375$$

We know that each firm produces 3.684 units in long-run equilibrium. Therefore, the new number of firms in the industry is

$$\frac{Q}{q} = \frac{671.375}{3.684} = 182.24 = 182 \text{ firms}$$

(Note that the fraction is rounded down to the nearest whole number. Any extra firms would bring price below the break-even level.)

QUESTIONS 6.3

1. If $y = 255 - x - x^2$ and $y = 180 + \frac{2}{3}x^2 - 21x$, find x and y.
2. Given the functions

$$2y + 4x^2 + 10x - 36 = 0$$

and

$$4y - 10x^2 + 24x = 24$$

find x and y.
3. A monopoly faces the marginal cost function

$$MC = 0.5q^2$$

and the marginal revenue function

$$MR = 200 - 4q$$

What output will maximize profits?
4. A price-discriminating monopoly sells in two markets whose demand schedules are

$$p_1 = 200 - 20q_1 \qquad p_2 = 120 - 5q_2$$

Its marginal cost schedule is

$$MC = 40 + 0.5q^2$$

where $q = q_1 + q_2$. How much should it sell in each market, and at what price, in order to maximize profit?
5. A firm's marginal cost schedule is

$$MC = 2.3 + 0.00012q^2$$

It sells its output in two separate markets with demand schedules

$$p_1 = 25 - 0.125q_1 \qquad p_2 = 12 - 0.05q_2$$

What prices and quantities will maximize profits if this firm is a price-discriminating monopoly?

6.6 POLYNOMIALS

Quadratic equations are a special case of polynomial equations. The general format of a polynomial function is

$$y = a_0 + a_1x + a_2x^2 + a_3x^3 + \ldots + a_nx^n$$

where n is any non-negative integer.

Linear equations are polynomials where $n = 1$. Quadratic equations are polynomials where $n = 2$.

When n is greater than 2 the solution of a polynomial equation by algebraic means becomes complex and time consuming. For practical purposes, how-

Quadratic equations

ever, you may use a spreadsheet to find a solution by the iterative method. This means calculating different values of an expression for different values of the unknown variable until a solution or a good approximation to it is found. As a spreadsheet can quickly perform the necessary calculations, it is an ideal tool for the calculation of polynomial solutions.

The format of the spreadsheet will depend on the problem tackled. Below are some examples of how problems can be approached.

Example 6.9

A firm's total costs (TC) are given by the function

$$TC = 420 + 32.5q - 6.25q^2 + 0.8q^3$$

where q is output level and TC is measured in pounds. If the firm's management is given a budget of £43,000, what output can it produce?

Solution

A spreadsheet needs to be constructed that will calculate TC for different values of q. You can then experiment with different ranges of q until the solution is found.

Using Lotus 1–2–3 format, a template can be constructed as follows. This template allows other values of the parameters of this cubic function to be entered in order to tackle other similar problems.

Enter labels in the cells shown and right justify (RJ) where indicated.

SOLUTION	cell	A1	
OF CUBIC	cell	B1	
POLYNOMIA	cell	C1	
L	cell	D1	
TC=a+bq+	cell	A3	(RJ)
cq^2+dq^3	cell	B3	
PARAMETER	cell	A5	
VALUES:	cell	B5	(RJ)
a =	cell	C5	(RJ)
b =	cell	C6	(RJ)
c =	cell	C7	(RJ)
d =	cell	C8	(RJ)
q	cell	A10	(RJ)
TC	cell	B10	(RJ)

Enter the parameter values for this particular problem

(a)	420	cell	D5
(b)	32.5	cell	D6

181

(c)	− 6.25	cell	D7
(d)	0.8	cell	D8

To get a 'ball park' idea of where the solution may lie, produce a range of values in jumps of 10 in the column headed q by entering

$$1 \qquad \text{in cell A11}$$

and the formula

$$+A11+10 \qquad \text{in cell A12}$$

Then copy this down the column for a few dozen rows.

To calculate the corresponding value of TC, enter the formula $+\$D\$5+\$D\$6*A11+\$D\$7*A11\^2+\$D\$8*A11\^3$ in cell B11 and copy it down the column. Format the range of the column to 2 decimal places.

You should find that when $q = 31$, TC = 19,254.05, and when $q = 41$, TC = 46,383.05. Therefore, TC = £43,000 lies somewhere between these values. To try to pinpoint the corresponding value of q, change the value in cell A11 to 31 and the formula in cell A12 to $+A11+1$ and copy down the column. Your spreadsheet should now look like Table 6.2. This shows that when $q = 40$, TC = 42,920.00. We already know from the 'ball park' run of this spreadsheet that when $q = 41$, TC = 46,383.05. Thus, if output is

Table 6.2

SOLUTION OF CUBIC POLYNOMIAL

TC=a+bq+ cq^2+dq^3

PARAMETER VALUES:

a =	420
b =	32.5
c =	−6.25
d =	0.8

q	TC
31	19254.05
32	21274.40
33	23435.85
34	25743.20
35	28201.25
36	30814.80
37	33588.65
38	36527.60
39	39636.45
40	42920.00

constrained to whole units, 40 is the maximum output that the firm's management can produce for a budget of £43,000.

Save your spreadsheet for use with other examples.

To solve other cubic polynomials, one simply enters the corresponding parameters into the spreadsheet set up for Example 6.9 above and adjusts the range of the independent variable (q) until the solution is found.

Example 6.10

If $TC = 880 + 72q - 14.5q^2 + 1.5q^3$, at what value of q will $TC = £9,889$?

Solution

Entering the new values for a, b, c and d into the spreadsheet and adjusting the range of q, one should get a spreadsheet similar to Table 6.3. This shows that $q = 21$ when $TC = £9,889$.

Table 6.3

SOLUTION OF CUBIC POLYNOMIAL		
TC=a+bq+ cq^2+dq^3		
PARAMETER VALUES:	a =	880
	b =	72
	c =	-14.5
	d =	1.5

q	TC
20	8520.00
21	9889.00
22	11418.00
23	13116.00
24	14992.00
25	17055.00
26	19314.00
27	21778.00
28	24456.00
29	27357.00

This spreadsheet can also be adjusted to cope with more complex polynomials. The crucial part of the spreadsheet is the formula in cell B11. This needs to be amended to calculate the value of the new polynomial function if more terms are added. Note, however, that large polynomial equations may have

several solutions. In particular, if there are both positive and negative coefficients, a polynomial function may equal zero at more than two values of the
independent variable. You can usually deduce the number of solutions from
the format of the equation, or get a plot from your spreadsheet to see how
many times the function crosses the horizontal axis. On the other hand, no
solutions may exist, in which case a graph will not cut the axis; e.g. there is no
positive value of x which will satisfy the equation $0 = 8 + 32x + 6x^2 + 0.9x^3$
although in this case there will be a negative solution.

Example 6.11

Is there a positive value of x for which

$$0 = -770,077.6 + 262x - 74x^2 + 12x^3 + 2x^4 - 0.05x^5?$$

Assume $x < 1,000.$

Solution

Call up the spreadsheet created for Example 6.9. Change the labels appro-
priately. There are two new parameters e and f as this polynomial is in the
form

$$y = a + bx + cx^2 + dx^3 + ex^4 + fx^5$$

You can fit these in by entering the labels

e =	cell	F5	(RJ)
f =	cell	F6	(RJ)

Table 6.4

SOLUTION OF 5th POWER   POLYNOMIAL

y=a+bx+ cx^2+dx^3+ex^4 +fx^5

PARAMETER	VALUES:				
		a =	-770077.6	e =	2
		b =	262	f =	-0.05
		c =	-74		
		d =	12		

x	y
30	-99817.60
31	-59993.15
32	-24823.20
33	4298.75
34	25835.20
35	38098.65
36	39245.60
37	27270.55
38	0.00
39	-44913.55

and putting the actual values 2 and 0.05 in cells G5 and G6 respectively. The new values for the parameters *a*, *b*, *c* and *d* obviously need to be entered in cells D5–D8. The formula in cell B11 now needs to be changed to

+$D$5+$D$6*A11+$D$7*A11^2+$D$8*A11^3+$G$5*A11^4+$G$6*A11^5

As well as copying this formula down the column, widen the B column to 15 places to cope with the large numbers that may be computed for *y*.
Using the same method as in the previous examples you should find the solution $x = 38$, as shown in Table 6.4.

QUESTIONS 6.4

(You will need a spreadsheet to tackle these.)
1. How much of *q* can be produced for £60,000 if the total cost function is

$$TC = 86 + 152q - 12q^2 + 0.6q^3?$$

2. What output can be produced for £150,000 if

$$TC = 130 + 62q - 3.5q^2 + 0.15q^3?$$

7

Series, time and investment

7.1 GROWTH AND DECLINE

In economics we come across many variables that grow, or decline, over time. A sum of money deposited in a building society will grow as interest accumulates on it. The amount of oil left in an oil well will decline as production continues over the years. In this chapter we explain how mathematics can help answer certain problems concerned with these variables that change over time. The main area of application is finance, where methods of appraising different forms of investment are explained. Other applications include the management of natural resources, where the implications of different depletion rates are analysed.

7.2 DISCRETE AND CONTINUOUS FUNCTIONS

The interest earned on money invested in a deposit account is normally paid at set regular intervals. Calculations of the return are therefore usually made with respect to specific time intervals. This point is made clearer in Figure 7.1(a) which shows the amount of money in a deposit account at any given moment in time assuming an initial deposit of £1,000 and interest credited at the end of each year at a rate of 10%. There is not a continuous relationship between time and the total sum in the deposit account. Instead there is a 'jump' at the end of each year when the interest on the account is paid. This is an example of a 'discrete' function. One knows exactly how much will be in the account at the end of every 12 months when the interest is added, but between these periods there is no change in the value of the account. A discrete function is one where the value of the dependent variable is known for specific values of the independent variable but does not continuously change between these values. Hence one gets a series of values rather than a continuum.

For example, teachers' salaries are based on scales with series of increments. A hypothetical salary scale might be as follows.

0 years of service	£15,000
1 years of service	£15,800

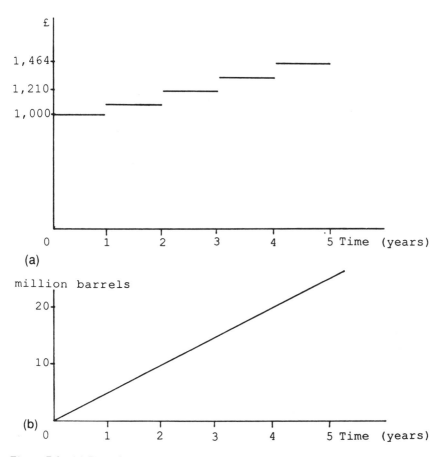

Figure 7.1 (a) Deposit account balance; (b) oil extraction

| 2 years of service | £16,600 |
| 3 years of service | £17,400 |

The relationship between salary and years of service is a discrete function. At any moment in time one knows what a teacher's salary will be but there is not a continuous relationship between time and salary level.

An example of a continuous function is illustrated in Figure 7.1(b) which shows the amount of oil extracted from a well, assuming an extraction rate of 5 million barrels per year. It is assumed that production is continuous and so there is a continuous smooth function showing the relationship between the amount of oil extracted and the time elapsed.

The distinction between these two sorts of function is not always so clearcut in practice. Although oil may physically flow continuously from an oil well, production figures are usually specified as a given amount per time period. Therefore problems that use annual production figures, for instance, are effectively dealing with a discrete function. On the other hand, interest may be credited to a deposit account, or charged on a loan, on a daily basis so that there is an almost continuous relationship over time between the value of the investment and the time elapsed. The point is, though, that there is a difference between discrete and continuous functions and that in this chapter we shall be dealing with a number of discrete-variable problems. These will mainly be concerned with series of values of different forms of investment over specified time intervals. The concept of continuous growth (and decay) is left until Chapter 14 where exponential functions are explained.

Because the problems in this chapter deal with discrete functions this makes them particularly suitable for solution by using a spreadsheet program, such as Lotus 1–2–3. Where a spreadsheet can help solve certain types of problem this will be made clear in each section, after the basic analysis has been explained. Most of the exercises can be solved without a spreadsheet, although in some cases manual solution would be rather time consuming. In a few instances a spreadsheet must be used to tackle certain problems unless you have a special financial calculator, but these can be skipped without missing out on an understanding of the analysis involved.

7.3 INTEREST

Time is money. If you borrow money you have to pay interest on it. If you invest money in a deposit account you expect to earn interest on it. In your economics course you should learn about the different reasons why interest is paid and what determines rates of interest in general, including how inflation is taken into account. In this chapter we shall assume interest rates to be given when different forms of investment are analysed.

From an investor's viewpoint the interest rate can be looked on as the 'opportunity cost of capital'. If a sum of money is tied up in a project for a year then the investor loses the interest that could have been earned by investing the money elsewhere, perhaps by putting it in a deposit account. This is the opportunity cost of this money being tied up in a project for 12 months. Although in practice interest rates can vary for different types of investment, and there is a profit margin between banks' lending and borrowing rates, only one rate of interest is assumed in most examples.

Simple interest is the interest that accrues on a given sum in a set time period. It is assumed that this interest is *not* reinvested along with the original capital. The amount of interest earned each time period will be the same if interest rates do not change and the total capital invested remains unaltered.

188

Example 7.1

An investor puts £20,000 into a deposit account and has the annual interest paid directly into a separate current account and then spends it. The deposit account pays 8.5% interest. How much interest is earned in the fifth year?

Solution

The interest paid each year will be 8.5% of £20,000. Thus in year 5 interest will be

$$0.085 \times £20,000 = £1,700$$

Most investment decisions, however, need to take into account the fact that any interest earned can be reinvested and so compound interest, explained below, is more relevant. The calculation of simple interest is such a basic arithmetic exercise that it will be assumed that all students can happily cope with such problems. The only mistake you are likely to make is to transform a percentage figure into a decimal fraction incorrectly.

Example 7.2

What is 0.5% of £400?

Solution

To convert any percentage figure to a decimal fraction you must divide it by 100. Therefore

$$0.5\% = \frac{0.5}{100} = 0.005$$

and so

$$0.5\% \text{ of } £400 = 0.005 \times £400 = £2$$

If you can remember that 1% = 0.01 then you should be able to transform any interest rate specified in percentage terms into a decimal fraction in your head. Try to do this for the following interest rates:

(i) 1.5%   (ii) 30%   (iii) 0.075%   (iv) 1.02%   (v) 5.5%

Now check your answers with a calculator. If you got any wrong you really ought to go back and revise Section 2.5 before proceeding. Converting decimal fractions back to percentage interest rates is, of course, simply a matter of multiplying by 100; e.g. 0.12 = 12%, 0.4 = 40%, 1.25 = 125%, 0.008 = 0.8%.

Compound interest is interest which is added to the original investment every time it accrues. The interest added in one time period will itself earn

189

interest in the following time period. The total value of an investment will therefore grow over time.

Example 7.3

If £600 is invested for 3 years at 8% interest compounded annually at the end of each year, what will the final value of the investment be?

Solution

	£
Initial sum invested	600.00
Interest at end of year $1 = 0.08 \times 600$	48.00
Total sum invested for year 2	648.00
Interest at end of year $2 = 0.08 \times 648$	51.84
Total sum invested for year 3	699.84
Interest at end of year $3 = 0.08 \times 699.84$	55.99
Final value of investment	755.83

Example 7.4

If £5,000 is invested at an annual rate of interest of 12% how much will the investment be worth after 2 years?

Solution

	£
Initial sum invested	5,000
Year 1 interest $= 0.12 \times 5,000$	600
Sum invested for year 2	5,600
Year 2 interest $= 0.12 \times 5,600$	672
Final value of investment	6,272

The above examples only involved the calculation of interest for a few years and so it did not take too long to solve the problem from first principles. How would you work out the final sum of an investment after a longer time period though, eg. 15 years?

Let us consider an investment at compound interest in algebraic terms. Assume that: A is the initial sum invested,

F is the final value of the investment,

i is the interest rate per time period (as a decimal fraction) and

n is the number of time periods.

The value of the investment at the end of each year will be $1 + i$ times the sum invested at the start of the year. In Example 7.3, above, the £648 at the start of year 2 is 1.08 times the initial investment of £600, for instance. The value of the investment at the start of year 3 is 1.08 times the value at the start of year 2, and so on. Thus

Value of investment after 1 year $= A(1+i)$

Value of investment after 2 years $= A(1+i)(1+i) = A(1+i)^2$

Value of investment after 3 years $= A(1+i)^2(1+i) = A(1+i)^3$ etc.

We can see that each value is A multiplied by $1+i$ to the power of the number of years the sum is invested. Thus, the value of the investment after n years is $A(1+i)^n$.

The formula for the final value F of an investment of £A for n time periods at interest rate i is therefore

$$F = A(1+i)^n$$

Let us rework Examples 7.3 and 7.4 using this formula just to check that we get the same answer.

Example 7.3 (reworked)

£600 invested for 3 years at 8%:

$$A = £300 \qquad n = 3 \qquad i = 0.08$$

$$F = A(1+i)^n = 600(1.08)^3 = 600(1.259712) = £755.83$$

Example 7.4 (reworked)

£5,000 invested for 2 years at 12%:

$$A = £5,000 \qquad n = 2 \qquad i = 0.12$$

$$F = A(1+i)^n = 5,000(1.12)^2 = 5,000(1.2544) = £6,272$$

Having satisfied ourselves that the formula works we can now tackle some more difficult problems.

Example 7.5

If £4,000 is invested for 10 years at an interest rate of 11% per annum what will the final value of the investment be?

Solution

$$A = £4,000 \qquad n = 10 \qquad i = 0.11$$

$$F = A(1 + i)^n = 4{,}000(1.11)^{10}$$
$$= 4{,}000(2.8394205)$$
$$= £11{,}357.68$$

The calculation of large powers of numbers, such as $(1.11)^{10}$, was explained in Chapter 2. This could be done on a basic pocket calculator but it is preferable to use a calculator with the function key $[y^x]$. Alternatively, if you can set up a spreadsheet that calculates the formula $F = A(1 + i)^n$ then you can simply enter different values of A, i and n to solve other similar problems.

Sometimes a compound interest problem may be specified in a rather different format, but the method of solution is still the same.

Example 7.6

You estimate that you will need £6,000 in 3 years' time to buy a new car, assuming a reasonable trade-in price for your old car. You have £4,000 which you can put into a fixed interest building society account earning 14%. Will you have enough to buy the car?

Solution

You need to work out the final value of your savings to see whether it will be greater than £6,000. Using the usual notation,

$$A = £4{,}000 \qquad n = 3 \qquad i = 0.14$$

$$F = A(1 + i)^n = 4{,}000(1.14)^3 = 4{,}000(1.481544) = £5{,}926.18$$

So the answer is 'almost'. You will have to find another £74 to get to £6,000 but perhaps you can get the car dealer to knock this off the price.

It is usually the custom to specify annual interest rates. However, some interest rates are set on a monthly basis, although the equivalent annual percentage rate (APR) should always be quoted. You are most likely to encounter monthly interest payments when taking out a loan rather than making an investment. The evaluation of an investment where interest is accrued monthly is done on exactly the same basis as when annual interest payments apply. As long as the interest rate i and the number of time periods n refer to the same period of time, then the formula $F = A(1 + i)^n$ can still be used.

Example 7.7

If interest is credited monthly at a monthly rate of 0.9% how much will £100 invested for 12 months accumulate to?

Solution

$$A = £100 \qquad n = 12 \qquad i = 0.9\% = 0.009$$

Therefore the final sum is

$$F = A(1+i)^n = 100(1.009)^{12} = 100(1.1135097) = £111.35$$

The APR is not simply 12 times the monthly interest rate. A final sum of £111.35 after investing £100 for one year corresponds to an annual rate of interest of 11.35%. This is greater than 12 times the monthly rate of 0.9%, since

$$12 \times 0.9\% = 10.8\%$$

Therefore to find the corresponding APR for any given monthly rate of interest i_m one should use the formula

$$\text{APR} = (1 + i_m)^{12} - 1$$

Because $(1 + i_m)^{12}$ gives the ratio of the final sum F to the initial amount A the -1 has to be added to the formula in order to get the proportional increase in F over A.

Example 7.8

If the monthly rate of interest is 1.75% what is the corresponding APR?

Solution

$$i_m = 1.75\% = 0.0175$$
$$\text{APR} = (1 + i_m)^{12} - 1$$
$$= (1.0175)^{12} - 1$$
$$= 1.2314393 - 1$$
$$= 0.2314 = 23.14\%$$

(In this example and in most others in this chapter the interest rate is rounded off to a maximum of 2 decimal places.)

If you have a credit card you can check out this formula by referring to the leaflet on interest rates that the credit card company should give you. For example, in June 1991 the TSB Trustcard had an interest rate per month of 2.2% and quoted the APR as 29.8%. We can check this using the formula

$$\text{APR} = (1 + i_m)^{12} - 1$$
$$= (1.022)^{12} - 1$$
$$= 1.2984067 - 1$$
$$= 0.2984 = 29.8\%$$

Changes in interest rates

What if interest rates are expected to change before the end of the investment period? The final sum can be calculated by slightly adjusting the usual formula.

Example 7.9

Interest rates are expected to be 14% for the next 2 years and then fall to 10% for the following 3 years. How much will £2,000 be worth if it is invested for 5 years?

Solution

After 2 years the final sum will be

$$F = A(1+i)^n = 2,000(1.14)^2 = 2,000(1.2996) = £2,599.20$$

If this sum is then invested for a further 3 years at the new interest rate of 10% then the final sum is

$$F = A(1+i)^n = 2,599.20(1.1)^3 = 2,599.20(1.331) = £3,459.54$$

This could have been worked out in one calculation by finding

$$F = 2,000(1.14)^2(1.1)^3 = £3,459.54$$

Therefore the formula for the final sum F that an initial sum A will accrue to after n time periods at interest rate i_n and q time periods at interest rate i_q is

$$F = A(1+i_n)^n(1+i_q)^q$$

If more than two interest rates are involved then the formula can be adapted along the same lines.

Example 7.10

What will £20,000 invested for 10 years be worth if the expected rate of interest is 12% for the first 3 years, 9% for the next 2 years and 8% thereafter?

Solution

$$\text{final sum } F = 20,000(1.12)^3(1.09)^2(1.08)^5 = £49,051.90$$

QUESTIONS 7.1

(Assume interest rates are annual unless specified otherwise.)
1. If £4,000 is invested at 5% interest for 3 years what will the final sum be?

2. How much will £200 invested at 12% be worth at the end of 4 years?

3. A parent invests £6,000 for a 7-year-old child in a fixed interest scheme which guarantees 8% interest. How much will the child have at the age of 21?

4. If £525 is invested in a deposit account that pays 6% interest for 6 years, what will the final sum be?

5. What will £24,000 invested at 11% be worth at the end of 5 years?

6. Interest rates are expected to be 10% for the next 3 years and then to fall to 8% for the following 3 years. How much will an investment of £3,000 be worth at the end of 6 years?

7. If £40,000 is invested at a monthly rate of 1% what will it be worth after 9 months? What is the corresponding APR?

8. A sum of £450,000 is invested at a monthly interest rate of 0.6%. What will the final sum be after 18 months? What is the corresponding APR?

9. Which is the better investment for someone wishing to invest a sum of money for 2 years:
   (a) an account which pays 0.9% monthly or
   (b) an account which pays 11% annually?

10. If £1,600 is invested at a quarterly rate of interest of 4.5% what will the final sum be after 18 months? What is the corresponding APR?

7.4 TIME PERIODS, INITIAL AMOUNTS AND INTEREST RATES

The formula for the final sum of an investment

$$F = A(1 + i)^n$$

contains the four variables F, A, i and n. So far we have only calculated F for given values of A, i and n. But, as you should know from your knowledge of algebra, if the values of any three of the variables in this equation are given then one can usually calculate the fourth.

Initial amount

Other than F, the easiest value to calculate is A, when F, i and n are given. One can simply substitute the given values into the formula above or the formula can be adjusted as follows. Since

$$F = A(1 + i)^n$$

then

$$\frac{F}{(1 + i)^n} = A$$

195

by dividing through by $(1 + i)^n$. Thus the initial sum formula is

$$A = F(1 + i)^{-n}$$

Example 7.11

How much money needs to be invested now in order to accumulate a final sum of £12,000 in 4 years' time at an annual rate of interest of 10%?

Solution

$$A = F(1 + i)^{-n}$$
$$= 12,000(1.1)^{-4}$$
$$= \frac{12,000}{1.4641} = £8,196.16$$

What we have actually done in the above example is find the sum of money that is equivalent to £12,000 in 4 years' time if interest rates are 10%. An investor would therefore be indifferent between (a) £8,196.16 now and (b) £12,000 in 4 years' time.

The £8,196.16 is therefore known as the 'present value' (PV) of the £12,000 in 4 years' time. We shall come back to this concept in the next few sections when methods of appraising different types of investment project are explained.

Time period

Calculating the time period is rather more tricky than the calculation of the initial amount. If

$$F = A(1 + i)^n$$

then

$$\frac{F}{A} = (1 + i)^n$$

If the values of F, A and i are given and one is trying to find n this means that one has to work out to what power $1 + i$ has to be raised to equal F/A. One way of doing this is via logarithms.

Example 7.12

For how many years must £1,000 be invested at 10% in order to accumulate £1,600?

Solution

$$A = £1,000 \qquad F = £1,600 \qquad i = 10\% = 0.1$$

Substituting these values into the formula

$$\frac{F}{A} = (1 + i)^n$$

we get

$$\frac{1,600}{1,000} = (1 + 0.1)^n$$

$$1.6 = (1.1)^n \tag{1}$$

If equation (1) is specified in logarithms then

$$\log 1.6 = n \log 1.1 \tag{2}$$

since to find the *n*th power of a number its logarithm must be multiplied by *n*. Finding logs, this means that (2) becomes

$$0.20412 = n\, 0.0413927$$

Therefore

$$n = \frac{0.20412}{0.0413927} = 4.93 \text{ years}$$

If investments must be made for whole years then the answer is 5 years. This answer can be checked using the formula

$$F = A(1 + i)^n = 1,000(1.1)^5 = 1,610.51$$

If the £1,000 is invested for a full 5 years then it accumulates to just over £1,600, which checks out with the answer above.

A general formula to solve for *n* can be derived as follows from the final sum formula:

$$F = A(1 + i)^n$$

$$\frac{F}{A} = (1 + i)^n$$

In log form

$$\log\left(\frac{F}{A}\right) = n \log(1 + i)$$

$$\frac{\log(F/A)}{\log(1 + i)} = n \tag{3}$$

Another approach would be to use the iterative method. This entails setting up equation (1) above

$$1.6 = (1.1)^n$$

and then computing $(1.1)^n$ for different values of n until the answer 1.6 is reached. This sort of exercise is obviously suited to spreadsheet analysis. If you have a suitable program then set up one column headed n with values from 1 to 10 in 0.1 increments. In the next column headed F/A, insert the formula $(1.1)^n$, where n is the value of the corresponding cell in the first column, and then copy the formula down the column. (If you use a Lotus 1–2–3 spreadsheet and the first value of n is in cell A6, say, then you would need to enter the formula $+1.1 \wedge A6$ in cell B6 to calculate F/A.) You should then be able to read off the value of n in the first column which corresponds to 1.6 (or as close as you can get to it) in the second column and get 4.9 as your answer.

If you do not want to have a column with such a large number of rows then you can just set up the column for n with increments of one unit and then get the program to plot a graph with F/A on the vertical axis and n on the horizontal axis. You will then be able to use a ruler to read off the value of n that corresponds to 1.6 on the F/A axis.

Although many students (or their lecturers) will prefer to use a spreadsheet to solve these types of problems the method of solution using logarithms is used in the other examples in this section for the benefit of students who do not have a spreadsheet program, and because logarithms will usually give a more accurate figure.

Example 7.13

How many years will £2,000 invested at 5% take to accumulate to £3,000?

Solution

$$A = 2,000 \qquad F = 3,000 \qquad i = 5\% = 0.05$$

Using the formula for the time period

$$n = \frac{\log(F/A)}{\log(1+i)}$$

$$= \frac{\log 1.5}{\log 1.05}$$

$$= \frac{0.1760913}{0.0211893} = 8.31 \text{ years}$$

198

Example 7.14

How long will any sum of money take to double its value if it is invested at 12.5%?

Solution

Let the initial sum be A. Therefore the final sum is

$$F = 2A$$

and

$$i = 12.5\% = 0.125$$

Substituting into the final sum formula

$$F = A(1+i)^n$$

we get

$$2A = A(1.125)^n$$

Cancelling the As,

$$2 = (1.125)^n$$

Therefore

$$\log 2 = n \log 1.125$$

$$n = \frac{\log 2}{\log 1.125} = \frac{0.30103}{0.0511525} = 5.9 \text{ years}$$

Interest rates

A method of calculating the interest rate on an investment is explained in the following example.

Example 7.15

If £4,000 invested for 10 years is to accumulate to £6,000 what must the interest rate be?

Solution

$$A = 4{,}000 \qquad F = 6{,}000 \qquad n = 10$$

Substituting these values into the formula

$$F = A(1+i)^n$$

199

we get

$$6{,}000 = 4{,}000(1 + i)^{10}$$
$$1.5 = (1 + i)^{10}$$
$$1 + i = {}^{10}\!\sqrt{(1.5)}$$
$$= 1.0413797$$

using the $[\sqrt[x]{y}]$ function key on a calculator. Therefore,

$$i = 0.0414 = 4.14\%$$

In general terms, if

$$F = A(1 + i)^n$$
$$\frac{F}{A} = (1 + i)^n$$
$$\sqrt[n]{(F/A)} = 1 + i$$
$$\sqrt[n]{(F/A)} - 1 = i \qquad (4)$$

This can also be written as

$$i = \left(\frac{F}{A}\right)^{1/n} - 1$$

Example 7.16

At what interest rate will £3,000 accumulate to £10,000 after 15 years?

Solution

Using the interest rate formula (4) above

$$i = \sqrt[n]{\left(\frac{F}{A}\right)} - 1 = \sqrt[15]{\left(\frac{10{,}000}{3{,}000}\right)} - 1 = {}^{15}\!\sqrt{(3.3333)} - 1 = 1.083574 - 1$$

$$i = 0.083574 = 8.36\%$$

Example 7.17

An initial investment of £50,000 increases to £56,711.25 after 2 years. What interest rate has been applied?

Solution

$$A = 50{,}000 \qquad F = 56{,}711.25 \qquad n = 2$$

Series, time and investment

Therefore

$$\frac{F}{A} = \frac{56,711.25}{50,000} = 1.134225$$

Substituting into the formula

$$i = \sqrt[n]{\left(\frac{F}{A}\right)} - 1 = \sqrt[2]{(1.13455)} - 1 = 1.065 - 1 = 0.065$$

$$i = 6.5\%$$

QUESTIONS 7.2

1. How much needs to be invested now in order to accumulate £10,000 in 6 years' time if the interest rate is 8%?
2. What sum invested now will be worth £500 in 3 years' time if it earns interest at 12%?
3. Do you need to invest more than £10,000 now if you wish to have £65,000 in 15 years' time and you have a deposit account which guarantees 14%?
4. You need to have £7,500 on 1 January next year. How much do you need to invest at 1.3% per month if your investment is made on 1 June?
5. How much do you need to invest now in order to earn £25,000 in 10 years' time if the interest rate is
   (a) 10%   (b) 8%   (c) 6.5%?
6. How many complete years must £2,400 be invested at 5% in order to accumulate a minimum of £3,000?
7. For how long must £5,000 be kept in a deposit account paying 8% interest before it accumulates to £7,500?
8. If it can earn 9.5% interest, how long would any given sum of money take to treble its value?
9. If one needs to have a final sum of £20,000 how many years must one wait if £12,500 is invested at 9%?
10. How long will £70,000 take to accumulate to £100,000 if it is invested at 11%?
11. If £6,000 is to accumulate to £10,000 after being invested for 5 years, what rate must it earn interest at?
12. What interest rate will turn £50,000 into £60,000 after 2 years?
13. At what interest rate will £3,000 accumulate to £4,000 after 4 years?
14. What monthly rate of interest must be paid on a sum of £2,800 if it is to accumulate to £3,000 after 8 months?
15. What rate of interest would turn £3,000 into £8,000 in 10 years?
16. At what rate of interest will £600 accumulate to £900 in 5 years?
17. Would you prefer (a) £5,000 now or (b) £8,000 in 4 years' time if money can be borrowed or lent at 11%?

7.5 INVESTMENT APPRAISAL: NET PRESENT VALUE

Assume that you have £10,000 which you wish to invest and that someone offers you the following proposal: pay £10,000 now and get £12,000 back in 12 months' time. Assume that the returns on this investment are guaranteed and there are no other costs involved. What would you do? Perhaps you would compare this return of 20% with the rate of interest your money could earn in a deposit account, say 8%. In a simple example like this the comparison of rates of return is perhaps the most intuitively obvious method of judging the proposal.

This is not the preferred method for investment appraisal, however. Economists consider that the net present value (NPV) method has several advantages over the method of comparing the project rate of return with the market interest rate, known as the internal rate of return (IRR) method. These advantages are explained more fully in the following section, but first it is necessary to understand what the NPV method involves.

We have already come across the concept of present value (PV) in Section 7.4. If a certain sum of money will be paid to you at some given time in the future its PV is the amount of money that would accumulate to this sum if it was invested now at the ruling rate of interest.

Example 7.18

What is the present value of £1,500 payable in 3 years' time if the relevant interest rate is 8%?

Solution

Using the initial amount investment formula, where

$$F = £1,500 \qquad i = 0.08 \qquad n = 3$$
$$A = F(1 + i)^{-n}$$
$$= 1,500(1.08)^{-3}$$
$$= \frac{1,500}{(1.08)^3}$$
$$= \frac{1,500}{1.259712} = £1,190.75$$

An investor would be indifferent between £1,190.75 now and £1,500 in 3 years' time. Thus £1,190.75 is the PV of £1,500 in 3 years' time.

In all the examples in this chapter it is assumed that future returns are assured with 100% certainty. Of course, in reality some people may place greater importance on earlier returns just because the future is thought to be more risky. If some form of measure of the degree of risk can be estimated then more advanced mathematical methods exist which can be used to adjust the investment appraisal methods explained in this chapter. Here we shall just assume that estimated future returns, and costs, are correct. One has to try to make the most rational decision based on whatever information is available.

The *net present value* (NPV) of an investment project is the PV of the future returns minus the cost of setting up the project.

Example 7.19

An investment project involves an initial outlay of £600 now and a return of £1,000 in 5 years' time. Money can be invested at 9%. What is the NPV?

Solution

The PV of £1,000 in 5 years' time at 9% is found using the initial amount formula

$$A = F(1+i)^{-n} = 1000(1.09)^{-5} = £649.93$$

Therefore NPV = £649.93 − £600 = £49.93.

The project in the example above is clearly worthwhile. The £1,000 in 5 years time is equivalent to £649.93 now and so the outlay required of only £600 makes it a bargain. In other words, one is being asked to pay £600 for something which is worth £649.93.

Another way of looking at the situation is to consider what alternative sum could be earned by the investor's £600. If £649.93 was invested for 5 years at 9% it would accumulate to £1,000. Therefore the lesser sum of £600 must obviously accumulate to a smaller sum. Using the usual investment formula this can be calculated as

$$F = A(1+i)^n = 600(1.09)^5 = 600(1.538624) = £923.17$$

The investor thus has the choice of

1 putting £600 into this investment project and securing £1,000 in 5 years' time, or
2 investing £600 at 9%, accumulating £923.17 in 5 years.

Option (1) is clearly the winner.

The basic criterion for deciding whether or not an investment project is worthwhile is therefore

$$NPV > 0$$

If the outlay is less than the PV of the future return it must be a profitable venture.

As well as deciding whether specific projects are profitable or not, an investor may have to decide how to allocate limited capital resources to competing investment projects. The rule for choosing between projects is that they should be ranked according to their NPV. If only one project is possible then the one with the largest NPV should be chosen.

Example 7.20

An investor can put money into any one of the following three ventures:

> Project A costs £2,000 now and pays back £3,000 in 4 years
> Project B costs £2,000 now and pays back £4,000 in 6 years
> Project C costs £3,000 now and pays back £4,800 in 5 years

The current interest rate is 10%. Which project should be chosen?

Solution

$$\text{NPV of project A} = 3,000(1.1)^{-4} - 2,000$$
$$= 2,049.04 - 2,000 = £49.04$$
$$\text{NPV of project B} = 4,000(1.1)^{-6} - 2,000$$
$$= 2,257.90 - 2,000 = £257.90$$
$$\text{NPV of project C} = 4,800(1.1)^{-5} - 3,000$$
$$= 2,980.42 - 3,000 = -£19.58$$

Project B has the largest NPV and is therefore the best investment. Project C has a negative NPV and so would not be worthwhile even if there was no competition.

The investment appraisal examples considered so far have only involved a single return payment at some given time in the future. However, most real investment projects involve a stream of returns occurring over several time periods. The same principle for calculating NPV is used to assess these projects, the initial outlay being subtracted from the sum of the PVs of the different future returns.

Example 7.21

An investment proposal involves an initial payment now of £40,000 and then returns of £10,000, £30,000 and £20,000 respectively in 1, 2 and 3 years' time. If money can be invested at 10% is this a worthwhile investment?

Solution

$$\text{PV of £10,000 in 1 year's time} = \frac{£10,000}{1.1} = £9,090.91$$

$$\text{PV of £30,000 in 2 years' time} = \frac{£30,000}{1.1^2} = £24,793.39$$

$$\text{PV of £20,000 in 3 years' time} = \frac{£20,000}{1.1^3} = £15,026.30$$

Total PV of future returns	£48,910.60
less Initial outlay	− £40,000
NPV of project	£ 8,910.60

This is greater than zero and so the project is worthwhile. At an interest rate of 10% one would need to invest a total of £48,910.60 to get back the stated returns and so £40,000 is clearly a bargain price.

The further into the future the expected return occurs the greater will be the discounting factor. This is made obvious in Example 7.22 below, where the returns are the same each time period. The PV of each successive year is smaller than that of the previous year because it is multiplied by $(1 + i)^{-1}$.

Example 7.22

An investment project requires an initial outlay of £7,500 and will pay back £2,000 at the end of the next 5 years. Is it worthwhile if capital can be invested elsewhere at 12%?

Solution

$$\text{PV of £2,000 in 1 year's time} = \frac{£2,000}{1.12} = £1,785.71$$

$$\text{PV of £2,000 in 2 years' time} = \frac{£2,000}{1.12^2} = £1,594.39$$

$$\text{PV of £2,000 in 3 years' time} = \frac{£2,000}{1.12^3} = £1,423.56$$

$$\text{PV of £2,000 in 4 years' time} = \frac{£2,000}{1.12^4} = £1,271.04$$

$$\text{PV of £2,000 in 5 years' time} = \frac{£2,000}{1.12^5} = £1,134.85$$

205

Total PV of future returns   £7,209.55

less Initial outlay   − £7,500.00

NPV of project   − £ 290.45

NPV < 0 and so this is not a worthwhile investment.

Investment appraisal using a spreadsheet

From the above examples one can see that the mathematics involved in calculating the NPV of a project can be quite time consuming. For this type of problem a spreadsheet program can be a great help. It is important, though, to understand what the program is doing.

To set up the general solution of an NPV calculation in algebraic terms assume that R_j is the net return in year j, i is the given rate of interest, n is the number of time periods in which returns occur and C is the initial cost of the project. Then

$$\text{NPV} = \frac{R_1}{1+i} + \frac{R_2}{(1+i)^2} + \dots + \frac{R_n}{(1+i)^n} - C$$

Using the Σ notation this becomes

$$\text{NPV} = \sum_{j=1}^{n} \frac{R_j}{(1+i)^j} - C \tag{1}$$

If the initial outlay C is considered as a negative return at time 0 (i.e. $R_0 = -C$) the formula can be more neatly stated as

$$\text{NPV} = \sum_{j=0}^{n} \frac{R_j}{(1+i)^j} \tag{2}$$

There will be no discounting of the initial outlay in the first term

$$\frac{R_0}{(1+i)^0}$$

since $(1+i)^0 = 1$. (Remember that $x^0 = 1$ whatever the value of x.)

The spreadsheet program Lotus 1–2–3 has an NPV function that allows the direct calculation of the NPV of a stream of returns for a given interest rate. One has to be careful in using this function, though, because it does not take into account the initial outlay.

To illustrate this point, the following example shows how a spreadsheet program can be used to work out the NPV of a project from first principles. The answer obtained is then compared with the Lotus 1–2–3 solution using the NPV function.

Example 7.23

An investment project requires an initial outlay of £25,000 with the following expected returns:

> £5,000   at the end of year 1
> £6,000   at the end of year 2
> £10,000 at the end of year 3
> £10,000 at the end of year 4
> £10,000 at the end of year 5

Is this a viable investment if money can be invested elsewhere at 15%?

Solution

Call up Lotus 1–2–3 on your computer. On your spreadsheet template set up the following labels and right justify (RJ) where indicated.

FILENAME	in cell A1	
NPV	in cell B1	
EX 756	in cell C1	
YEAR	in cell A3	(RJ)
RETURN	in cell B3	(RJ)
PV RET	in cell C3	(RJ)
INTEREST	in cell C2	
TOTAL PV	in cell B11	
LOTUS NPV	in cell B12	
NPV	in cell B13	

These should appear as shown in Table 7.1.

Table 7.1

FILENAME	NPV	EX 756	
		INTEREST	15.00%
YEAR	RETURN	PV RET	
0	−25000	−25000	
1	5000	4347.826	
2	6000	4536.862	
3	10000	6575.162	
4	10000	5717.532	
5	10000	4971.767	
	TOTAL PV	1149.150	
	LOTUSNPV	26149.15	
	NPV	1149.150	

In cell A4 enter the figure 0. In cell A5 enter the formula +A4+1 and then copy this formula down to cell A9 using the copy command/C. You should get the years 0 to 5 appearing. In cell D2 enter the interest rate 0.15 and then reformat this into the percentage format 15%.

In cell B4 enter − 25000. This is the initial outlay and is hence a negative return in year 0.

In cells B5–B9 enter the expected returns for each year as given in the question.

To obtain the PV of returns in any year j, one needs to use the discounting formula

$$PV = \frac{R_j}{(1+i)^j}$$

In cell C4 on the spreadsheet you therefore need to enter the formula +B4/(1+$D$2)^A4. The B4 refers to the actual return in that year, $D$2 refers to the given interest rate in cell D2 and A4 refers to the power to which $1 + i$ is raised. Given that there is a zero in cell A4 one should now get the same − 25000 appearing in cell C4.

Now copy the formula in cell C4 into cells C5–C9. The values appearing in these cells will be the PVs of the returns in cells B5–B9. The first can easily be checked using a pocket calculator as

$$PV \text{ of £5,000 in 1 year's time} = \frac{5,000}{1.15} = 4,347.826$$

The NPV of the project as a whole will therefore be the sum of the PVs in cells C4–C9, which includes the negative return of the initial outlay.

To obtain this sum enter the formula @SUM (C4.C9) in cell C11. This gives the project NPV of £1,149.15 which is positive and hence means the project is a viable investment opportunity.

This can be compared with the answer obtained using the Lotus 1–2–3 built-in NPV function. In cell C12 enter @NPV(D2,B5.B9). This will calculate the PV of the returns in years 1–5 using the interest rate you have entered in D2. The Lotus 1–2–3 NPV function always treats the number in the first cell of the range (i.e. B5) as the return at the end of year 1. The computed answer of 26,149.15 is the total PV of the returns in years 1–5. To get the overall NPV of the project one has to subtract the initial outlay. This can be done by entering the formula +C12+B4 in cell C13. (Note that the contents of B4 are already negative and are therefore added, not subtracted.)

The answer in cell C13 should be identical with that in cell C11, at 1,149.15. If it is, then you have demonstrated how a spreadsheet can calculate the NPV of an investment project from first principles and get the same answer as that computed using the built-in Lotus 1–2–3 NPV function after allowing for the initial cost of setting up the project. Now save your worksheet onto a disk.

The question asks if the project is worthwhile. Given a positive NPV of £1,149,15, the answer is clearly 'yes'.

The worksheet you have created can be used to work out the NPV for other projects. The initial cost and returns need to be entered in cells B4–B9 and the new interest rate goes in D2. Obviously if there are more (or less) years when returns occur then rows will need to be added (or deleted or left blank).

As investment appraisal involves the comparison of different projects, as well as the assessment of the financial viability of individual projects, a spreadsheet can be adapted to work out the NPV for more than one project.

The following example shows how the spreadsheet created for Example 7.23 can be extended so that two projects can be compared.

Example 7.24

An investor has to choose between two projects A and B whose outlay and returns are set out in Table 7.2. Which is the better investment if the going rate of interest is 10%?

Table 7.2

	Project A	Project B
Initial outlay	30,000	30,000
Return in 1 year's time	6,000	8,000
Return in 2 years' time	10,000	8,000
Return in 3 years' time	10,000	8,000
Return in 4 years' time	10,000	8,000
Return in 5 years' time	8,000	8,000

All values are in pounds

Solution

Call up the worksheet in Table 7.1 which you created for Example 7.23 and make the following label changes:

Change contents of cell C1 to	EX757	
In cell D3 insert	RETURN	(RJ)
In cell E3 insert	PV RET	(RJ)
In cell B10 insert	PROJECTA	
In cell D10 insert	PROJECTB	
Put the new interest rate in cell	D2, i.e. enter 0.1	

In cells B4–B9 enter the (negative) outlay and returns for project A. The NPV should automatically be calculated as £3,029.66, as shown in Table 7.3. In cells D4–D9 enter the outlay and returns for project B. To calculate PVs enter the formula

209

Table 7.3

FILENAME NPV		EX 757		
		INTEREST	10.00%	
YEAR	RETURN	PV RET	RETURN	PV RET
0	-30000	-30000	-30000	-30000
1	6000	5454.545	8000	7272.727
2	10000	8264.462	8000	6611.570
3	10000	7513.148	8000	6010.518
4	10000	6830.134	8000	5464.107
5	8000	4967.370	8000	4967.370
	PROJECTA		PROJECTB	
	TOTAL PV	3029.661		326.2941
	LOTUSNPV	33029.66		30326.29
	NPV	3029.661		326.2941

$$+D4/(1+\$D\$2)^\wedge A4$$

in cell E4 and then copy into cells E5–E9.

The total NPV (both versions) for project B is then found by entering

@SUM(E4.E9)	in cell E11
@NPV(D2,D5.D9)	in cell E12
+E12+D4	in cell E13

The computed NPV for project B is £326.29 compared with £3,029.66 for project A. Therefore, although both projects are financially viable, the better investment is project A because it has the greater NPV.

If you do not have access to a spreadsheet program then you may have to work out the NPV of different projects from first principles. However, there are now available financial calculators with an NPV function which may be a cheaper alternative than a computer. To assist students without a spreadsheet program a set of discounting factors is reproduced in Table 7.4. Although the actual monetary returns will differ from project to project the discounting factor will be the same for a given time period and a given rate of interest. For example, the PV of a sum of money £x payable in 8 years' time when the interest rate is 7% will be

$$\frac{£x}{(1.07)^8} \quad \text{or} \quad £x(1.07)^{-8}$$

The value of $(1.07)^{-8}$ can be read off from Table 7.4 by looking at the column headed 7% and the row corresponding to year 8, giving a figure of 0.582009. If £x was £525 then the PV would be

$$£525(0.582009) = £305.55$$

Table 7.4 Discounting factors for net present value

Rate of interest i	6.00%	7.00%	8.00%	9.00%	10.00%	11.00%
Year 0	1	1	1	1	1	1
1	0.943396	0.934579	0.925925	0.917431	0.909090	0.900900
2	0.889996	0.873438	0.857338	0.841679	0.826446	0.811622
3	0.839619	0.816297	0.793832	0.772183	0.751314	0.731191
4	0.792093	0.762895	0.735029	0.708425	0.683013	0.658730
5	0.747258	0.712986	0.680583	0.649931	0.620921	0.593451
6	0.704960	0.666342	0.630169	0.596267	0.564473	0.534640
7	0.665057	0.622749	0.583490	0.547034	0.513158	0.481658
8	0.627412	0.582009	0.540268	0.501866	0.466507	0.433926
9	0.591898	0.543933	0.500248	0.460427	0.424097	0.390924
10	0.558394	0.508349	0.463193	0.422410	0.385543	0.352184
11	0.526787	0.475092	0.428882	0.387532	0.350493	0.317283
12	0.496969	0.444011	0.397113	0.355534	0.318630	0.285840

Rate of interest i	12.00%	13.00%	14.00%	15.00%	16.00%	17.00%
Year 0	1	1	1	1	1	1
1	0.892857	0.884955	0.877192	0.869565	0.862068	0.854700
2	0.797193	0.783146	0.769467	0.756143	0.743162	0.730513
3	0.711780	0.693050	0.674971	0.657516	0.640657	0.624370
4	0.635518	0.613318	0.592080	0.571753	0.552291	0.533650
5	0.567426	0.542759	0.519368	0.497176	0.476113	0.456111
6	0.506631	0.480318	0.455586	0.432327	0.410442	0.389838
7	0.452349	0.425060	0.399637	0.375937	0.353829	0.333195
8	0.403883	0.376159	0.350559	0.326901	0.305025	0.284782
9	0.360610	0.332884	0.307507	0.284262	0.262952	0.243403
10	0.321973	0.294588	0.269743	0.247184	0.226683	0.208037
11	0.287476	0.260697	0.236617	0.214943	0.195416	0.177809
12	0.256675	0.230705	0.207559	0.186907	0.168462	0.151974

The appearance on the market of pocket calculators with a $[y^x]$ function key has made tables like this largely redundant, however. To calculate $£525(1.07)^{-8}$ one only has to enter

$$525[\div]1.07[y^x]8[=]$$

or

$$525[\times]1.07[y^x]8[+/-][=]$$

The following examples show how Table 7.4 can be used to calculate the NPV for different projects. You can, of course, check these figures on your spreadsheet.

Example 7.25

An investor can put money into one of the two projects whose returns at the end of each year are shown in Table 7.5. Both require an initial outlay

Basic mathematics for economists

of £50,000 and the given rate of interest is 6%. Which is the better investment?

Table 7.5

	Project A	Project B
Year 1	4,500	3,000
Year 2	6,000	3,000
Year 3	7,500	6,000
Year 4	9,000	6,000
Year 5	10,500	12,000
Year 6	12,000	12,000
Year 7	12,000	15,000
Year 8	12,000	15,000

All values are given in pounds.

Solution

Reading the discount factors for years 1–8 for 6% from Table 7.4 the project NPVs can be computed as in Table 7.6. Therefore, at 6% interest, A is the better investment.

Table 7.6

		Project A		Project B	
Year	Discount factor	Return	NPV	Return	NPV
1	0.943396	4,500	4,245.28	3,000	2,830.188
2	0.889996	6,000	5,339.98	3,000	2,669.989
3	0.839619	7,500	6,297.14	6,000	5,037.715
4	0.792093	9,000	7,128.84	6,000	4,752.561
5	0.747258	10,500	7,846.21	12,000	8,967.098
6	0.704960	12,000	8,459.53	12,000	8,459.526
7	0.665057	12,000	7,980.69	15,000	9,975.856
8	0.627412	12,000	7,528.95	15,000	9,411.185
Total PV of future returns			54,826.62		52,104.12
less Initial outlay			− 50,000.00		− 50,000.00
Project NPV			4,826.62		2,104.12

Example 7.26

The returns from two projects A and B are as set out in Table 7.7. If each project requires an initial outlay of £30,000 and the market rate of interest is 9%, which is the better investment?

Solution

From Table 7.8, when $i = 9\%$, B is the better investment.

212

Series, time and investment

Table 7.7

Year	Project A	Project B
1	2,500	4,000
2	3,000	4,000
3	6,000	6,000
4	7,500	6,000
5	8,500	8,000
6	9,000	8,000
7	9,000	10,000
8	4,000	5,000

All values are given in pounds.

Table 7.8

		Project A		Project B	
Year	Discount factor	Return	NPV	Return	NPV
1	0.917431	2,500	2,293.58	4,000	3,669.72
2	0.841679	3,000	2,525.04	4,000	3,366.72
3	0.772183	6,000	4,633.10	6,000	4,633.10
4	0.708425	7,500	5,313.19	6,000	4,250.55
5	0.649931	8,500	5,524.42	8,000	5,199.45
6	0.596267	9,000	5,366.41	8,000	4,770.14
7	0.547034	9,000	4,923.31	10,000	5,470.34
8	0.501866	4,000	2,007.47	5,000	2,509.33
Total PV of future returns			32,586.50		33,869.36
less Initial outlay			− 30,000.00		− 30,000.00
Project NPV			2,586.50		3,869.36

QUESTIONS 7.3

1. The following investment projects all involve an outlay now and a single return at some point in the future. Calculate the NPV and say whether or not each is a worthwhile investment:

   (a) £1,100 outlay, £1,500 return after 3 years, interest rate 8%
   (b) £750 outlay, £1,000 return after 5 years, interest rate 9%
   (c) £10,000 outlay, £12,000 return after 3 years, interest rate 8%
   (d) £50,000 outlay, £75,000 return after 3 years, interest rate 14%
   (e) £50,000 outlay, £100,000 return after 5 years, interest rate 14%
   (f) £5,000 outlay, £7,000 return after 3 years, interest rate 6%
   (g) £5,000 outlay, £7,750 return after 5 years, interest rate 6%
   (h) £5,000 outlay, £8,500 return after 6 years, interest rate 6%

2. An investor has to choose between the following three projects:

   Project A requires an outlay of £35,000 and returns £60,000 after 4 years
   Project B requires an outlay of £40,000 and returns £75,000 after 5 years
   Project C requires an outlay of £25,000 and returns £50,000 after 6 years

213

Which project would you advise this investor to put money into if the cost of capital is 10%?

3. A firm has a choice between three investment projects, all of which involve an initial outlay of £36,000. The returns at the end of the next 4 years are given in Table 7.9. If the interest rate is 15%, say (a) whether each project is viable or not, and (b) which is the best investment.

Table 7.9

Year	Project A	Project B	Project C
1	15,000	5,000	20,000
2	15,000	10,000	15,000
3	15,000	20,000	10,000
4	15,000	25,000	5,000

All values are given in pounds.

4. If money can be invested elsewhere at 6%, is the following project worthwhile?

Initial outlay	£100,000
Return at end of year 1	£10,000
Return at end of year 2	£12,000
Return at end of year 3	£15,000
Return at end of year 4	£18,000
Return at end of year 5	£20,000
Return at end of year 6	£20,000
Return at end of year 7	£20,000
Return at end of year 8	£15,000
Return at end of year 9	£10,000
Return at end of year 10	£5,000

5. Would you put £40,000 into a project which pays back nothing in the first year but then brings annual net returns of £12,000 from the end of year 2 until the end of year 6, assuming an interest rate of 8%?

6. A project requires an initial outlay of £20,000 and will pay back the following returns (in £):

1,000 at the end of year 1
1,000 at the end of year 2
2,000 at the end of year 3
2,000 at the end of year 4
5,000 at the end of year 5
5,000 at the end of year 6
5,000 at the end of year 7
5,000 at the end of year 8
5,000 at the end of year 9
5,000 at the end of year 10

Is this project a worthwhile investment if the going rate of interest is (a) 9%, (b) 10%?

7. Which of the three projects shown in Table 7.10 is the best investment if the interest rate is 20%?

Table 7.10

	Project A	Project B	Project C
Outlay now	85,000	40,000	40,000
Return after year 1	20,000	15,000	10,000
Return after year 2	24,000	20,000	12,000
Return after year 3	30,000	25,000	12,000
Return after year 4	30,000	0	12,000
Return after year 5	25,000	0	15,000
Return after year 6	20,000	0	15,000

All values are given in pounds.

7.6 THE INTERNAL RATE OF RETURN

The internal rate of return (IRR) method of investment appraisal involves finding the rate of return (r) on a project and comparing it with the market rate of interest (i). If $r > i$ then the project is viable. Alternative projects can be ranked according to the magnitude of the different rates of return.

Example 7.27

Find the IRR for the three projects in Table 7.11, decide whether they are viable if the market rate of interest is 7%, and then rank them in order of profitability according to the IRR method.

Table 7.11

	Project A	Project B	Project C
Initial outlay	£5,000	£4,000	£8,000
Return after 1 year	£5,750	£4,300	£8,500

Solution

In this simple example it is obvious from basic arithmetic that

$$\text{IRR for A} = r_A = \frac{750}{5,000} = 0.15 = 15\%$$

$$\text{IRR for B} = r_B = \frac{300}{4,000} = 0.075 = 7.5\%$$

$$\text{IRR for C} = r_C = \frac{500}{8,000} = 0.0625 = 6.25\%$$

215

Basic mathematics for economists

Only projects A and B produce an IRR of more than the market rate of interest of 7% and so C is not viable.

A is preferred to B because $r_A > r_B$.

From the above example one can see that the IRR is the rate of interest which, if applied to the initial outlay, gives the return in year 1. Put another way, r is the rate of interest at which the PV of the future return equals the initial outlay, thus making the NPV of the whole project zero. This principle can be utilized to help calculate the IRR for more complex problems.

Example 7.28

Use the IRR method to evaluate the following project given a market rate of interest of 11%.

Initial outlay	£75,000
Return at end of year 1	£15,000
Return at end of year 2	£20,000
Return at end of year 3	£20,000
Return at end of year 4	£25,000
Return at end of year 5	£25,000
Return at end of year 6	£12,000

Solution

One needs to find the value of r for which

$$0 = -75,000 + 15,000(1+r)^{-1} + 20,000(1+r)^{-2} + 20,000(1+r)^{-3}$$
$$+ 25,000(1+r)^{-4} + 25,000(1+r)^{-5} + 12,000(1+r)^{-6}$$

The algebraic method of solution is far too complex and time consuming to consider using here. The most practical method is to use the Lotus 1–2–3 spreadsheet program. This has an IRR function which can immediately calculate r. You can also use the NPV function to check that the value of r is indeed that for which the NPV is zero.

A worksheet that solves this problem is shown in Table 7.12. This is constructed as follows. Enter the labels

IRR-NPV	in cell A1	
EXAMPLE	in cell B1	
762	in cell C1	
IRR	in cell B2	(RJ)
YEAR	in cell A3	
RETURN	in cell B3	

216

INTEREST	in cell C3
LOTUSNPV	in cell E3
NPV	in cell F3

Reduce the width of column D to 2 units to make the spreadsheet easier to read. Enter 0 in cell A4 and the formula +A4+1 in cell A5 and copy this to cells A6–A10.

Enter the initial outlay as a negative number in cell B4 and the annual returns in cells B5–B10.

To use the Lotus 1–2–3 built-in IRR function enter

@IRR(0.1, B4.B10)   in cell C2

The 0.1 is a guess at the value of r. (The closer the initial guess, the quicker the calculation using this function.) Note that the range over which the IRR is calculated includes the initial outlay in cell B4. Format cell C2 to per cent to 2 decimal places and you should get 14.14%, which is the actual IRR for this project.

To check this against the NPV calculation, first enter a range of interest rates in column C. To do this enter, say, 0.05 in cell C4 and then the formula +C4+0.01 in cell C5. Copy this down the column and put in per cent format.

In cell E4 enter @NPV(C4,$B$5.$B$10). This will calculate the PV of the future returns using the interest rate in C4. Copy this formula down the column.

Table 7.12

IRR–NPV	EXAMPLE	762		
	IRR	14.14%		
YEAR	RETURN	INTEREST	LOTUSNPV	NPV
0	−75000	5.00%	98813.35	23813.35
1	15000	6.00%	95686.58	20686.58
2	20000	7.00%	92706.56	17706.56
3	20000	8.00%	89864.67	14864.67
4	25000	9.00%	87152.86	12152.86
5	25000	10.00%	84563.64	9563.641
6	12000	11.00%	82090.03	7090.037
		12.00%	79725.53	4725.536
		13.00%	77464.06	2464.062
		14.00%	75299.93	299.9377
		15.00%	73227.85	−1772.14
		16.00%	71242.85	−3757.14
		17.00%	69340.28	−5659.71
		18.00%	67515.80	−7484.19
		19.00%	65765.31	−9234.68
		20.00%	64085.00	−10914.9

You will remember that the Lotus 1–2–3 NPV function does not take into account the initial outlay. The true NPV of the project is therefore calculated in column F by entering the formula +E4+$B$4 in cell F4 (to subtract the initial outlay in cell B4) and then copying this formula down the column.

The IRR will be the rate of interest that corresponds to an NPV of zero. From the spreadsheet in Table 7.12 one can see that this will occur somewhere between the interest rates of 14% and 15%, which checks out with the IRR of 14.14% computed in cell C2.

To illustrate this result, the spreadsheet you have constructed can be used to obtain a plot with the interest rate on the horizontal axis and the NPV (as calculated in column F) on the vertical axis using the /G command. This will cut the interest rate axis at 14.14% as Figure 7.2 shows.

The given market rate of interest is 11% and so, as the calculated IRR of 14.14% exceeds this, the project is worthwhile according to this criterion.

Deficiencies of the IRR criterion

Although the IRR method may appear to be the most obvious and easily understood criterion for deciding on investment projects, and is still frequently used, it has several deficiencies which make it less useful than the NPV method.

First, it ignores the total value of the profit, as illustrated below.

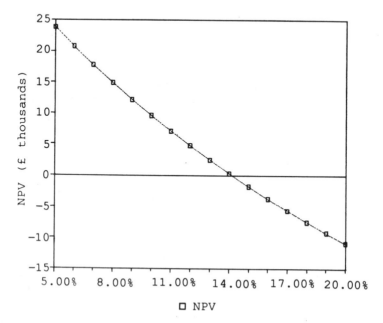

Figure 7.2

Example 7.29

A firm has to choose between projects A and B. The interest rate is 6%. Project A involves an initial outlay of £18,000 and a return in 1 year's time of £20,000. Project B involves an initial outlay of £2,000 and a return in 1 year's time of £2,500.

Solution

The IRR method ranks B as the best investment opportunity, since for A

$$r_A = \frac{20,000}{18,000} - 1 = 1.11 - 1 = 0.11 = 11\%$$

and for B

$$r_B = \frac{2,500}{2,000} - 1 = 1.25 - 1 = 0.25 = 25\%$$

The NPV method, however, would rank A as the better investment since

$$NPV_A = -18,000 + \frac{20,000}{1.06} = -18,000 + 18,867.92 = £867.92$$

$$NPV_B = -2,000 + \frac{2,500}{1.06} = -2,000 + 2,358.49 = £358.49$$

If the firm has a straightforward choice between A and B, then A is clearly the better investment. (The possibility of the firm using its initial £18,000 for investing in nine separate projects all with the same returns as B is ruled out.)

Some students may still not be convinced that the IRR method is faulty in the above example as one so often sees the rate of return used as a measure of the success of an investment in the press and other sources. Let us therefore work from first principles and consider the total assets of the firm after one year.

Assume that the firm has up to £18,000 at its disposal. If it puts this all into project A, then at the end of the year total assets will be £20,000.

If its puts £2,000 into project B, then it can also invest the remaining £16,000 elsewhere at the going rate of interest of 6%. Its total assets will therefore be as follows:

Return on project B	£ 2,500
plus £16,000 invested at 6% = 16,000 × 1.06	£16,960
Total assets	£19,460

Thus the firm is in a better financial position overall at the end of the year if it chooses project A, which is what the NPV method recommends but what the IRR method advises against.

The second drawback of the IRR method is that there may not be one unique solution for *r* when there are several negative terms in the polynomial to be solved. This point was made in Chapter 6 when the solution of polynomial equations was discussed and is illustrated in the example below.

Example 7.30

How should a firm choose between projects A and B in Table 7.13 if the interest rate is 19%?

Table 7.13

	Project A	Project B
Year 0 Net outlay	– 40,000	– 40,000
Year 1 Net return	15,000	15,000
Year 2 Net return	28,000	28,000
Year 3 Net return	– 25,000	– 25,000
Year 4 Net return	80,000	90,000
Year 5 Net return	30,000	30,000
Year 6 Net return	– 20,000	– 20,000

All values are given in pounds.

Solution

Calculating the IRR in projects with variable streams of future returns such as these involves solving for *r* in a polynomial equation.

In project A, for example, one needs to find the value of *r* for which

$$40,000 = \frac{15,000}{1+r} + \frac{28,000}{(1+r)^2} - \frac{25,000}{(1+r)^3} + \frac{80,000}{(1+r)^4} + \frac{30,000}{(1+r)^5} - \frac{20,000}{(1+r)^6} \quad (1)$$

The algebraic solution to this problem is complex and time consuming. However, if you have access to the spreadsheet package Lotus 1–2–3 or other financial computer packages then you can simply use the IRR function as already explained in Example 7.28 above. Before you do this, take a close look at the figures. Projects A and B are almost identical except that project B has a £10,000 greater return than project A in year 4. There are negative figures in year 3 because a further investment is required, and in year 6 because the company has to pay to dismantle the site and return it to an environmentally acceptable state.

Before doing any calculations, one would think that B must obviously be the best choice whatever the criterion used, because of the extra return in year 4. However, putting these figures into a revised version of the spreadsheet which was developed for Example 7.28 (so that the NPV and IRR for

220

Table 7.14

IRR-NPV	EXAMPLE	764		IRR A	IRR B
				39.53%	-225.50%
YEAR	RETURN A	RETURN B	INTEREST	NPV A	NPV B
0	-40000	-40000	19.00%	22964.39	27951.08
1	15000	15000	20.00%	21415.46	26237.99
2	28000	28000	21.00%	19923.49	24588.56
3	-25000	-25000	22.00%	18485.90	22999.89
4	80000	90000	23.00%	17100.23	21469.20
5	30000	30000	24.00%	15764.17	19993.90
6	-20000	-20000	25.00%	14475.52	18571.52
			26.00%	13232.18	17199.69
			27.00%	12032.19	15876.21
			28.00%	10873.66	14598.95
			29.00%	9754.805	13365.91
			30.00%	8673.927	12175.20
			31.00%	7629.417	11025.00
			32.00%	6619.743	9913.597
			33.00%	5643.450	8839.351
			34.00%	4699.151	7800.715
			35.00%	3785.529	6796.211
			36.00%	2901.329	5824.434
			37.00%	2045.357	4884.046
			38.00%	1216.475	3973.773
			39.00%	413.5995	3092.402
			40.00%	-364.303	2238.778
			41.00%	-1118.21	1411.797
			42.00%	-1849.08	610.4120

two projects can be computed) gives the result $IRR_A = 39.53\%$, a handsome return, but $IRR_B = -225.5\%$, as Table 7.14 shows.

Something is clearly amiss! Why should project B give a negative IRR when it shows a better set of figures than project A? The reason is that when there are several instances of negative figures in the future stream of returns there are several solutions for r in the relevant polynomial, i.e. (1) above. The computer has thrown up one of the negative mathematical solutions which clearly does not give a true reflection of the project's financial attractiveness.

Again, the NPV method can easily cope with a stream of returns which switch from positive to negative and back again without producing any spurious answers. In this example you can check (from the spreadsheet in Table 7.14 or calculating from first principles) that when $i = 19\%$

$$NPV_A = £22,964.39$$

$$NPV_B = £27,951.08$$

This suggests that B is the better investment, which is the solution that one would expect from the figures in this example with B's higher return in year 4.

Table 7.15

Interest	NPV$_A$	NPV$_B$
39.00%	413.60	3,092.40
40.00%	– 364.30	2,238.78
41.00%	– 1,118.21	1,411.80
42.00%	– 1,849.08	610.41
43.00%	– 2,557.80	– 166.38

One can always get an estimate of the relevant mathematical solution for r_B (IRR for B) by calculating the NPV of project B for various interest rates and noting the value at which it becomes zero. From Table 7.14 you can read off the values given in Table 7.15. This means that r_A must lie between 39% and 40%, which checks out with the computed answer of 39.53% using the IRR function, and that r_B is between 42% and 43%. This means that B is preferred to A, giving the same ranking as the NPV method. The point of this example, though, is that a programmed IRR function can give misleading answers if the stream of returns alternates between positive and negative values.

A third advantage that the NPV investment appraisal method has over the IRR method is that it can easily cope with forecasts of variable interest rates. The IRR method just involves comparing the computed IRR from an investment project with one given interest rate and so it could not be applied to Example 7.31 below.

Example 7.31

An investment project involves an initial outlay of £25,000 and net annual returns as follows:

£6,000	at the end of year 1
£8,000	at the end of year 2
£8,000	at the end of year 3
£10,000	at the end of year 4
£6,000	at the end of year 5

Interest rates are currently 15% but are forecast to fall to 12% next year and 10% the following year. They will then rise by 1 percentage point each year. Is the project worthwhile?

Solution

Unfortunately this variation in interest rates means that one cannot simply use the NPV function on a spreadsheet program and so the answer is computed manually. The discount rate will be different for each year and so

$$NPV = -25,000 + \frac{6,000}{1.15} + \frac{8,000}{1.15 \times 1.12} + \frac{8,000}{1.15 \times 1.12 \times 1.1}$$

$$+ \frac{10,000}{1.15 \times 1.12 \times 1.1 \times 1.11} + \frac{6,000}{1.15 \times 1.12 \times 1.1 \times 1.11 \times 1.12}$$

$$= -25,000 + \frac{6,000}{1.15} + \frac{8,000}{1.288} + \frac{8,000}{1.4168} + \frac{10,000}{1.572648} + \frac{6,000}{1.7613657}$$

$$= -25,000 + 5,217.39 + 6,211.18 + 5,646.53 + 6,358.70 + 3,406.45$$

$$= -25,000 + 26,840.25$$

$$= £1,840.25$$

This is positive and so the investment is worthwhile.

Although this is not a straightforward NPV calculation, spreadsheet programs can be adapted to do the calculations. One suggested format for solving Example 7.31 is shown in Table 7.16 using the Lotus 1–2–3 notation. This should produce the figures shown in Table 7.17 which confirm that NPV is £1,840.25.

Table 7.16

	A	B	C	D	E	F
1	NPV	WITH	VARIABLE	INTEREST	RATES	
2						
3	YEAR	i	DISCOUNT	FACTOR	RETURN	PV
4	0	0	+1/(1+B4)	1	−25000	+D4*E4
5	1	0.15	+1/(1+B5)	+D4*C5	6000	+D5*E5
6	2	0.12	+1/(1+B6)	+D5*C6	8000	+D6*E6
7	3	0.10	+1/(1+B7)	+D6*C7	8000	+D7*E7
8	4	0.11	+1/(1+B8)	+D7*C8	10000	+D8*E8
9	5	0.12	+1/(1+B9)	+D8*C9	6000	+D9*E9
10						
11				TOTAL	NPV	@SUM(F4.F9)

Table 7.17

NPV	WITH	VARIABLE	INTEREST	RATES	
YEAR		i DISCOUNT	FACTOR	RETURN	PV
0	0	1	1	−25000	−25000
1	0.15	0.869565	0.869565	6000	5217.391
2	0.12	0.892857	0.776397	8000	6211.180
3	0.1	0.909090	0.705815	8000	5646.527
4	0.11	0.900900	0.635870	10000	6358.702
5	0.12	0.892857	0.567741	6000	3406.447
			TOTAL	NPV	1840.248

Basic mathematics for economists

QUESTIONS 7.4

1. Calculate the IRR for the projects in Table 7.18 and then say whether or not the IRR ranking is consistent with the NPV ranking for these projects if the market rate of interest is 15%.

Table 7.18

	Project A	Project B	Project C	Project D
Outlay now	£20,000	£6,000	£25,000	£10,000
Return after 1 year	£24,000	£8,500	£30,000	£12,000

2. Two projects A and B each involve an initial outlay of £40,000 and guarantee the returns (in £) given in Table 7.19. The market rate of interest is 18%. Which is the better investment according to (a) the IRR criterion, (b) the NPV criterion?

Table 7.19

	Project A	Project B
End of year 1	15,000	10,000
End of year 2	20,000	12,000
End of year 3	25,000	12,000
End of year 4	0	12,000
End of year 5	0	15,000
End of year 6	0	15,000

3. Using the spreadsheet format as set out in Table 7.12 for Example 7.28, find the IRR and show that the NPV of the following project is zero when the discount rate used is approximately equal to this IRR.

Outlay now: £25,000
Annual returns: (1) £4,000;   (2) £6,000;   (3) £7,500;
   (4) £7,500;   (5) £10,000;   (6) £10,000

7.7 GEOMETRIC SERIES AND ANNUITIES

You may have noted that in some of the examples in Sections 7.5 and 7.6 above the return on the investment was the same in each time period. Although not many physical investment projects give such a constant stream of returns there are other forms of financial investments which are designed to. These are called 'annuities'. For example, someone might pay a fixed sum for a quaranteed pension payment of £8,000 a year for the next 5 years. The present value of a steady stream of a fixed return of £a per year for the next n years when interest rates are i% will therefore be

$$a(1+i)^{-1} + a(1+i)^{-2} + \dots + a(1+i)^{-n}$$

224

This sequence of terms is a special case of what is known as a 'geometric series'. There exists a mathematical formula for the sum of such sequences of numbers which means that one does not have to calculate each of the terms separately before summing. This would obviously be useful if you did not have access to a computer with an NPV program, or had to write your own program, but is this formula of any use otherwise? Well, there are some forms of annuities that are called 'perpetual annuities' which promise a fixed annual monetary return forever. For example, a bond that pays a fixed 6% return on a nominal price of £100 is a perpetual annuity of £6. The NPV of such an annuity at a rate of interest i would be

$$\frac{6}{1+i} + \frac{6}{(1+i)^2} + \ldots + \frac{6}{(1+i)^n} + \ldots$$

as n continues to infinity. Each successive term gets smaller and smaller but the sum of this sequence continues to grow as n gets bigger. You cannot sum such an infinite series of numbers without using the formula for the sum of an infinite geometric series. The next section deals with infinite geometric series and perpetual annuities but first, in this section, we look at some more general features of geometric series and the appraisal of investments in annuities with a finite lifespan.

Geometric series

A geometric series is a sequence of terms where each successive term is the previous term multiplied by a common ratio. The series starts with a given initial term. Any number of terms may be in a series.

Example 7.32

If the given initial term is 24 and the common ratio is 5, what is the corresponding geometric series? (Find up to six terms.)

Solution

The series will be

$$24 \qquad 24 \times 5 \qquad 24 \times 5^2 \qquad 24 \times 5^3 \qquad 24 \times 5^4 \qquad 24 \times 5^5$$

Example 7.33

A firm's sales revenue is initially £40,000 and then grows by 20% each successive year. What is the pattern of sales revenue over 5 years?

Solution

Each year's sales are 120% of the previous year's. The time profile of sales revenue is therefore a geometric series with an initial term of £40,000 and a common ratio of 1.2. Thus (in £) the series is

40,000   40,000 × 1.2   40,000 × 1.2^2 40,000 × 1.2^3 40,000 × 1.2^4

If we use the algebraic notation a for initial term, k for the common ratio and n for the number of terms then the general form of a geometric series will be

$$a, \ ak, \ ak^2, \ \ldots, \ ak^{n-1}$$

Note that the last (nth) term is ak^{n-1} and not ak^n, because the initial term a is not multiplied by k.

Sum of a geometric series

The sum of a geometric series can be found by simply adding all the terms together. This is easy enough to do for Example 7.32 above using a pocket calculator. More complex series are more time consuming to sum in this way, however, and so we need to derive the formula for the sum.

The general form for the sum of a geometric series with n terms will be

$$GP_n = a + ak + ak^2 + \ldots + ak^{n-1} \tag{1}$$

Multiplying each term by k gives

$$kGP_n = ak + ak^2 + \ldots + ak^{n-1} + ak^n$$

Subtracting (1) $GP_n = a + ak + ak^2 + \ldots + ak^{n-1}$

gives $(k-1)GP_n = -a + ak^n$

Therefore

$$GP_n = \frac{-a + ak^n}{k-1} = \frac{-a(1-k^n)}{k-1} = \frac{(-1)a(1-k^n)}{(-1)(1-k)} = \frac{a(1-k^n)}{1-k}$$

Thus the formula for the sum of a geometric series is

$$GP_n = \frac{a(1-k^n)}{1-k}$$

The examples below illustrate how this can be used to sum some simple numerical sequences of numbers.

Example 7.34

Sum the geometric series

$$15 \quad 45 \quad 135 \quad 405 \quad 1215 \quad 3645$$

Solution

In this geometric series with six terms, each number except the first is 3 times the previous one. Thus

$$a = 15 \quad k = 3 \quad n = 6$$

Substituting these values into the formula for the sum of a geometric series we get

$$GP_n = \frac{a(1-k^n)}{1-k} = \frac{15(1-3^6)}{-2}$$

$$= \frac{15(1-729)}{-2}$$

$$= \frac{15(-728)}{-2}$$

$$= 15 \times 364$$

$$= 5,460$$

You can check that the formula gives the same answer as that found by using a pocket calculator to sum the numbers and in this simple example using the calculator may in fact be the quicker method. In other more complex cases the formula will provide the quickest method of solution.

Example 7.35

A firm expects its sales to grow by 12% per month. If its January sales figure is 9,200 units per month what will its expected total annual sales be?

Solution

The firm's total annual sales will be the sum of the geometric series

$$9,200 + 9,200(1.12) + 9,200(1.12)^2 + \ldots + 9,200(1.12)^{11}$$

In this example $a = 9,200$, $k = 1.12$ and $n = 12$. Therefore, the sum is

$$GP_n = \frac{a(1-k^n)}{1-k} = \frac{9,200(1-1.12^{12})}{1-1.12} = \frac{9,200(1-3.896)}{1-1.12} = \frac{-26,642.979}{-0.12} = £222,024.83$$

Example 7.36

A 5 year saving scheme requires investors to pay in £5,000 now followed by £5,000 at 12 month intervals. Interest is credited at 14% at the end of each year of the investment. What will the final sum be at the end of the fifth year?

Solution

The £5,000 invested at the start of year 5 will be worth 5,000(1.14). The £5,000 invested at the start of year 4 will be worth $5,000(1.14)^2$ etc. Therefore the final sum will be

$$5,000(1.14) + 5,000(1.14)^2 + \ldots + 5,000(1.14)^5$$

This is a geometric series with $a = 5,000(1.14)$, $k = 1.14$ and $n = 5$. The sum will be

$$GP_n = \frac{a(1 - k^n)}{1 - k}$$

$$= \frac{5,000(1.14)(1 - 1.14^5)}{1 - 1.14}$$

$$= \frac{5,700(1 - 1.9254146)}{-0.14}$$

$$= \frac{-5,274.8626}{-0.14}$$

$$= £37,677.59$$

The formula for the sum of a geometric series can be used in NPV calculations for a constant stream of returns but one has to be very careful not to get the algebraic terminology mixed up because, as in the previous example, the initial term includes the constant ratio.

Example 7.37

An annuity will pay £8,000 at the end of each year for 5 successive years, the first payment being 12 months from the initial purchase date. What is the maximum price any rational investor would pay for such an annuity if the opportunity cost of capital is 10%?

Solution

The maximum purchase price will be the NPV of the stream of returns, using 10% as the discount rate. Therefore (in £):

$$NPV = \frac{8,000}{1.1} + \frac{8,000}{1.1^2} + \frac{8,000}{1.1^3} + \frac{8,000}{1.1^4} + \frac{8,000}{1.1^5}$$

This is a geometric series with five terms. The first term a is $8,000/1.1$ (*not* 8,000). The constant ratio k is $1/1.1$. Therefore

$$NPV = \frac{a(1 - k^n)}{1 - k}$$

$$= \frac{\frac{8,000}{1.1}\left[1 - \left(\frac{1}{1.1}\right)^5\right]}{1 - \frac{1}{1.1}}$$

$$= \frac{8,000(1 - 0.6209211)}{1.1(1 - 1/1.1)}$$

$$= \frac{8,000(0.3790789)}{1.1 - 1} = \frac{3,032.6312}{0.1}$$

$$= £30,326.31$$

In the above example some of the terms cancelled out. The same terms will cancel in any annuity NPV calculations and so a simplified general formula for the NPV of an annuity can be derived.

Assuming an annual payment of R for n years and an interest rate of i, then

$$NPV = \frac{R}{1 + i} + \frac{R}{(1 + i)^2} + \ldots + \frac{R}{(1 + i)^n}$$

In this geometric series

$$a = \frac{R}{1 + i}$$

$$k = \frac{1}{1 + i}$$

Therefore

$$NPV = \frac{a(1 - k^n)}{1 - k}$$

$$= \frac{\frac{R}{1 + i}\left[1 - \left(\frac{1}{1 + i}\right)^n\right]}{1 - \frac{1}{1 + i}}$$

$$= \frac{R\left[1-\left(\frac{1}{1+i}\right)^{n}\right]}{1+i-1}$$

$$= \frac{R\left[1-\frac{1}{(1+i)^{n}}\right]}{1+i-1} = \frac{R[1-(1+i)^{-n}]}{i}$$

Thus for any annuity

$$\text{NPV} = \frac{R[1-(1+i)^{-n}]}{i}$$

We can use this formula to check the answer to Example 7.37 above. Given $R = 8,000$, $i = 0.1$ and $n = 5$, then

$$\text{NPV} = \frac{8,000[1-(1.1)^{-5}]}{0.1} = £30,326.30$$

This is the same answer as that derived from first principles.

Example 7.38

An annuity will pay £2,000 a year for the next 5 years, with the first payment in 12 months' time. Capital can be invested elsewhere at an interest rate of 14%. Is £6,000 a reasonable price to pay for this annuity?

Solution

For this annuity (in £)

$$\text{NPV} = 2,000(1.14)^{-1} + 2,000(1.14)^{-2} + 2,000(1.14)^{-3}$$
$$+ 2,000(1.14)^{-4} + 2,000(1.14)^{-5}$$

To sum this geometric series we use the formula

$$\text{NPV} = \frac{R[1-(1+i)^{-n}]}{i}$$

where in this example $R = 2,000$, $i = 0.14$ and $n = 5$. Therefore

$$\text{NPV} = \frac{2,000[1-(1.14)^{-5}]}{0.14}$$

$$= \frac{2,000(1-0.5193686)}{0.14}$$

$$= £6,866.16$$

The NPV of this annuity is greater than its purchase price and so it is clearly a worthwhile investment.

Example 7.39

What would you pay for an annuity that promises to pay £450 a year for the next 10 years given an interest rate of 8%?

Solution

In this example $R = £450$, $i = 8\% = 0.08$ and $n = 10$. The present value of the stream of returns will be

$$\text{NPV} = \frac{R[1 - (1 + i)^{-n}]}{i}$$

$$= \frac{450[1 - (1.08)^{-10}]}{0.08}$$

$$= \frac{450(1 - 0.4631935)}{0.08}$$

$$= \frac{241.56293}{0.08}$$

$$= £3,019.54$$

Thus any price less than or equal to £2,985.79 would make this annuity worthwhile investing in.

QUESTIONS 7.5

1. In the geometric series below (i) identify the constant ratio, (ii) say what the sixth term will be and (iii) calculate the sum of each series up to ten terms using the formula for summation of a geometric series.

   (a) 8, 20, 50, . . .
   (b) 0.5, 1.5, 4.5, . . .
   (c) 2, 2.8, 3.92, . . .
   (d) 60, 48, 38.4, . . .
   (e) 2.4, 1.8, 1.35, . . .

2. A firm starts producing a new product. It sells 420 units in January and then sales increase by 10% each month. What will total demand be in the last 6 months of the year?

3. What would be the maximum price you would pay for the following annuities if money can be invested elsewhere at 8%?

231

Annuity A pays £200 a year for the next 8 years
Annuity B pays £900 a year for the next 4 years
Annuity C pays £6,000 a year for the next 12 years

4. Would you pay £3,500 for an annuity which guarantees to pay you £750 annually for the next 7 years if you can invest money elsewhere at 9%?
5. What would be a reasonable price to pay for a pension plan which guarantees to pay £200 a month for the next 2 years if you can earn 1.2% a month on your bank deposit account?

7.8 PERPETUAL ANNUITIES

We now return to the problem of how to calculate the worth of an annuity that promises to pay a fixed annual return indefinitely. The NPV of the stream of returns from a perpetual annuity is an infinite geometric progression. Whether or not one can find the sum of an infinite geometric progression depends on whether the progression is convergent or divergent. Before looking at the formal mathematical conditions for convergence or divergence these concepts are illustrated with some simple examples.

When you were at school you may have come across the teaser about the frog jumping across a pond, which goes something like this.

'A frog is sitting on a leaf in the middle of a circular pond. The pond is 10 metre in radius and the frog jumps 5 metres with its first jump. Its second jump is 2.5 m, its third jump 1.25 m and so on. How many jumps will it take for the frog to reach the edge of the pond? Assume that each time it jumps it lands on a leaf.'

The right answer is, of course, 'never'. Each time the frog manages to jump half of the remaining distance to the edge of the pond. The total distance the frog travels in n jumps is given by the sum of the geometric series

$$5 + (0.5)5 + (0.5)^2 5 + \ldots + (0.5)^{n-1} 5$$

As n gets larger the sum of this series continues to increase but never actually reaches 10 metres. Only if an infinite number of jumps can be made will the total distance travelled be 10 metres. Thus in this example we have a geometric series which converges on 10 metres.

A divergent geometric series is illustrated below.

Example 7.40

| 40 | 60 | 90 | 135 | . . . |

This sequence can be written as the geometric series

| 40 | 40(1.5) | $40(1.5)^2$ | $40(1.5)^3$ | . . . | $40(1.5)^n$ |

It is intuitively obvious that each successive term is larger than the previous one and that as the number of terms approaches infinity the sum of the series will also become infinitely large. In this case even the last term $40(1.5)^n$ will become infinitely large. There is no set quantity towards which the sum of the series converges.

Most of you will probably have already guessed by now that the way to distinguish a convergent and a divergent geometric series is to look at the common ratio k. If k is greater than 1, and positive, successive terms become larger and larger and the series diverges. If k is less than 1, and positive, then successive terms become smaller and smaller and the series converges. The 'and positive' condition is included because it is mathematically possible to have a negative common ratio, in which case the conditions for convergence are that $|k| < 1$.

To find the sum of a convergent geometric series (such as the case of a perpetual annuity) let us look again at the general formula for the sum of a geometric series:

$$GP_n = \frac{a(1 - k^n)}{1 - k}$$

This can be rewritten as

$$GP_n = \frac{a}{1 - k} - \left(\frac{a}{1 - k}\right)k^n \tag{1}$$

If $-1 < k < 1$ then the value of k^n approaches zero as n approaches infinity (i.e. $k^n \to 0$ as $n \to \infty$) and so the second term in (1) will disappear and the sum to infinity will be

$$GP_n = \frac{a}{1 - k} \tag{2}$$

We can now use formula (2) for the frog example. The total distance jumped is

$$\sum_{n=0}^{\infty} 5(0.5)^n$$

In this geometric series $k = 0.5$ and $a = 5$. The sum for an infinite number of terms will thus be

$$\frac{a}{1 - k} = \frac{5}{1 - 0.5} = \frac{5}{0.5} = 10 \text{ metres}$$

The NPV of a perpetual annuity can also be found using this formula although care must be taken to include the discounting factor in the initial term, as explained in the following example.

Example 7.41

What is the NPV of an annuity which will pay £6 a year *ad infinitum*, with the first payment due in 12 months' time? Assume that capital can be invested elsewhere at 15%.

Solution

$$NPV = \frac{6}{1.15} + \frac{6}{1.15^2} + \ldots + \frac{6}{1.15^n}$$

where $n \to \infty$. In this geometric series $a = 6/1.15$ and $k = 1/1.15$. This is clearly convergent as $-1 < k < 1$. Summing to infinity:

$$NPV = \frac{a}{1-k}$$

$$= \frac{\frac{6}{1.15}}{1 - \frac{1}{1.15}}$$

$$= \frac{6}{1.15\left(1 - \frac{1}{1.15}\right)}$$

$$= \frac{6}{1.15 - 1} = \frac{6}{0.15} = £40$$

A simplified formula for the NPV of a perpetual annuity can be derived as certain terms will always cancel out, as Example 7.41 above illustrates.

Assume that an annuity pays a fixed return R each year, starting in 12 months' time, and the opportunity cost of capital is $i\%$. For this annuity

$$NPV = R(1+i)^{-1} + R(1+i)^{-2} + \ldots + R(1+i)^{-n}$$

where $n \to \infty$. This is a geometric series with $a = R(1+i)^{-1}$ and $k = (1+i)^{-1}$. Using the formula for the sum of an infinite converging geometric series

$$NPV = \frac{a}{1-k} = \frac{R(1+i)^{-1}}{1-(1+i)^{-1}} = \frac{R}{1+i-1}$$

(multiplying through by $1+i$). Thus the formula for the NPV of a perpetual annuity is

$$NPV = \frac{R}{i}$$

Reworking Example 7.41 above using this formula we get

$$NPV = \frac{6}{0.15} = £40$$

which is identical to the answer derived from first principles, although the formula obviously makes the calculations much easier.

Example 7.42

An investment opportunity involves an initial outlay of £50,000 and gives a £5,000 annual return, starting in 12 months' time and continuing indefinitely. Capital can be invested elsewhere at 8%. Is this worth considering?

Solution

The NPV of the annual income stream can be calculated using the formula for the NPV of a perpetual annuity.

$$NPV = \frac{R}{i} = \frac{5,000}{0.08} = £62,500$$

This is greater than the initial outlay and so this investment is clearly an attractive proposition.

Example 7.43

What would be the maximum price you would pay for a perpetual annuity that will pay £900 per annum, starting in 12 months' time, given an interest rate of 15%?

Solution

For this stream of returns

$$NPV = \frac{R}{i} = \frac{900}{0.15} = £6,000$$

This is the maximum price a rational investor would pay for this annuity.

QUESTIONS 7.6

1. Identify which of the following geometric series are convergent and then calculate the sum to which these series converge as the number of terms approaches infinity:

   (a) 4, 6, 9, . . .
   (b) 120, 96, 76.8, . . .

(c) 0.8, − 1.2, 1.8, . . .

(d) 36, 12, 4, . . .

(e) 500, 500(0.48), 500(0.48)^2, . . .

(f) 850, 850(1.2)$^{-1}$, 850(1.2)$^{-2}$, . . .

2. What is the maximum price you would pay for a perpetual annuity that will commence annual payment of £400 in 12 months' time if the market rate of interest is 13%?

3. Is it worth paying £10,000 for a perpetual annuity of £1,500 per annum, commencing payments in 12 months' time, if money can be invested elsewhere at 14%?

4. What would you calculate the price of an annuity paying £12,000 per annum (starting in 12 months' time) to be if the market rate of interest is

(a) 5%, (b) 10%, (c) 15%, (d) 20%?

5. A government bond guarantees an annual payment of £14; what will it be priced at, given a market rate of interest of 9%?

7.9 LOAN REPAYMENTS

If someone takes out a loan now, to be paid off in regular equal instalments over a given time period, how can these payments be calculated? The following example shows how the formula for calculating the NPV of an annuity can be adapted for this type of problem. As most loans are paid off monthly we shall mainly use monthly rates of interest in this section.

Example 7.44

If a £2,000 loan is taken out now to be paid back over the next 12 months at a monthly interest rate of 2% what will the monthly payments be?

Solution

From the lender's viewpoint the repayments can be viewed as a monthly annuity which pays £R per month for the following 12 months, where R is the monthly repayment. If the lender is willing to exchange the loan of £2,000 for this stream of payments then this must be the NPV of this 'annuity'.

The formula for the NPV of an annuity is (from Section 7.7)

$$\text{NPV} = \frac{R[1 - (1 + i)^{-n}]}{i} \tag{1}$$

However, this time instead of calculating NPV we need to work out the value of R for a given size of loan (L).

Substituting L for NPV in (1), we get

$$\frac{R[1 - (1 + i)^{-n}]}{i} = L$$

$$R[1 - (1 + i)^{-n}] = iL$$

$$R = \frac{iL}{1 - (1 + i)^{-n}}$$

which is the general formula for calculating loan repayments.

Substituting the values of i, n and L for this example, which are $L = 2,000$, $i = 2\% = 0.02$ and $n = 12$, into the loan repayment formula gives

Table 7.20 Powers for loan repayments: Calculation of $(1 + i)^{-n}$

i (%)	$n = 12$	$n = 24$	$n = 36$	$n = 60$	$n = 240$	$n = 300$
0.50	0.941905	0.887185	0.835644	0.741372	0.302096	0.223965
0.55	0.936300	0.876658	0.820815	0.719574	0.268103	0.192920
0.60	0.930731	0.866260	0.806255	0.698427	0.237949	0.166190
0.65	0.925197	0.855991	0.791961	0.677911	0.211199	0.143174
0.70	0.919700	0.845848	0.777927	0.658008	0.187467	0.123355
0.75	0.914238	0.835831	0.764148	0.638699	0.166412	0.106287
0.80	0.908811	0.825937	0.750621	0.619966	0.147731	0.091588
0.85	0.903418	0.816165	0.737339	0.601791	0.131154	0.078927
0.90	0.898061	0.806514	0.724299	0.584157	0.116444	0.068022
0.95	0.892738	0.796981	0.711495	0.567049	0.103390	0.058627
1.00	0.887449	0.787566	0.698924	0.550449	0.091805	0.050534
1.05	0.882194	0.778266	0.686582	0.534343	0.081523	0.043561
1.10	0.876972	0.769081	0.674463	0.518717	0.072397	0.037553
1.15	0.871784	0.760008	0.662564	0.503554	0.064296	0.032376
1.20	0.866630	0.751048	0.650880	0.488842	0.057105	0.027915
1.25	0.861508	0.742197	0.639409	0.474567	0.050721	0.024070
1.30	0.856419	0.733454	0.628145	0.460715	0.045053	0.020757
1.35	0.851363	0.724819	0.617084	0.447275	0.040022	0.017900
1.40	0.846339	0.716290	0.606224	0.434232	0.035554	0.015438
1.45	0.841347	0.707865	0.595560	0.421577	0.031586	0.013316
1.50	0.836387	0.699543	0.585089	0.409295	0.028064	0.011486
1.55	0.831459	0.691324	0.574807	0.397378	0.024935	0.009908
1.60	0.826562	0.683204	0.564711	0.385813	0.022156	0.008584
1.65	0.821696	0.675185	0.554797	0.374590	0.019689	0.007375
1.70	0.816861	0.667263	0.545061	0.363699	0.017497	0.006363
1.75	0.812057	0.659438	0.535501	0.353130	0.015550	0.005491
1.80	0.807284	0.651708	0.526114	0.342873	0.013820	0.004738
1.85	0.802541	0.644073	0.516895	0.332918	0.012284	0.004089
1.90	0.797828	0.636531	0.507842	0.323257	0.010919	0.003529
1.95	0.793146	0.629080	0.498953	0.313881	0.009706	0.003046
2.00	0.788493	0.621721	0.490223	0.304782	0.008628	0.002629

$$R = \frac{0.02 \times 2{,}000}{1 - (1.02)^{-12}}$$

$$= \frac{40}{1 - 0.7884934}$$

$$= £189.12 \text{ per month}$$

To work out loan repayments using the above formula, one really needs a calculator with a power function. Table 7.20 shows some calculated values that may be useful for this type of problem for those of you who do not have such a calculator to hand.

Example 7.45

What will be the monthly repayments on a loan of £6,000 taken out for 5 years at a monthly interest rate of 0.7%?

Solution

$$L = £6{,}000 \qquad i = 0.7\% = 0.007 \qquad n = 5 \times 12 = 60$$

Using the repayment formula

$$R = \frac{iL}{1 - (1+i)^{-n}}$$

$$= \frac{0.007 \times 6{,}000}{1 - (1.007)^{-60}}$$

$$= \frac{0.007 \times 6{,}000}{1 - 0.658008}$$

$$= \frac{42}{0.342}$$

$$= £122.81 \text{ monthly repayment}$$

(Note: Table 7.20 is used to find the value of 1.007^{-60} in the above example. One just reads along the row for 0.7% until one gets to the column for 60.)

Example 7.46

What are the monthly payments on a repayment mortgage of £60,000 taken out for 25 years if the monthly rate of interest is 0.75%?

Solution

$$L = 60{,}000 \qquad i = 0.75\% = 0.0075 \qquad n = 25 \times 12 = 300 \text{ months}$$

Using the repayment formula

$$R = \frac{iL}{1-(1+i)^{-n}}$$

$$= \frac{0.0075 \times 60{,}000}{1-(1.0075)^{-300}}$$

$$= \frac{450}{1-0.106287}$$

$$= 503.52$$

If only the APR for a loan is quoted, then it will be necessary to calculate the equivalent monthly interest rate before working out monthly repayments.

Example 7.47

If a loan of £4,200 is taken out over a period of 3 years at an APR of 6.8% what will the monthly repayments be?

Solution

First we need to convert the APR of 6.8% to its equivalent monthly rate i_m. We know that

$$1 + \text{APR} = (1 + i_m)^{12}$$

and so

$$\sqrt[12]{(1 + \text{APR})} = 1 + i_m$$

$$\sqrt[12]{(1 + \text{APR})} - 1 = i_m$$

Substituting in the value APR $= 6.8\% = 0.068$

$$i_m = \sqrt[12]{(1 + 0.068)} - 1$$

$$= \sqrt[12]{(1.068)} - 1$$

$$= 1.0054974 - 1$$

$$= 0.0055 \text{ (to 2 significant dp)}$$

The values to be entered into the loan repayment formula are therefore

$$L = 4{,}200 \qquad i = 0.0055 \qquad n = 3 \times 12 = 36$$

giving

239

$$R = \frac{iL}{1 - (1 + i)^{-n}}$$

$$= \frac{0.0055 \times 4,200}{1 - (1.0055)^{-36}}$$

$$= \frac{23.1}{1 - 0.820815}$$

$$= £128.92 \text{ monthly payment}$$

From an individual consumer's viewpoint you may be more interested in finding out the interest rate you have to pay on a loan. All lenders now have to quote their APR by law, but you may still wish to check this.

Example 7.48

A car dealer offers you a £12,000 car for a £4,000 deposit now followed by 24 monthly payments of £400. What is the APR on this effective loan of £8,000?

Solution

As in the examples above, treat the stream of repayments as an annuity for the lender. Referring again to the formula for loan repayments

$$R = \frac{iL}{1 - (1 + i)^{-n}}$$

we can see that even if we know that $L = 8,000$, $R = 400$ and $n = 24$ this still leaves us with the awkward equation

$$400 = \frac{i \times 8,000}{1 - (1 + i)^{-24}}$$

to solve for i (the monthly interest rate) which can then be used to calculate the APR.

The only practical way for you to solve this is to use a spreadsheet program. In one column insert a series of monthly interest rates i (say from 0.5% to 2.0% in 0.05% intervals), then in another column calculate the corresponding APR, which will be $(1 + i)^{12} - 1$, and then in a third column calculate the repayment. One suggested format for a Lotus 1–2–3 worksheet to tackle this sort of problem can be constructed as follows.

Enter the following labels, right justifying (RJ) where indicated:

LOAN	in cell A1	
APR	in cell B1	(RJ)

EX 795	in cell C1
MONTHLY i	in cell A3
APR	in cell C3 (RJ)
PAYMENT	in cell E3
LOAN=	in cell F1
N MONTHS=	in cell F2

Enter 0.005 in cell A4 and then the formula +A4+0.0005 in cell A5 and copy down the column for about 40 rows. Format this column to percentage, to 2 decimal places. To calculate the APR corresponding to the monthly rates in column A, enter the formula +(1+A4)^12 − 1 in cell C4 and copy down the column and put in percentage format.

Enter 8000 in cell G1   (the loan value)
Enter   24 in cell G2   (the repayment period in months)

In cell E4 enter the formula

$$+A4*\$G\$1/(1-(1+A4)^{\wedge}-\$G\$2)$$

and copy it down the column. This will calculate the loan repayment R using the formula

$$R = \frac{iL}{1 - (1+i)^{-n}}$$

where i is the monthly interest rate in column A, L is the loan (entered in cell G1) and n is the number of months (entered in cell G2). Your spreadsheet should look like Table 7.21.

You should be able to read off the following values about three-quarters of the way down the spreadsheet:

MONTHLY i	APR	PAYMENT
1.5%	19.56%	399.39
1.55%	20.27%	401.72

Therefore a £400 per month payment corresponds approximately to a 20% APR.

This spreadsheet format can be used to solve similar types of problems. One only needs to change the total loan figure in cell G1 and the time period in cell G2 to compute a new set of repayment figures for a range of monthly interest rates.

Example 7.49

A loan company will require 36 monthly payments of £426 in return for a loan of £12,500. What APR is it charging?

Solution

Using the spreadsheet constructed for Example 7.48 above the new values for the loan and time period can be entered in cells G1 and G2. One can then read off the values:

MONTHLY i	APR	PAYMENT
1.15%	14.71%	426.0071

The APR is therefore approximately 14.71%.

Table 7.21

LOAN	APR	EX 795		LOAN =	8000
				N MONTHS=	24
MONTHLY i		APR	PAYMENT		
0.50%		6.17%	354.5648		
0.55%		6.80%	356.7316		
0.60%		7.44%	358.9064		
0.65%		8.08%	361.0890		
0.70%		8.73%	363.2795		
0.75%		9.38%	365.4779		
0.80%		10.03%	367.6842		
0.85%		10.69%	369.8983		
0.90%		11.35%	372.1203		
0.95%		12.01%	374.3501		
1.00%		12.68%	376.5877		
1.05%		13.35%	378.8332		
1.10%		14.03%	381.0865		
1.15%		14.71%	383.3476		
1.20%		15.39%	385.6165		
1.25%		16.08%	387.8931		
1.30%		16.77%	390.1776		
1.35%		17.46%	392.4698		
1.40%		18.16%	394.7697		
1.45%		18.86%	397.0774		
1.50%		19.56%	399.3928		
1.55%		20.27%	401.7159		
1.60%		20.98%	404.0467		
1.65%		21.70%	406.3852		
1.70%		22.42%	408.7313		
1.75%		23.14%	411.0852		
1.80%		23.87%	413.4466		
1.85%		24.60%	415.8157		
1.90%		25.34%	418.1925		
1.95%		26.08%	420.5768		
2.00%		26.82%	422.9687		
2.05%		27.57%	425.3682		
2.10%		28.32%	427.7753		
2.15%		29.08%	430.1899		
2.20%		29.84%	432.6121		
2.25%		30.60%	435.0418		
2.30%		31.37%	437.4790		

QUESTIONS 7.7

1. What will be the monthly repayments on a loan of £6,500 taken out over 5 years at a monthly interest rate of 1.2%?
2. You wish to buy a car priced at £6,000 by putting down a cash deposit of £2,000 and borrowing £4,000, the loan being paid back in monthly instalments over 2 years. How much will you have to budget to pay out of your salary if the monthly interest rate is 1.4%?
3. A loan company will lend you £5,000, repayable over the next 3 years in monthly payments. What will these payments be if the APR on the loan is 24.6%?
4. What will be the monthly payments on a repayment mortgage of £75,000 taken out over 20 years if the interest rate is fixed at 0.95% per month?
5. A loan of £800 is taken out. What APR is being charged if the monthly payments are

   (a) £27.00 over 36 months?
   (b) £31.13 over 36 months?
   (c) £25.00 over 48 months?
   (d) £21.78 over 48 months?

6. A car dealer has on offer a special '0% finance' deal on the advertised price of £8,671 for a particular model. This requires an initial deposit of £1,734 followed by 24 monthly payments of £289.00. If you could get the price reduced to £8,095 if you paid cash and can earn 9% per annum on money invested in a building society, which method would you use to purchase this car?

7.10 OTHER APPLICATIONS OF GROWTH AND DECLINE

In the previous sections of this chapter, various mathematical methods have been explained in order to solve a variety of problems concerned with finance and investment. Rather than introducing even more mathematical methods, this section now considers how the techniques already explained can be adapted to non-financial problems. A series of different types of problem are presented and the most appropriate method of solution is explained. Note that in all these examples growth and decline are treated as discrete processes.

Example 7.50

There are limited world reserves of mineral M. The current rate of extraction is 45 million tonnes a year, with all mined material being used up by manufacturing industry. This is expected to increase at 3% per annum. Total estimated reserves are 1200 million tonnes. When will they be expected to run out if this 3% growth rate in consumption continues?

Solution

The annual pattern of consumption will be (in millions of tonnes) the geometric series

$$45, \quad 45(1.03), \quad 45(1.03)^2, \quad \ldots, \quad 45(1.03)^{n-1}$$

where n is the number of years that mining continues, $a = 45$ is the initial term and $k = 1.03$ is the common ratio. The sum of this geometric series is

$$\frac{a(1 - k^n)}{1 - k} = \frac{45(1 - 1.03^n)}{1 - 1.03}$$

which must sum to 1,200 if all reserves are used up. Therefore

$$1{,}200 = \frac{45(1 - 1.03^n)}{-0.03}$$

$$-\frac{36}{45} = 1 - 1.03^n$$

$$1.03^n = 1 + 0.8$$

$$1.03^n = 1.8$$

Putting this in logarithmic form we get

$$n \log 1.03 = \log 1.8$$

$$n = \frac{\log 1.8}{\log 1.03} = \frac{0.2552725}{0.0128372} \doteq 19.885$$

Therefore mineral M is expected to run out within 20 years at the current rate of extraction.

Example 7.51

A developing country currently produces 3,600 million units of food per annum and this rate of production is expected to increase by 4% a year. Its population is currently 25 million and expected to grow by 6% per annum. The minimum recommended intake of food is 1,200 units of food per person per year, on average. This country is clearly in trouble if this situation persists. Assuming no changes in growth rate of production or population growth, no imports and exports and no foreign aid, when will food production fall below the subsistence level?

Solution

The demand for food is $1{,}200 \times$ population. Thus initial demand is $1{,}200 \times 25$ million $= 3{,}000$ million units.

The rate of growth of the population is 6% and so total demand for food after n years will be $3,000(1.06)^n$ million units. Initial production is 3,600 million units of food. The rate of growth of production is 4% and so total production after n years will be $3,600(1.04)^n$ million units.

The subsistence level is reached in year n when

$$\text{food demand} = \text{food production}$$

$$3,000(1.06)^n = 3,600(1.04)^n$$

$$\left(\frac{1.06}{1.04}\right)^n = \frac{3,600}{3,000}$$

$$(1.0192307)^n = 1.2$$

putting in log form,

$$n \log 1.01923 = \log 1.2$$

$$n = \frac{\log 1.2}{\log 1.01923} = \frac{0.0791812}{0.0082722} = 9.572$$

Therefore food production will fall below the subsistence level in 10 years' time.

Example 7.52

Estimated reserves of an oil field are 84 billion barrels. What annual growth in the rate of extraction will exhaust the oil in 12 years given that this year's production is 6 billion barrels?

Solution

Total oil extraction will be the sum of the geometric series with 12 terms:

$$6 + 6(1 + r) + 6(1 + r)^2 + \ldots + 6(1 + r)^{11}$$

where r is the growth in the extraction rate.

Using the usual terminology, the initial term $a = 6$ and the constant ratio $k = 1 + r$. The sum of a geometric series is

$$GP_n = \frac{a(1 - k^n)}{1 - k}$$

Therefore

$$84 = \frac{6[1 - (1 + r)^{12}]}{1 - (1 + r)} = \frac{6[1 - (1 + r)^{12}]}{-r}$$

Basic mathematics for economists

The only practical way to solve this equation for r is to set up a spreadsheet with this formula and calculate different values of GP_n for different values of r. The closest you get to 84 will be the solution.

A Lotus 1–2–3 spreadsheet to solve this problem can be constructed as follows.

Enter the labels below, right justifying (RJ) where indicated.

OIL	in cell A1	
EX 7103	in cell B1	
GROWTH	in cell A3	(RJ)
USAGE	in cell C3	(RJ)
INITIAL	in cell E1	
AMOUNT =	in cell E2	

Table 7.22

OIL EX 7103		INITIAL	
		AMOUNT =	6
GROWTH	USAGE		
1.00%	76.09501		
1.10%	76.51974		
1.20%	76.94731		
1.30%	77.37774		
1.40%	77.81105		
1.50%	78.24726		
1.60%	78.68640		
1.70%	79.12847		
1.80%	79.57351		
1.90%	80.02152		
2.00%	80.47253		
2.10%	80.92657		
2.20%	81.38364		
2.30%	81.84378		
2.40%	82.30699		
2.50%	82.77331		
2.60%	83.24275		
2.70%	83.71534		
2.80%	84.19109		
2.90%	84.67003		
3.00%	85.15217		
3.10%	85.63755		
3.20%	86.12617		
3.30%	86.61807		
3.40%	87.11326		
3.50%	87.61176		
3.60%	88.11361		
3.70%	88.61882		
3.80%	89.12741		
3.90%	89.63940		
4.00%	90.15483		

Series, time and investment

Enter 0.01 in cell A4. Put the initial amount 6 in cell F2. Enter the formula +A4+0.001 in cell A5 and copy down the page and put in percentage format to get a series of percentage figures increasing by 0.1% each row.

In cell C4 enter the formula +$F$2*(1−(1+A4)^12)/−A4 and copy down the page. Your spreadsheet should now look like Table 7.22 and you should be able to read off the following values:

GROWTH	USAGE
2.70%	83.71534
2.80%	84.19109

Therefore a growth rate of approximately 2.76% in the annual level of extraction will exhaust the oil reserves in 12 years.

If you wish to pinpoint the exact growth rate then you can alter the graduations in the growth column to finer divisions (e.g. 0.01%). This spreadsheet may be used to solve similar problems. One only needs to alter the initial amount in cell F2 and the power to which $1+A4$ is raised in the formula in cell C4. Alternatively, you could enter the number of time periods in cell F5, with the label n= in cell E5, and change the formula in cell C4 to $F$2*(1−(1+A4)^$F$5)/−A4.

Example 7.53

Although the UK's population is not growing very much, the usage of water is rising rapidly as people use more appliances such as washing machines and dishwashers, take more baths, use garden sprinklers etc. In a water authority's area the current river flows allow a maximum extraction rate of 100 million gallons per day and current usage is 25 million gallons per day. When will a crisis point be reached if consumption grows by 4% per annum? What rate of growth would allow current supply sources to be sufficient for the next 100 years?

Solution

Consumption rate in n years' time will be (in millions of gallons) $25(1.04)^n$.
The crisis point will be reached when

$$100 = 25(1.04)^n$$
$$4 = 1.04^n$$

Putting in log form this gives

$$\log 4 = n \log 1.04$$
$$n = \frac{\log 4}{\log 1.04} = \frac{0.60206}{0.017033} = 35.346$$

247

Therefore a growth rate of 4% can be sustained for 35 years.

If current water supplies are to last another 100 years at growth rate r, then the current maximum supply rate of 100 million gallons per day will equal demand when

$$100 = 25(1 + r)^{100}$$

$$4 = (1 + r)^{100}$$

$$\sqrt[100]{4} = 1 + r$$

$$1.0139595 = 1 + r$$

$$0.0139595 = r$$

Therefore current water supplies will be sufficient for the next 100 years with a growth rate of 1.4%.

Example 7.54

A retailer has to order stock of seasonal good X in one batch at the start of the season. The first week's sales are expected to be 200 units of X. Past years' sales suggest that demand will grow by 5% a week for the next 14 weeks and then fall by 10% a week for the remaining 10 weeks of the season. How much stock needs to be ordered to meet the anticipated sales for the whole 25 week season?

Solution

This problem involves the summing of two separate geometric series.

Sales over the first 15 weeks are expected to be

$$200 + 200(1.05) + 200(1.05)^2 + \ldots + 200(1.05)^{14}$$

In this geometric series

$$a = 200 \qquad k = 1.05 \qquad n = 15$$

The sum of this series will be

$$\frac{a(1 - k^n)}{1 - k} = \frac{200(1 - 1.05^{15})}{1 - 1.05}$$

$$= \frac{200(1 - 2.0789282)}{-0.05}$$

$$= \frac{-215.78564}{-0.05}$$

$$= 4,315.7128$$

Thus, to the nearest whole unit over the first 15 weeks, sales will be 4,316. In the fifteenth week the sales will be

$$200(1.05)^{14} = 395.98632$$
$$= 396 \text{ (to nearest whole unit)}$$

Over the remaining 10 weeks the total sales will be

$$396(0.9) + 396(0.9)^2 + \ldots + 396(0.9)^{10}$$

In this geometric series $a = 396(0.9)$, $k = 0.9$ and $n = 10$. The sum of this series will be

$$\frac{a(1-k^n)}{1-k} = \frac{396(0.9)(1-0.9^{10})}{1-0.9}$$
$$= \frac{356.4(0.6513216)}{0.1}$$
$$= \frac{232.131}{0.1} = 2,321.31$$

Thus to the nearest whole unit total sales will be 2,321.
Over the whole season expected sales will be

$$4,316 + 2,321 = 6,637 \text{ units}$$

Example 7.55

Average annual income in a developing country is $420, and average annual expenditure on food is $280. If average income rises at 3% per annum and income elasticity of demand for food is 0.8, when will average expenditure on food reach $340?

Solution

For every 1% rise in average income Y, there will be a 0.8% rise in food consumption F, given an income elasticity of demand of 0.8. Therefore a 3% growth in Y will mean a 2.4% growth in F. The mathematical problem then becomes 'how long will it take $280 to grow to $340 at a growth rate of 2.4%?' and we need to solve for n in the equation

$$340 = 280(1.024)^n$$
$$\frac{340}{280} = 1.024^n$$
$$1.2142857 = 1.024^n$$

Putting in log form

$$\log 1.2142857 = n \log 1.024$$

$$n = \frac{\log 1.2142857}{\log 1.024} = \frac{0.0843209}{0.0103} = 8.1864938$$

Therefore, average food expenditure will reach $340 after approximately 8.19 years.

QUESTIONS 7.8

1. Total reserves of mineral z are 140 million tonnes. Current annual consumption is 18 million tonnes. If consumption is expected to grow by 4.5% a year, how long will these reserves last?
2. A country's gross national product (GNP) is forecast to grow at 3% per annum and its population is expected to expand at 1.5% per annum. GNP is currently $12,000 million and the population is 15 million, giving a GNP per capita of $800. When will GNP per capita reach $1,000?
3. What rate of growth of consumption will allow the current reserves of a natural resource to last for the next 50 years if this year's consumption is forecast to be 8 million tonnes and total reserves are 1220 million tonnes?
4. World annual usage of mineral M is actually declining by 5% a year. The current rate of extraction is 65 million tonnes per year. Total reserves in existence amount to 1,500 million tonnes. Will they last forever if the 5% per annum decline persists?
5. The estimated reserves of resource R are 1,650 million tonnes. Current annual consumption is 80 million tonnes. What percentage reduction in the annual consumption rate will ensure that the resource never runs out, assuming that consumption falls each year by the same percentage?

8

Introduction to calculus

8.1 THE DIFFERENTIAL CALCULUS

This chapter explains some of the basic techniques of calculus and their application to economic problems. We shall be concerned here with what is known as the 'differential calculus'.

Differentiation is a method used to find the slope of a function at any point. Although this is a useful tool in itself, it also forms the basis for some very powerful techniques for solving optimization problems, which are explained in this and the following chapters.

The basic technique of differentiation is quite straightforward and easy to apply. We shall therefore first consider some applications and later on (in Section 8.3) look at the fundamental principles on which differential calculus is based.

Consider a simple example of the function

$$y = f(x)$$

which only has one term:

$$y = 6x^2$$

To derive an expression for the slope of this function the basic rule of differentiation is: multiply the whole term by the value of the power of x in the term and then deduct 1 from the power of x. In this example there is a term in x^2 and so the power of x is 2. Using the above rule the expression for the slope of this function becomes

$$2(6x) = 12x$$

This is known as the derivative of y with respect to x, and is usually written as dy/dx, which is read as 'dy by dx'.

We can check that this is approximately correct by looking at the graph of the function $y = 6x^2$ in Figure 8.1. Any term in x^2 will rise at an ever increasing rate as x is increased. In other words, the slope of this function must increase as x increases. The slope is the derivative of the function with respect to x, which we have just worked out to be $12x$. As x increases the term $12x$ will

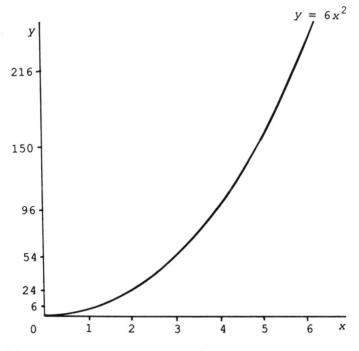

$y = 6x^2$

Figure 8.1

also obviously increase and so we can confirm that the formula derived for the slope of this function does behave in the expected fashion.

To determine the actual value of the slope of the function $y = 6x^2$ for any given value of x, one simply enters the given value of x into the formula

$$\text{slope} = 12x$$

When $x = 4$, then slope $= 48$; when $x = 5$, then slope $= 60$ etc.

Example 8.1

What is the slope of the function $y = 4x^2$ when x is 8?

Solution

By differentiating y we get slope $= dy/dx = 8x$ for any value of x. When $x = 8$, then slope $= 8(8) = 64$.

Example 8.2

Find a formula that gives the slope of the function $y = 6x^3$ for any value of x.

252

Solution

Slope $= dy/dx = 18x^2$ for any value of x.

Example 8.3

What is the slope of the function $y = 45x^4$ when $x = 10$?

Solution

Slope $= dy/dx = 180x^3$. When $x = 10$, then slope $= 180(1,000) = 180,000$.

QUESTIONS 8.1

1. Derive an expression for the slope of the function $y = 12x^3$.
2. What is the slope of the function $y = 6x^4$ when $x = 2$?
3. What is the slope of the function $y = 0.2x^4$ when $x = 3$?
4. Derive an expression for the slope of the function $y = 52x^5$.
5. Make up your own single-term function and then differentiate it.

8.2 RULES FOR DIFFERENTIATION

The rule for differentiation can be formally stated as:
if $y = ax^n$ where a and n are given parameters then
$dy/dx = nax^{n-1}$.
When there are several terms in x added together or subtracted in a function then this rule for differentiation is applied to each term individually. (The special rules for differentiating functions where terms are multiplied or divided are explained in Chapter 12.)

Example 8.4

Differentiate the function $y = 3x^2 + 10x^3 - 0.2x^4$.

Solution

$$\frac{dy}{dx} = 6x + 30x^2 - 0.8x^3$$

Example 8.5

Find the slope of the function $y = 6x^2 - 0.5x^3$ when $x = 10$.

Solution

Slope $= dy/dx = 12x - 1.5x^2$.
When $x = 10$, slope $= 120 - 1.5(100) = 120 - 150 = -30$.

Example 8.6

Derive an expression for the slope of the function $y = 4x^2 + 2x^3 - x^4 + 0.1x^5$ for any value of x.

Solution

$$\text{slope} = \frac{dy}{dx} = 8x + 6x^2 - 4x^3 + 0.5x^4$$

In using the formula for differentiation, one has to remember that $x^1 = x$ and $x^0 = 1$.

Example 8.7

Differentiate the function $y = 8x$.

Solution

$$y = 8x = 8x^1$$
$$\frac{dy}{dx} = 8x^{1-1} = 8x^0 = 8$$

Example 8.8

Derive an expression for the slope of the function $y = 30x - 0.5x^2$ for any value of x.

Solution

$$\text{slope} = \frac{dy}{dx} = 30x^0 - 2(0.5)x = 30 - x$$

Example 8.9

Differentiate the function $y = 14x$.

Solution

$$\frac{dy}{dx} = 14x^{1-1} = 14x^0 = 14$$

In practice all we need to know is that the derivative of any term in x (to the power of 1) is simply the value of the parameter that x is multiplied by, e.g. 14 in Example 8.9 above.

If the term is a constant, such as in the function $y = 5$, it can be written as being multiplied by x^0, e.g. $y = 5x^0$. Thus

$$\frac{dy}{dx} = 0(5x^{-1}) = 0$$

Therefore any constant terms always disappear when a function is differentiated.

Example 8.10

Differentiate the function $y = 20 + 4x - 0.5x^2 + 0.01x^3$.

Solution

$$\frac{dy}{dx} = 4 - x + 0.03x^2$$

Example 8.11

Derive an expression for the slope of the function $y = 6 + 3x - 0.1x^2$.

Solution

$$\text{slope} = \frac{dy}{dx} = 3 - 0.2x$$

Even when the power of x in a function is negative or not a whole number, the same rule for differentiation still applies.

Example 8.12

What is the slope of the function $y = 4x^{0.5}$ when $x = 4$?

Solution

Slope $= dy/dx = 2x^{-0.5}$. When $x = 4$, slope $= 2/4^{0.5} = 2/2 = 1$.

Example 8.13

Differentiate the function $y = x^{-1} + x^{0.5}$.

Solution

$$\frac{dy}{dx} = -x^{-2} + 0.5x^{-0.5}$$

QUESTIONS 8.2

1. Differentiate the function $y = x^3 + 60x$.
2. What is the slope of the function $y = 12 + 0.5x^4$ when $x = 5$?
3. Derive a formula for the slope of the function $y = 4 + 4x^{-1} - 4x$.
4. What is the slope of $y = 4x^{0.5}$ when $x = 4$?
5. Differentiate the function $y = 25 - 0.1x^{-2} + 2x^{0.3}$.
6. Make up your own function with at least three different terms in x and then differentiate it.

8.3 MARGINAL REVENUE AND TOTAL REVENUE

What differentiation actually does is look at the effect of an infinitely small change in the independent variable x on the dependent variable y in a function $y = f(x)$. This may seem a strange concept, and the rest of this section tries to explain how it works, but first consider the following example which shows how a function can be differentiated from first principles.

Example 8.14

Differentiate the function $y = 6x + 2x^2$ from first principles.

Solution

Assume that x is increased by the small amount Δx (Δ is the Greek letter 'delta' which usually signifies a change in a variable). This will produce a small change Δy in y.

Given the original function

$$y = 6x + 2x^2 \tag{1}$$

the new value of y (i.e. $y + \Delta y$) can be found by substituting the new value of x (i.e. $x + \Delta x$) into the function. Thus

$$y + \Delta y = 6(x + \Delta x) + 2(x + \Delta x)^2$$
$$y + \Delta y = 6x + 6\Delta x + 2x^2 + 4x\Delta x + 2(\Delta x)^2$$

Subtracting (1) $\quad \underline{y \qquad = 6x \quad + \quad 2x^2}$

gives $\quad \Delta y \qquad = 6\Delta x + 4x\Delta x + 2(\Delta x)^2$

Dividing through by Δx,

$$\frac{\Delta y}{\Delta x} = 6 + 4x + 2\Delta x$$

If Δx becomes infinitely small, then the last term disappears and

256

$$\frac{\Delta y}{\Delta x} = 6 + 4x \qquad (2)$$

By definition, dy/dx is the effect of an infinitely small change in x on y. Thus, from (2), $dy/dx = 6 + 4x$.

This is the same result for dy/dx that would be obtained using the basic rules for differentiation explained in Section 8.2. It is obviously quicker to use these rules than to differentiate from first principles. However, Example 8.14 can now help you to understand how differential calculus can be applied to economics.

Up to this point we have been using the usual algebraic notation for a single variable function, assuming that y is dependent on x. Changing the notation so that we can look at some economic applications does not alter the rule for differentiation as long as functions are specified in a form where one variable is dependent on another.

In introductory economics texts, marginal revenue (MR) is sometimes defined as the increase in the total revenue (TR) received from sales caused by an increase in output by 1 unit. This is not a precise definition though. It only gives an approximate value for marginal revenue between the two given outputs and will also vary if the units that output is measured in are changed. A more precise definition of marginal revenue is that it is the rate of change of total revenue relative to increases in output.

In Figure 8.2 the rate of change of total revenue between points B and A is

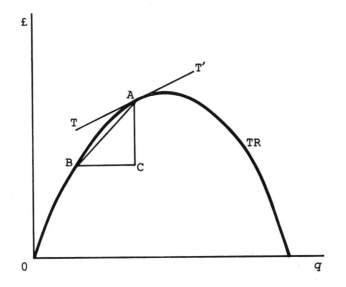

Figure 8.2

257

$$\frac{\Delta TR}{\Delta Q} = \frac{AC}{BC} = \text{the slope of the line AB}$$

which is an approximate value for marginal revenue over this output range.

Now suppose that the distance between B and A gets smaller. As point B moves along TR towards A the slope of the line AB gets closer to the value of the slope of the tangent TT′ through A. Thus for a very small change in output, MR will be almost equal to the slope of TR at A. If the change becomes infinitesimally small, then the slope of AB will exactly equal the slope of TT′. Therefore, MR will be equal to the slope of the TR function at any given output.

The idea of an 'infinitesimally small' change is not always easy to grasp. If the change is infinitesimally small then, one may argue, this means that it is zero and so there is not really any change at all. This is true, but it does not alter the fact that we have used the concept of an infinitesimally small change to derive the result that at any given output MR is equal to the slope of the TR function. We know that the slope of a function can be found by differentiation and so it must be the case that

$$MR = \frac{dTR}{dq}$$

Example 8.15

Given that $TR = 80q - 2q^2$, derive the MR function.

Solution

$$MR = \frac{dTR}{dq} = 80 - 4q$$

This result helps explain some of the properties of the relationship between TR and MR. The demand schedule D in Figure 8.3 represents the function $p = 80 - 2q$. We know that $TR = pq$: therefore

$$TR = (80 - 2q)q = 80q - 2q^2$$

which is the same as the TR function in Example 8.15 above.

This function is plotted in the lower section of Figure 8.3 and the function for MR, already derived, is plotted in the top section.

One can see that when TR is rising, MR is positive, as one would expect, and when TR is falling, MR is negative. As the rate of increase of TR gets smaller so does the value of MR. When TR is at its maximum, MR is zero.

With the function for MR derived above it is very straightforward to find the value of the output at which TR is a maximum. We know that TR is at its maximum when MR is zero. Thus

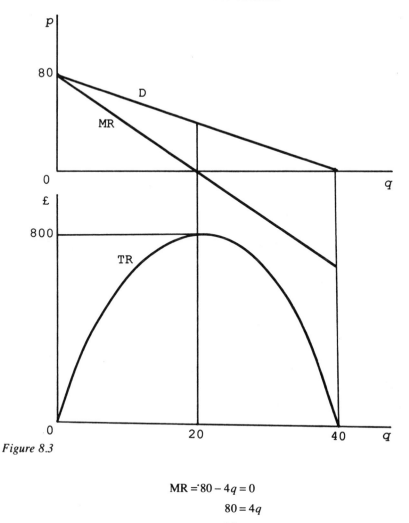

Figure 8.3

$$MR = 80 - 4q = 0$$
$$80 = 4q$$
$$20 = q$$

when TR is a maximum.

One can also see that the MR function has the same intercept on the vertical axis as this straight line demand schedule, but twice its slope. This result holds for any linear downward-sloping demand schedule.

As a proof assume that $p = a - bq$ represents a linear demand schedule where a and b are given parameters. Therefore

$$TR = pq = (a - bq)q = aq - bq^2$$

$$MR = \frac{dTR}{dq} = a - 2bq$$

However, it should also be noted that this result does not hold for non-linear demand schedules.

Example 8.16

Given the demand schedule $p = 80 - q^{0.5}$ derive the corresponding MR function.

Solution

$$TR = pq = (80 - q^{0.5})q = 80q - q^{1.5}$$

$$MR = \frac{dTR}{dq} = 80 - 1.5q^{0.5}$$

In this non-linear case the intercept on the price axis is still 80 but the slope of MR is 1.5 times the slope of the demand function. (Note that it is implicitly assumed that only the positive value of $q^{0.5}$ is relevant to this demand function. The same assumption applies to all other demand functions in this format that are used here.)

For those of you who are still not convinced that the idea of looking at an 'infinitesimally small' change can help find the rate of change of a function at a point, Example 8.17 below shows how a spreadsheet can be used to calculate rates of change for very small increments. This example is for illustrative purposes only though. The main reason for using calculus in the first place is to enable the immediate calculation of rates of change at any point of a function.

Example 8.17

For the total revenue function

$$TR = 500q - 2q^2$$

find the value of MR when $q = 80$ (i) using calculus, and (ii) using a spreadsheet and looking at small increments in q above the given value of 80.
    Compare the two answers.

Solution

(i) $$MR = \frac{dTR}{dq} = 500 - 4q$$

Thus when $q = 80$

$$MR = 500 - 4(80) = 500 - 320 = 180$$

(ii) Set up a spreadsheet using Lotus 1–2–3 as follows. Enter the labels given below and right justify (RJ) where indicated:

CALCULUS	in cell A1	
MR=dTR/dq	in cell C1	
GIVEN	in cell A3	
FUNCTION	in cell B3	
TR=	in cell C3	(RJ)
$500q-2q^2$	in cell D3	
BASE q=	in cell A5	(RJ)
BASE TR=	in cell D5	(RJ)
DELTA q	in cell A8	(RJ)
q	in cell B8	(RJ)
TR	in cell C8	(RJ)
DELTA TR	in cell D8	(RJ)
dTR/dq	in cell E8	(RJ)

Enter the given (base) value of q of 80 in cell B5. To calculate the corresponding (base) value of TR, enter the formula

$$+500*B5-2*B5^2 \qquad \text{in cell E5}$$

To produce a column of smaller and smaller increments in q in the DELTA q column, enter the number 10 in cell A9 and enter the formula +A9/10 in cell A10 and copy it down to cell A19. Then widen columns to 12 units.

To calculate the new value of q (which is BASE q plus DELTA q), enter the formula

$$+\$B\$5+A9 \qquad \text{in cell B9}$$

Copy this down to cell B19 and widen columns to 14 units.

To calculate the corresponding values of TR enter the formula

$$+500*B9-2*B9^2 \qquad \text{in cell C9}$$

Copy this down to cell C19 and widen the column to 14 units.

The increments in TR (from its base value) can then be calculated by entering the formula

$$+C9-\$E\$5 \qquad \text{in cell D9}$$

Copy this down the DELTA TR column and increase column width to 12 units.

Finally, to calculate MR over the increment range (i.e. dTR/dq) enter the formula

$$+D9/A9 \qquad \text{in cell E9}$$

Copy down the column to cell E19 and widen the column to 12 units.

Table 8.1

CALCULUS		MR=dTR/dq		
GIVEN	FUNCTION	TR = 500q - 2q^2		
BASE q =	80	BASE TR =		27200
DELTA q	q	TR	DELTA TR	dTR/dq
10	90	28800	1600	160
1	81	27378	178	178
0.1	80.1	27217.98	17.98	179.8
0.01	80.01	27201.7998	1.7998	179.98
0.001	80.001	27200.179998	0.179998	179.998
0.0001	80.0001	27200.018	0.01799998	179.9998000
0.00001	80.00001	27200.0018	0.001799999	179.9999801
0.000001	80.000001	27200.00018	0.00018	179.9999990
0.0000001	80.0000001	27200.000018	0.000018	179.9999881
0.00000001	80.00000001	27200.000002	0.0000018	179.9999154
0.000000001	80.000000001	27200	0.00000018	179.9999154

You should now have a spreadsheet like that shown in Table 8.1. This shows that as increments in q become smaller and smaller, the value of MR (i.e. dTR/dq) approaches 180.

This is consistent with the answer obtained by calculus in (i).

QUESTIONS 8.3

1. Given the demand schedule $p = 120 - 3q$ derive a function for MR and find the output at which TR is a maximum.
2. For the demand schedule $p = 40 - 0.5q$ find the value of MR when $q = 15$.
3. Find the output at which MR is zero when $p = 720 - 4q^{0.5}$ describes the demand schedule.
4. A firm knows that the demand function for its output is $p = 400 - 0.5q$. What price should it charge to maximize sales revenue?
5. Make up your own demand function and then derive the corresponding MR function and find the output level which corresponds to zero marginal revenue.

8.4 MARGINAL COST AND TOTAL COST

Just as MR can be shown to be the rate of change of the TR function, so the marginal cost (MC) is the rate of change of the total cost (TC) function. In fact, in nearly all situations where one is dealing with the concept of a marginal increase in a function, the marginal function is equal to the rate of change of the original function, i.e. to derive the marginal function one just differentiates the original function.

Example 8.18

Given $TC = 6 + 4q^2$ derive the MC function.

Solution

$$MC = \frac{dTC}{dq} = 8q$$

The example above is somewhat unrealistic in that it assumes an MC function that is a straight line. This is because the TC function is given as a simple quadratic function, whereas one normally expects a TC function to have a shape similar to that shown in Figure 8.4. This represents a cubic function with certain properties to ensure that the rate of change of TC first falls and then rises, and that the TC function never actually falls as output increases (i.e. MC is never negative). The flattest point of this TC schedule is at M, which corresponds to the minimum value of MC.

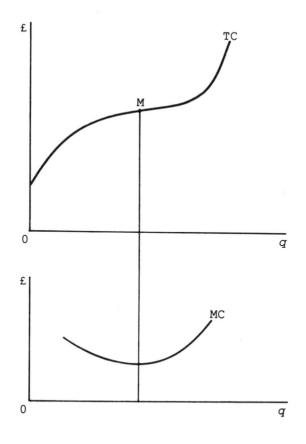

Figure 8.4

A cubic total cost function has these properties if $TC = aq^3 + bq^2 + cq + d$ where a, b, c and d are parameters such that a, c, $d > 0$, $b < 0$ and $b^2 < 3ac$. This applies to the TC functions in the examples below.

Example 8.19

If $TC = 2.5q^3 - 13q^2 + 50q + 12$ derive the MC function.

Solution

$$MC = \frac{dTC}{dq} = 7.5q^2 - 26q + 50$$

Example 8.20

If a firm faces the TC function

$$TC = 40 + 82q - 6q^2 + 0.2q^3$$

when will the average variable cost be at its minimum value?

Solution

The theory of costs tells us that MC will cut the minimum point of both the average cost (AC) and the average variable cost (AVC) functions. We therefore need to derive the MC and AVC functions and find where they intersect.

It is obvious from this TC function that total fixed costs (TFC) = 40 and total variable costs (TVC) = $82q - 6q^2 + 0.2q^3$. Therefore,

$$AVC = \frac{TVC}{q} = 82 - 6q + 0.2q^2$$

and

$$MC = \frac{dTC}{dq} = 82 - 12q + 0.6q^2$$

Setting MC = AVC

$$82 - 12q + 0.6q^2 = 82 - 6q + 0.2q^2$$
$$0.4q^2 = 6q$$
$$q = \frac{6}{0.4} = 15$$

at the minimum point of AVC.

(When you have covered the next chapter, come back to this example and see if you can think of another way of solving it.)

QUESTIONS 8.4

1. If $TC = 65 + q^{1.5}$ what is MC when $q = 25$?
2. Derive a formula for MC if $TC = 4q^3 - 20q^2 + 60q + 40$.
3. If $TC = 0.5q^3 - 3q^2 + 25q + 20$ derive functions for (a) MC, (b) AC, (c) the slope of AC.
4. What is special about MC if $TC = 25 + 0.8q$?
5. Make up your own TC function and then derive the corresponding MC function.

8.5 PROFIT MAXIMIZATION

We are now ready to see how calculus can help a firm to maximize profits, as the following examples illustrate. At this stage we shall just use the $MC = MR$ rule for profit maximization. The second condition (MC cuts MR from below) will be dealt with in the next chapter.

Example 8.21

A monopoly faces the demand schedule $p = 460 - 2q$ and the cost schedule $TC = 20 + 0.5q^2$. How much should it sell to maximize profit and what will this maximum profit be? (All costs and prices are in pounds.)

Solution

To find the output where $MC = MR$ we first need to derive the MC and MR functions. Given

$$TC = 20 + 0.5q^2$$

then

$$MC = \frac{dTC}{dq} = q \qquad (1)$$

As

$$TR = pq = (460 - 2q)q = 460q - 2q^2$$

then

$$MR = \frac{dTR}{dq} = 460 - 4q \qquad (2)$$

To maximize profit $MR = MC$. Therefore, from (1) and (2),

$$460 - 4q = q$$
$$460 = 5q$$
$$92 = q$$

The actual maximum profit when the output is 92 will be

$$TR - TC = (460q - 2q^2) - (20 + 0.5q^2)$$
$$= 460q - 2q^2 - 20 - 0.5q^2$$
$$= 460q - 2.5q^2 - 20$$
$$= 460(92) - 2.5(8,464) - 20$$
$$= 42,320 - 21,160 - 20 = £21,140$$

Example 8.22

A firm faces the demand schedule $p = 184 - 4q$ and the TC function $TC = q^3 - 21q^2 + 160q + 40$. What output will maximize profit?

Solution

Given

$$TR = pq = (184 - 4q)q = 184q - 4q^2$$

then

$$MR = \frac{dTR}{dq} = 184 - 8q$$

Given

$$TC = q^3 - 21q^2 + 160q + 40$$

then

$$MC = \frac{dTC}{dq} = 3q^2 - 42q + 160$$

To maximize profits MC = MR. Therefore,

$$3q^2 - 42q + 160 = 184 - 8q$$
$$3q^2 - 34q - 24 = 0$$
$$(q - 12)(3q + 2) = 0$$
$$q - 12 = 0 \quad \text{or} \quad 3q + 2 = 0$$
$$q = 12 \quad \text{or} \quad q = -2/3$$

One cannot produce a negative quantity and so the firm must produce 12 units of output in order to maximize profits.

QUESTIONS 8.5

1. A monopoly faces the following TR and TC schedules:

$$TR = 300q - 2q^2$$

Introduction to calculus

$$TC = 12q^3 - 44q^2 + 60q + 30$$

What output should it sell to maximize profit?
2. A firm faces the demand function $p = 190 - 0.6q$ and the total cost function $TC = 40 + 30q + 0.4q^2$.

(a) What output will maximize profit?
(b) What output will maximize total revenue?
(c) What will the output be if the firm makes a profit of £4,760?

3. A firm's total revenue and total cost functions are

$$TR = 52q - q^2$$

$$TC = \frac{q^3}{3} - 2.5q^2 + 34q + 4$$

At what output will profit be maximized?

8.6 RESPECIFYING FUNCTIONS

Sometimes demand schedules are specified in the form $q = f(p)$. To derive the relationship between MR and q, one first has to rearrange this sort of function to get the inverse function in the form $p = f(q)$.

Example 8.23

Derive the MR function for the demand schedule $q = 400 - 0.1p$.

Solution

Given

$$q = 400 - 0.1p$$

then

$$10q = 4,000 - p$$
$$p = 4,000 - 10q$$

Then, proceeding in the usual fashion,

$$TR = pq = (4,000 - 10q)q = 4,000q - 10q^2$$

$$MR = \frac{dTR}{dq} = 4,000 - 20q$$

Basic mathematics for economists

Example 8.24

A firm faces the demand schedule $q = 200 - 4p$ and the cost schedule $TC = 0.1q^3 - 0.5q^2 + 2q + 8$. What price will maximize profit?

Solution

$q = 200 - 4p$ can be rewritten as $p = 50 - 0.25q$. This is a linear demand schedule and so MR has the same intercept and twice the slope. Thus

$$MR = 50 - 0.5q$$

$$MC = \frac{dTC}{dq} = 0.3q^2 - q + 2$$

To maximize profits MC = MR. Therefore,

$$0.3q^2 - q + 2 = 50 - 0.5q$$
$$0.3q^2 - 0.5q - 48 = 0$$

Using the formula for the solution of quadratic equations

$$q = \frac{-(-0.5) \pm \sqrt{(-0.5)^2 - 4 \times 0.3 \times (-48)}}{2 \times 0.3}$$

$$= \frac{0.5 \pm \sqrt{0.25 + 57.6}}{0.6}$$

$$= \frac{0.5 \pm \sqrt{57.85}}{0.6}$$

Disregarding the negative solution as output cannot be negative

$$q = \frac{0.5 + 7.6}{0.6} = \frac{8.1}{0.6} = 13.5$$

Substituting this output into the demand function

$$p = 50 - 0.25q = 50 - 3.375 = 46.625$$

QUESTIONS 8.6

1. Given the demand function $q = 150 - 3p$ derive a function for MR.
2. A firm faces the demand schedule $q = 40 - p^{0.5}$ (where $p^{0.5} \geqslant 0$, $q \leqslant 40$) and the cost schedule $TC = q^3 - 2.5q^2 + 50q + 16$. What price should it charge to maximize profit?
3. Find the MR function corresponding to the demand schedule $q = (60 - 2.5p)^{0.5}$.

8.7 POINT ELASTICITY OF DEMAND

Price elasticity of demand is defined as

$$e = (-1) \frac{\text{percentage change in quantity}}{\text{percentage change in price}}$$

However, looking at the changes in price and quantity between points A and B on the demand schedule D in Figure 8.5, the question one may ask is 'percentage of what'? Clearly the change in quantity Δq is a much larger percentage of q_1 than q_2 for example. If we consider elasticity of demand at a point on the demand schedule this problem can be avoided. In Chapter 4 some simple examples of point elasticity based on linear demand schedules were considered. Now we can look at how point elasticity can also be derived for non-linear demand schedules.

If the movement along D from A to B is very small then

$$e = (-1) \frac{\Delta q / q}{\Delta p / p} = (-1) \frac{p}{q} \frac{1}{\Delta p / \Delta q} \tag{1}$$

The very small movement effectively means that we can assume $p_1 = p_2 = p$ and $q_1 = q_2 = q$. As B gets nearer to A the value of $\Delta p / \Delta q$, which is the slope of the line AB, gets closer to the slope of the tangent TT′ at A. (Note that, as price falls, Δp is negative, giving a negative value for the relevant slopes.) Thus for an infinitesimally small movement from A

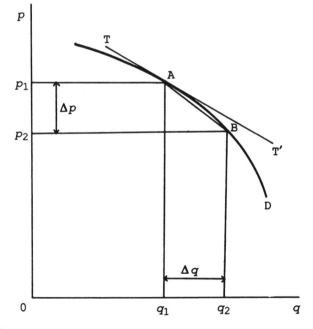

Figure 8.5

Basic mathematics for economists

$$\frac{\Delta p}{\Delta q} = \frac{dp}{dq} = \text{slope of D at A}$$

Thus, substituting this result into (1) above, the formula for point elasticity of demand becomes

$$e = (-1)\frac{p}{q}\frac{1}{dp/dq}$$

Example 8.25

What is the point elasticity when price is 12 for the demand function $p = 60 - 3q$?

Solution

$$\frac{dp}{dq} = -3$$

Given $p = 60 - 3q$, then

$$3q = 60 - p$$

$$q = \frac{60 - p}{3}$$

When $p = 12$, then

$$q = \frac{60 - 12}{3} = \frac{48}{3} = 16$$

Therefore,

$$e = (-1)\frac{p}{q}\frac{1}{dp/dq} = (-1)\frac{12}{16}\frac{1}{-3} = 0.25$$

Example 8.26

If a firm's demand schedule is represented by the function $q = 60 - 2p^{0.5}$ (where $p^{0.5} \geq 0$, $q \leq 60$), what is the elasticity of demand when the quantity is 8?

Solution

Deriving the inverse of the function

$$q = 60 - 2p^{0.5}$$
$$2p^{0.5} = 60 - q$$
$$p^{0.5} = 30 - 0.5q$$

270

$$p = (30 - 0.5q)^2 = 900 - 30q + 0.25q^2$$

Therefore,

$$\frac{dp}{dq} = -30 + 0.5q$$

When $q = 8$,

$$\frac{dp}{dq} = -30 + 0.5(8) = -30 + 4 = -26$$

Also, when $q = 8$,

$$p = 900 - 240 + 16 = 676$$

Thus

$$e = (-1)\frac{676}{8}\frac{1}{-26} = 3.25$$

QUESTIONS 8.7

1. What is the point elasticity of demand when price is 20 for the demand schedule $p = 45 - 1.5q$?
2. Explain why the point elasticity of demand decreases in value as one moves down a straight line demand schedule.
3. Given the demand function $p = 600 - 0.5q^2$, what is the elasticity of demand when the quantity is 30?
4. Explain why the demand function $p = 265q^{-1}$ will have the same point elasticity of demand at all prices and say what its value is.

8.8 TAX YIELD

Elementary supply and demand analysis tells us that the effect of a per-unit tax t on a good sold in a competitive market can be illustrated by shifting the supply schedule vertically by the amount of the tax, as shown in Figure 8.6. This will cause the price paid by consumers to rise and the quantity bought to fall as equilibrium moves from A to B. The change in total revenue spent by consumers will depend on the price elasticity of demand.

The Chancellor of the Exchequer, however, is more interested in the total amount of tax raised for the government, or the tax yield (TY), than total consumer expenditure. If a per-unit tax is increased, the quantity bought will fall. The question, however, is whether this fall in quantity will outweigh the effect on TY of the increase in the tax raised on each unit. To answer this we need to know the rate of change of the tax yield with respect to increases in the per-unit tax.

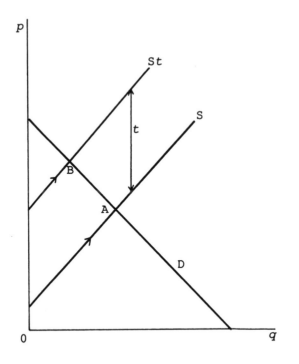

Figure 8.6

Example 8.27

A market has the demand schedule $p = 92 - 2q$ and the supply schedule $p = 12 + 3q$. What per unit tax will raise the maximum tax revenue for the government? (All prices are in pounds.)

Solution

Let the per-unit tax be t. This changes the supply schedule to $p = 12 + t + 3q$, i.e. the intercept on the price axis shifts vertically upwards by the amount t.

In equilibrium,

$$\text{supply } p = \text{demand } p$$

Therefore,

$$12 + 3q + t = 92 - 2q$$
$$5q = 80 - t$$
$$q = 16 - 0.2t$$

272

Introduction to calculus

The tax yield is (amount sold) × (per-unit tax). Therefore,

$$TY = qt = (16 - 0.2t)t = 16t - 0.2t^2$$

The rate of change of TY with respect to t is

$$\frac{dTY}{dt} = 16 - 0.4t$$

If $dTY/dt > 0$, an increase in t will increase TY. From the formula for dTY/dt derived above, one can see that as t is increased the value of dTY/dt falls. Therefore in order to maximize TY, t should be increased until $dTY/dt = 0$.
Thus

$$\frac{dTY}{dt} = 16 - 0.4t = 0$$

$$16 = 0.4t$$

$$40 = t$$

Therefore a per-unit tax of £40 will maximize TY.

Rather than working from first principles, as in the above example, a general formula can be derived for the rate of change of the tax yield with respect to a per-unit tax if it is assumed that both supply and demand schedules are linear.

Assume that these schedules are as follows:

$$\text{demand } p = a + bq$$
$$\text{supply } p = c + dq$$

where a, b, c and d are parameters (note that we expect $b < 0$). With a per-unit tax of t, the supply schedule becomes

$$p = c + dq + t$$

Setting supply price equal to demand price

$$c + dq + t = a + bq$$
$$q(d - b) = a - c - t$$
$$q = \frac{a - c}{d - b} - \frac{t}{d - b}$$
$$TY = qt = \frac{a - c}{d - b}t - \frac{t^2}{d - b}$$
$$\frac{dTY}{dt} = \frac{a - c}{d - b} - \frac{2t}{d - b} \tag{1}$$

273

We can check this formula using the figures from Example 8.27 above. Given the demand schedule $p = 92 - 2q$ and the supply schedule $p = 12 + 3q$, then

$$a = 92 \qquad b = -2 \qquad c = 12 \qquad d = 3$$

and so substituting into (1) above

$$\frac{dTY}{dt} = \frac{92 - 12}{3 - (-2)} - \frac{2t}{5} = \frac{80}{5} - \frac{2t}{5} = 16 - 0.4t$$

This is the same as the function for dTY/dt that was derived from first principles in Example 8.27.

QUESTIONS 8.8

1. Given the demand schedule $p = 180 - 8q$ and the supply schedule $p = 25 + 2q$ what level of per-unit tax would maximize the government's tax yield?
2. Change one of the parameters in Question 1 above and work out the new answer.
3. Assume a market has the demand schedule $q = 40 - 0.5p$ and the supply schedule $q = 2p + 4$. The government currently imposes a per-unit tax of £3. If this tax is slightly increased will TY rise or fall?

8.9 THE KEYNESIAN MULTIPLIER

In a simple Keynesian macroeconomic model with no government sector and no foreign trade, it is assumed that

$$Y = C + I \tag{1}$$

where Y is the national income, C is consumption and I is investment, exogenously fixed, and

$$C = a + bY \tag{2}$$

where a and b are parameters.

The marginal propensity to consume (MPC) is the rate of change of consumption as the national income increases, which is equal to $dC/dY = b$. The multiplier is the rate of change of national income in response to an increase in exogenously determined investment, i.e. dY/dI. The result that the multiplier is equal to

$$\frac{1}{1 - \text{MPC}}$$

can be easily derived by differentiation.

Substituting (2) into (1) we get

$$Y = a + bY + I$$

Therefore,

$$Y(1 - b) = a + I$$

$$Y = \frac{a+I}{1-b} = \frac{a}{1-b} + \frac{I}{1-b}$$

Thus

$$\frac{dY}{dI} = \frac{1}{1-b}$$

which is the formula for the multiplier.

This can be used to calculate the increase in investment necessary to achieve any specified increase in national income.

Example 8.28

In a basic Keynesian macroeconomic model it is assumed that $Y = C + I$ where $I = 250$ and $C = 0.75Y$. What is the equilibrium level of Y? What increase in I would be needed to cause Y to increase to 1,200?

Solution

$$Y = C + I = 0.75Y + 250$$

$$0.25Y = 250$$

So the equilibrium level of Y is 1,000.

For any increase in I (ΔI) the resulting increase in Y (ΔY) will be determined by the formula

$$\Delta Y = K\Delta I \tag{1}$$

where K is the multiplier. We know that

$$K = \frac{1}{1 - \text{MPC}}$$

In this example, MPC $= dC/dY = 0.75$. Therefore,

$$K = \frac{1}{1 - 0.75} = \frac{1}{0.25} = 4 \tag{2}$$

The required change in Y is

$$\Delta Y = 1{,}200 - 1{,}000 = 200 \tag{3}$$

275

Therefore, substituting (2) and (3) into (1),

$$200 = 4\Delta I$$

So the required increase in I is $\Delta I = 50$.

Multipliers for other exogenous variables in more complex macroeconomic models can be derived using the same method. However, for differentiation with respect to one exogenous variable the other variables must remain constant and so we shall return to this topic in Chapter 10 where partial differentiation is explained.

QUESTIONS 8.9

1. In a basic Keynesian macroeconomic model it is assumed that $Y = C + I$ where $I = 820$ and $C = 60 + 0.8Y$.

   (a) What is the marginal propensity to consume?
   (b) What is the equilibrium level of Y?
   (c) What is the value of the multiplier?
   (d) What increase in I is required to increase Y to 5,000?
   (e) If this increase takes place will savings $(Y - C)$ still equal I?

9

Unconstrained optimization

9.1 MAXIMIZATION AND ZERO SLOPE

Consider the demand function

$$p = 60 - 0.2q$$

The corresponding total revenue function is

$$TR = pq = (60 - 0.2q)q = 60q - 0.2q^2$$

This will take an inverted U-shape similar to that shown in Figure 9.1.

If we ask the question 'when is TR at its maximum?' the answer is obviously at M, which is the highest point on the curve. At this maximum position the

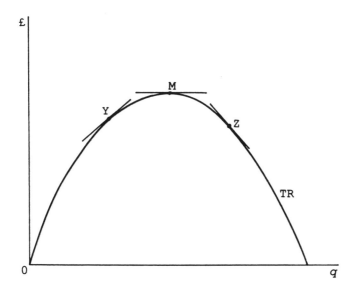

Figure 9.1

TR schedule is flat. To the left of M, TR is rising and has a positive slope, and to the right of M, the TR schedule is falling and has a negative slope. At M itself the slope is zero.

We can therefore say that for a function *of this shape* the maximum point will be where its slope is zero. In Chapter 8 we learned that the slope of a function can be obtained by differentiation. So, for the function

$$TR = 60q - 0.2q^2$$

$$slope = \frac{dTR}{dq} = 60 - 0.4q$$

The slope is zero when

$$60 - 0.4q = 0$$
$$60 = 0.4q$$
$$150 = q$$

Therefore TR is maximized when quantity is 150.

QUESTIONS 9.1

1. What output will maximize total revenue if $TR = 250q - 2q^2$?
2. If a firm faces the demand schedule $p = 90 - 0.3q$ how much does it have to sell to maximize sales revenue?
3. A firm faces the total revenue schedule $TR = 600q - 0.5q^2$

   (a) What is the marginal revenue when q is 100?
   (b) When is the total revenue at its maximum?
   (c) What price should the firm charge to achieve this maximum TR?

4. For the non-linear demand schedule $p = 750 - 0.1q^2$ what output will maximize the sales revenue?

9.2 SECOND-ORDER CONDITION FOR A MAXIMUM

In the example in Section 9.1 above, it was obvious that the TR function was a maximum when its slope was zero because we knew the shape of the function. However, consider the functions in Figure 9.2. The function in Figure 9.2(a) has a slope of zero at N, but this is its minimum point not its maximum. In the case of the function in Figure 9.2(b) the slope is zero at I, but this is neither a maximum nor a minimum point.

The examples in Figure 9.2 clearly illustrate that although a zero slope is *necessary* for a function to be at its maximum it is not a *sufficient* condition. A zero slope just means that the function is at what is known as a 'stationary point', i.e. its slope is neither increasing nor decreasing. Some stationary

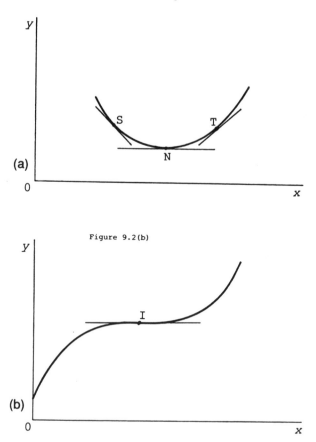

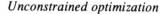

Figure 9.2(b)

Figure 9.2

points will be turning points, i.e. the slope changes from positive to negative (or vice versa) at these points, and will be maximum (or minimum) points of the function.

In order to find out whether a function is at a maximum or a minimum or a point of inflexion (as in Figure 9.2(b)) when its slope is zero we have to consider what are known as the *second-order* conditions. (The first-order condition for any of the three forms of stationary point is that the slope of the function is zero.)

The second-order conditions tell us what is happening to the rate of change of the slope of the function. If the rate of change of the slope is negative it means that the slope decreases as the variable on the horizontal axis is increased. If the slope is decreasing and one is at a point where the actual slope is zero this means that the slope of the function is positive slightly to the left and negative slightly to the right of this point. This is the case in Figure 9.1

279

above. The slope is positive at Y, zero at M and negative at Z. Thus, if the rate of change of the slope of a function is negative at the point where the actual slope is zero then that point is a maximum.

This is the second-order condition for a maximum. Until now, we have just assumed that a function is maximized when its slope is zero if a sketch graph suggests that it takes an inverted U-shape. From now on we shall make a more rigorous check of the second-order conditions to confirm whether or not a function is maximized (or minimized) at any stationary points.

It is a straightforward exercise to find the rate of change of the slope of a function by differentiation. We know that the slope of a function $y = f(x)$ can be found by differentiation. Therefore if we differentiate the function for the slope of the original function, i.e. dy/dx, we get the rate of change of the slope. This is known as the second-order derivative and is written d^2y/dx^2.

Example 9.1

Show that the function $TR = 60q - 0.2q^2$ satisfies the second-order condition for a maximum when $q = 150$.

Solution

The slope of TR (i.e. the marginal revenue) is

$$\frac{dTR}{dq} = 60 - 0.4q = 0 \text{ when } q = 150$$

The rate of change of the slope is

$$\frac{d^2TR}{dq^2} = -0.4$$

Whatever the value of q, -0.4 is always negative.
Therefore, when q is 150, TR must be a maximum.

In the example above the second-order derivative did not depend on the value of q at the function's turning point, but for other functions one may need to substitute the turning point value into the second-order derivative to check the second-order conditions.

Example 9.2

For the non-linear demand function

$$p = 194.4 - 0.2q^2$$

show that TR is a maximum when q is 18.

Solution

$$TR = pq = (194.4 - 0.2q^2)q = 194.4q - 0.2q^3$$

For a turning point

$$\frac{dTR}{dq} = 194.4 - 0.6q^2 = 0$$

$$194.4 = 0.6q^2$$

$$324 = q^2$$

$$18 = q$$

When $q = 18$ then

$$\frac{d^2TR}{dq^2} = -1.2q = -1.2(18) = -21.6 < 0$$

Therefore, second-order conditions are satisfied and TR is a maximum when q is 18.

QUESTIONS 9.2

Find stationary points for the following functions and say whether or not they are at their maximum at these points.

1. $TR = 720q - 0.3q^2$
2. $TR = 225q - 0.12q^3$
3. $TR = 96q - q^{1.5}$
4. $AC = 51.2q^{-1} + 0.4q^2$

9.3 SECOND-ORDER CONDITION FOR A MINIMUM

By the same reasoning as that set out in Section 9.2 above, if the rate of change of the slope of a function is positive at the point when the slope is zero then the function is at a minimum. This is illustrated in Figure 9.2(a). The slope of the function is negative at S, zero at N and positive at T, i.e. the rate of change of the slope is positive at the stationary point N.

Example 9.3

Find the minimum point of the average cost function

$$AC = 25q^{-1} + 0.1q^2$$

Solution

$$\text{slope} = \frac{dAC}{dq} = -25q^{-2} + 0.2q = 0$$

$$0.2q = 25q^{-2}$$

$$q^3 = 125$$

Therefore AC is at a stationary point when $q = 5$. The rate of change of the slope is

$$\frac{d^2AC}{dq^2} = 50q^{-3} + 0.2$$

$$= \frac{50}{125} + 0.2 \text{ when } q = 5$$

$$= 0.4 + 0.2 = 0.6 > 0$$

Therefore the second-order condition for a minimum value of AC is satisfied when q is 5.

The actual value of AC at its minimum point will be

$$25q^{-1} + 0.1q^2 = \frac{25}{5} + 0.1 \times 25 = 5 + 2.5 = 7.5$$

QUESTIONS 9.3

Find whether any stationary points exist for the following functions for positive values of q, and say whether or not the stationary points are at the minimum values of the function.

1. $AC = 345.6q^{-1} + 0.8q^2$
2. $AC = 600q^{-1} + 0.5q^{1.5}$
3. $MC = 30 + 0.4q^2$
4. $TC = 15 + 27q - 9q^2 + q^3$
5. $MC = 8.25q$

9.4 SUMMARY OF SECOND-ORDER CONDITIONS

If $y = f(x)$ and there is a stationary point where $dy/dx = 0$, then

1 if $d^2y/dx^2 < 0$, this point is a maximum
2 if $d^2y/dx^2 > 0$, this point is a minimum

Note that, strictly speaking, (1) and (2) are conditions for *local* maximums and minimums. It is possible that a function may take the shape shown in

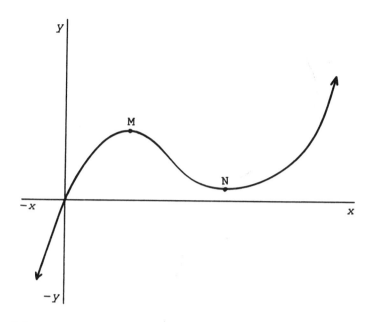

Figure 9.3

Figure 9.3. There is no true global maximum or minimum, as values of y continue towards plus and minus infinity as shown by the arrows. Points M and N, which satisfy the above second-order conditions for maximum and minimum, are therefore just local maximum and minimum points.

If $d^2y/dx^2 = 0$, there may be an inflexion point that is neither a maximum nor a minimum, such as I in Figure 9.2(b). To check if this is so one really needs to investigate further, looking at the third, fourth and possibly higher order derivatives, but we will not go into these conditions here. In all the economic applications given in this text, it will be obvious whether or not a function is at a maximum or minimum at any stationary points.

It should be noted that many functions do not have maximum or minimum points. Linear functions are an obvious example. Such cases cannot satisfy the first-order conditions for a turning point, i.e. that $dy/dx = 0$, except when horizontal.

Example 9.4

In Chapter 5 we considered an example of a break-even chart where a firm was assumed to have the total cost function $TC = 18q$ and the total revenue function $TR = 240 + 14q$. Show that the profit-maximizing output cannot be determined for this firm.

Solution

Profit is

$$\pi = TR - TC$$
$$= 240 + 14q - 18q$$
$$= 240 - 4q$$

Thus

$$\frac{d\pi}{dq} = -4$$

There is obviously no output level at which $-4 = 0$ and so no turning point exists. Therefore the profit-maximizing output cannot be determined.

End-Point Solutions

There are some possible exceptions to these first- and second-order conditions for maximum and minimum values of functions. If the domain of a function is restricted, then a maximum or minimum point may be determined by this restriction, giving what is known as an 'end-point' or 'corner' solution. In such cases, the usual rules for optimization set out in this chapter will not apply. For example, suppose a firm faces the total cost function (in £)

$$TC = 45 + 18q - 5q^2 + q^3$$

Its slope will be

$$\frac{dTC}{dq} = 18 - 10q + 3q^2 = 0 \tag{1}$$

for a stationary point
However, if we try using the quadratic equation formula to find a value of q for which (1) holds we see that

$$q = \frac{10 \pm \sqrt{100 - 216}}{6} = \frac{10 \pm \sqrt{-116}}{6}$$

and so no solution exists. There is no turning point as no value of q corresponds to a zero slope for this function.

However, if the domain of q is restricted to non-negative values then TC will be at its minimum value of £45 when $q = 0$. Mathematically the conditions for minimization are not met at this point but, from a practical viewpoint, the minimum cost that this firm can ever face is the £45 it must pay even if nothing is produced. This is an example of an end-point solution.

Therefore, when tackling problems concerned with the minimization or maximization of economic variables, you need to ask whether or not there are restrictions on the domain of the variable in question which may give an end-point solution.

QUESTIONS 9.4

1. A firm faces the demand schedule $p = 200 - 2q$ and the total cost schedule

$$TC = \tfrac{2}{3}q^3 - 14q^2 + 222q + 50$$

Derive expressions for the following functions and find out whether they have maximum or minimum points. If they do, say what value of q this occurs at and calculate the actual value of the function at this output.

(a) Marginal cost
(b) Average variable cost
(c) Average fixed cost
(d) Total revenue
(e) Marginal revenue
(f) Profit

2. Construct your own example of a function that has a turning point. Explain whether this turning point is a maximum or a minimum.
3. A firm attempting to expand output in the short-run faces the total product of labour schedule $TP = 24L^2 - L^3$. At what levels of L will

(a) TP_L (b) MP_L (c) AP_L

be at their maximum levels?

Do the following functions have maximum or minimum points?
4. $TC = 12 + 62q - 10q^2 + 1.2q^3$
5. $TC = 6 + 2.5q$
6. $y = 60 + 3.8x^2 - 0.04x^3$

9.5 PROFIT MAXIMIZATION

We have already encountered some problems involving the maximization of a profit function. As profit maximization is perhaps the most common optimization problem that you will encounter in economics, in this section we shall carefully work through the second-order condition for profit maximization and see how it relates to the different intersection points of a firm's MC and MR schedules.

Consider the firm whose marginal cost and marginal revenue schedules are shown by MC and MR in Figure 9.4. At what output will profit be maximized?

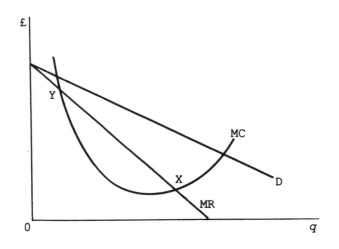

Figure 9.4

The first rule for profit maximization is that profits are at a maximum when MC = MR. However, you will note that in Figure 9.4 there are two points, X and Y, where MC = MR. Only X satisfies the second rule for profit maximization, which is that MC cuts MR from below at the point of intersection. This corresponds to the second-order condition for a maximum required by the differential calculus, as illustrated in the following example.

Example 9.5

Find the profit-maximizing output for a firm with the total cost function $TC = 4 + 97q - 8.5q^2 + \frac{1}{3}q^3$ and the total revenue function $TR = 58q - 0.5q^2$.

Solution

First let us find where MC and MR intersect.

$$MC = \frac{dTC}{dq} = 97 - 17q + q^2 \qquad (1)$$

$$MR = \frac{dTR}{dq} = 58 - q \qquad (2)$$

Therefore when MC = MR

$$97 - 17q + q^2 = 58 - q$$
$$39 - 16q + q^2 = 0 \qquad (3)$$
$$(3 - q)(13 - q) = 0$$

286

Thus $q = 3$ or $q = 13$. These are the two outputs at which the MC and MR schedules intersect, but which one satisfies the second rule for profit maximization?

To answer this question, the problem can be reformulated by deriving a function for profit and then trying to find its maximum. Thus, profit will be

$$\pi = TR - TC = 58q - 0.5q^2 - (4 + 97q - 8.5q^2 + \tfrac{1}{3}q^3)$$
$$= 58q - 0.5q^2 - 4 - 97q + 8.5q^2 - \tfrac{1}{3}q^3$$
$$= -39q + 8q^2 - 4 - \tfrac{1}{3}q^3$$

Differentiating

$$\frac{d\pi}{dq} = -39 + 16q - q^2 = 0 \text{ for a stationary point} \tag{4}$$

Equation (4) is the same as (3) above and has the same two solutions, i.e. $q = 3$ or $q = 13$. However, using this method we can also explore the second-order conditions. From (4) we can derive

$$\frac{d^2\pi}{dq^2} = 16 - 2q$$

When $q = 3$ then $d^2\pi/dq^2 = 16 - 6 = 10$ and therefore π is a minimum. When $q = 13$ then $d^2\pi/dq^2 = 16 - 26 = -10$ and therefore π is a maximum. Thus only one of the intersection points of MR and MC corresponds to the profit-maximizing output. This will be where MC cuts MR from below.

We can prove that this must be so as follows:

slope of $MC = \dfrac{dMC}{dq} = -17 + 2q$ from (1).

slope of $MR = \dfrac{dMR}{dq} = -1$ from (2).

When $q = 3$, then the slope of MC is

$-17 + 2(3) = -17 + 6 = -11 <$ slope of MR (i.e. steeper negative slope).

When $q = 13$, then the slope of MC is

$-17 + 2(13) = 9 >$ slope of MR

Thus, when $q = 3$, the MC schedule has a steeper negative slope than MR and so must cut it from above. When $\dot{q} = 13$, MC has a positive slope and so must cut MR from below.

QUESTIONS 9.5

1. A monopoly faces the total revenue schedule

$$TR = 300q - 2q^2$$

and the total cost schedule

$$TC = 12q^3 - 44q^2 + 60q + 30$$

Are there two output levels at which $MC = MR$? Which is the profit-maximizing output?

2. If a firm faces the demand schedule $p = 120 - 3q$ and the total cost schedule $TC = 120 + 36q + 1.2q^2$, what output levels, if any, will

(a) maximize profit   (b) minimize profit?

3. Explain why a firm which is a monopoly seller in a market with the demand schedule $p = 66.8 - 0.4q$ and which faces the total cost schedule

$$TC = 220 + 120q - 12q^2 + 0.5q^3$$

can never make a positive profit.

4. What is the maximum profit a firm can make if it faces the demand schedule $p = 660 - 3q$ and the total cost schedule $TC = 25 + 240q - 72q^2 + 6q^3$?

5. If a firm faces the demand schedule $p = 53.5 - 0.7q$ and the total cost schedule $TC = 400 + 35q - 6q^2 + 0.1q^3$ what *price* will maximize profits?

9.6 INVENTORY CONTROL

In Chapter 8 we considered a few applications of differentiation, such as tax yield maximization, without taking second-order conditions into account. We can now look at some more applications where it is not obvious that a function is maximized (or minimized) when its slope is zero and where second-order conditions must be fully investigated. Here we look at how the optimum order size for a firm wishing to minimize ordering and storage costs can be calculated.

A firm which uses a particular component has to take into account costs other than the actual purchase price of the component. These include the following.

1 Reorder costs: each order for a consignment of components will involve administration work, delivery, unloading etc.
2 Storage costs: The more components a firm has in storage the more space will be needed in the firm's premises. There is also the opportunity cost of the firm's capital which will be tied up in the components it has paid for.

If a firm only makes a few large orders its storage costs will be high but, on the other hand, if it makes lots of small orders its reorder costs will be high. How then can it decide on the optimum order size?

Unconstrained optimization

Assume that the total annual demand for components is evenly spread over the year. Let this total annual demand be defined as Q and let the size of each order for components be q.

Assume that each order is of equal size and that inventory levels are run down to zero before the next consignment arrives. Also assume that F is the fixed cost for each order and S is the storage costs per unit per year. If each consignment of size q is run down at a constant rate then the average amount of stock held will be $q/2$. (This is illustrated in Figure 9.5 where t represents the time interval between orders.) Thus storage costs for the year will be $(q/2)S$. The number of orders made in a year will be Q/q. Thus the total order costs for the year will be $(Q/q)F$.

The firm will wish to choose the order size that minimizes the total of order costs plus storage costs, defined as TC. The mathematical problem is therefore to find the value of q that minimizes

$$TC = (Q/q)F + (q/2)S$$

Differentiating with respect to q

$$\frac{dTC}{dq} = \frac{-QF}{q^2} + \frac{S}{2} \tag{1}$$

For a stationary point

$$0 = -\frac{QF}{q^2} + \frac{S}{2}$$

$$\frac{QF}{q^2} = \frac{S}{2}$$

$$\frac{2QF}{S} = q^2$$

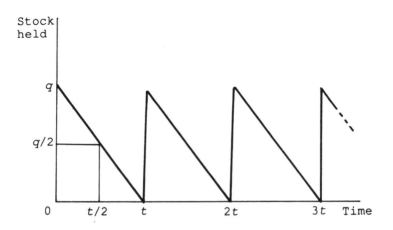

Figure 9.5

Therefore the optimal order size is

$$q = \sqrt{\left(\frac{2QF}{S}\right)} \qquad (2)$$

As F and S will usually be exogenously given, then q depends on the square root of the total annual demand.

The second-order conditions need to be inspected to check that this turning point is a minimum. If (1) above is rewritten as

$$\frac{dTC}{dq} = -QFq^{-2} + \frac{S}{2}$$

then we can see that

$$\frac{d^2TC}{dq^2} = \frac{2QF}{q^3} > 0$$

as Q, F and q must all be positive quantities.

Thus any positive value of q that satisfies the first-order condition (2) above must also satisfy the second-order condition for a minimum value of TC.

Example 9.6

A firm uses 200,000 units of a component in a year, with demand evenly spread over the year. Each order for a batch of components costs the fixed amount of £80 per order in addition to the purchase price. Each unit held in stock over the year costs £8. What is the optimum order size?

Solution

The optimum order size is q and so the average stock held is $q/2$. The number of orders is

$$\frac{Q}{q} = \frac{200,000}{q}$$

at £80 each.

$$TC = \text{order} + \text{stock holding costs}$$

$$= \frac{200,000(80)}{q} + \frac{8q}{2}$$

$$= 16,000,000q^{-1} + 4q$$

For a stationary point

$$\frac{dTC}{dq} = -16,000,000q^{-2} + 4 = 0$$

$$q^2 = \frac{16,000,000}{4} = 4,000,000$$

$$q = \sqrt{(4,000,000)} = 2,000$$

$$\frac{d^2TC}{dq^2} = 32,000,000q^{-3} > 0 \quad \text{for any } q > 0$$

and so the second-order condition for a minimum is met for the stationary point when q is 2,000. Therefore the optimum order size is 2,000 units.

We could, of course, have solved this problem by just substituting the given values into the formula for optimal order size (2) derived earlier. Thus

$$q = \sqrt{\left(\frac{2QF}{S}\right)} = \sqrt{\left(\frac{2 \times 200,000 \times 80}{8}\right)} = 2,000$$

QUESTIONS 9.6

In all the questions below assume that demand is spread evenly over the year and stock is run down to zero before a new order is placed.

1. A firm uses 6,000 tonnes of commodity X every year. The fixed transaction costs involved with each order are £80. Each tonne of X held in stock costs £6 per annum. How many separate orders for X should the firm make during the year?

2. If each order for a batch of components costs £700 to make, storage costs per annum per component are £20 and annual usage is 4,480 components, what is the optimal order size?

3. A firm uses 1,280 units of a component each year. The cost of making an order is £540 and each component held in stock for a year costs the firm £6. What average order size would you advise the firm to make? Assume that the demand for this component is steady from year to year and that the same number of orders do not have to be made within each 12 month period.

4. A firm uses 1,400 units per year of component G. Each order costs £350 to make and average storage costs per unit of G are £20. There is also an extra 'capacity' cost given that the firm has to provide warehousing capable of storing a full order size of q even though this warehousing space will be under-utilized most of the time. This 'capacity' cost will be £15 per unit of G in an order. Adapt the optimal order size formula to include this extra cost and then find the optimal order size for this firm.

10

Partial differentiation

10.1 PARTIAL DIFFERENTIATION AND THE MARGINAL PRODUCT

Some functions have more than one independent variable. For example, the production function $Q = f(K, L)$ has the two independent variables L and K.

The value of the function will change if one of the independent variables is increased whilst the other is held constant. If K is held constant and L is increased then we will trace out the total product of labour (TP_L) schedule (TP_L is the same thing as Q). This will typically take a shape similar to that shown in Figure 10.1.

In your introductory microeconomics course the marginal product of L (MP_L) was probably defined as the increase in TP_L caused by a one-unit increment in L, assuming K to be fixed at some given level. A more precise definition, however, is that MP_L is the rate of change of TP_L with respect to L. For any given value of L this is the slope of the TP_L function. (Refer back to Section 8.3 if you do not understand why.) Thus the MP_L schedule in Figure 10.1 is at its maximum when the TP_L schedule is at its steepest, at M, and is zero when TP_L is at its maximum, at N.

Partial differentiation is a technique for deriving the rate of change of a function with respect to increases in one variable when all other variables in the function are held constant. Therefore, if the production function $Q = f(K, L)$ is differentiated with respect to L, with K held constant, we get the rate of change of total product with respect to L, in other words MP_L.

The basic rule for partial differentiation is that all variables, other than the one the function is being differentiated with respect to, are treated as constants. Apart from this, partial differentiation follows the normal differentiation rules explained in Chapter 8. If it is assumed that $Q = f(K, L)$ then the usual notation for the partial derivative of this function with respect to L is $\partial Q/\partial L$. The curved ∂ in a partial derivative distinguishes it from the derivative of a single variable function where a normal letter 'd' is used.

Partial differentiation

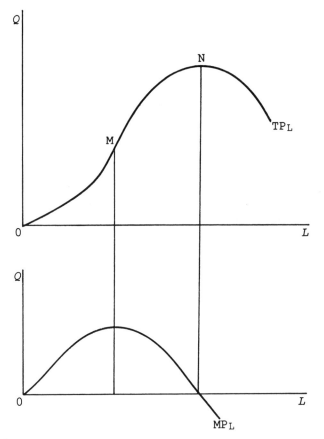

Figure 10.1

Example 10.1

If $y = 14x + 3z^2$, find the partial derivatives of this function with respect to x and z.

Solution

$$\frac{\partial y}{\partial x} = 14$$

(The $3z^2$ disappears as it is treated as a constant. One then just differentiates the term $14x$ with respect to x.)

$$\frac{\partial y}{\partial z} = 6z$$

(The $14x$ is treated as a constant and disappears. One then just differentiates the term $3z^2$ with respect to z.)

Example 10.2

Find the partial derivatives of the function $y = 6x^2z$.

Solution

In this function the variable held constant does not disappear as it is multiplied by the other variable.

$$\frac{\partial y}{\partial x} = 12xz$$

treating z as a constant.

$$\frac{\partial y}{\partial z} = 6x^2$$

treating x^2 as a constant.

Example 10.3

For the production function $Q = 20K^{0.5}L^{0.5}$

(i) derive a function for MP_L, and
(ii) show that MP_L decreases as one moves along an isoquant.

Solution

(i) $MP_L = \dfrac{\partial Q}{\partial L} = 10K^{0.5}L^{-0.5} = \dfrac{10K^{0.5}}{L^{0.5}}$

Note that this MP_L function will continuously slope downward, unlike the MP_L function illustrated in Figure 10.1.

(ii) If the function for MP_L above is multiplied top and bottom by $2L^{0.5}$, then we get

$$MP_L = \frac{20K^{0.5}L^{0.5}}{2L} = \frac{Q}{2L} \tag{1}$$

An isoquant joins combinations of K and L that yield the same output level. Thus if Q is held constant and L is increased then the function (1) shows us that MP_L will decrease.

We can now see that for any Cobb–Douglas production function in the format $Q = AK^\alpha L^\beta$ the law of diminishing marginal productivity holds for each

Partial differentiation

input as long as $0 < \alpha$, $\beta < 1$. Making the usual assumption that K is fixed and L is variable, the marginal product of L is found in the usual way by partial differentiation. Thus

$$MP_L = \frac{\partial Q}{\partial L} = \beta AK^\alpha L^{\beta-1} = \frac{\beta AK^\alpha}{L^{1-\beta}}$$

If K is held constant then, as L is increased, $L^{1-\beta}$ increases (given $0 < \beta < 1$) and so the whole term for MP_L decreases in value, i.e. the marginal product falls as L is increased.

Similarly, if K was increased while L was held constant,

$$MP_K = \frac{\partial Q}{\partial K} = \alpha AK^{\alpha-1}L^\beta = \frac{\alpha AL^\beta}{K^{1-\alpha}}$$

which falls as K increases.

When there are more than two inputs in a production function, the same principles still apply. For example if $Q = AX_1^a X_2^b X_3^c X_4^d$ where X_1, X_2, X_3 and X_4 are inputs, then the marginal product of X_3 will be

$$\frac{\partial Q}{\partial X_3} = cAX_1^a X_2^b X_3^{c-1} X_4^d = \frac{cAX_1^a X_2^b X_4^d}{X_3^{1-c}}$$

which decreases as X_3 increases, *ceteris paribus*.

Note that the concept of diminishing marginal productivity is a short-run concept which explains what happens to the marginal product of one input if that input is increased whilst all other input levels are held constant. It should not be confused with the long-run concept of 'diminishing returns to scale' which applies when *all* inputs are increased by the same proportion (see Section 4.9).

We can also see that for a production function in the usual Cobb–Douglas format the marginal product functions will continuously decline towards zero and will never 'bottom out' for finite values of L; i.e. they will never reach a minimum point where the slope is zero. If, for example,

$$Q = 25K^{0.4}L^{0.5}$$

then

$$MP_L = \frac{\partial Q}{\partial L} = 12.5K^{0.4}L^{-0.5}$$

The first-order condition for a minimum

$$\frac{12.5K^{0.4}}{L^{0.5}} = 0$$

is satisfied only if $12.5K^{0.4} = 0$, which only holds if $K = 0$ and hence $Q = 0$, or if L becomes infinitely large.

Since, for finite values of L, MP_L will still remain positive however large L becomes, this means that on the isoquant map for a two-input Cobb–Douglas production function the isoquants will never 'bend back'; i.e. there will not be an uneconomic region.

Other possible formats for production functions are possible though. For example, if

$$Q = 4.6K^2 + 3.5L^2 - 0.012K^3L^3$$

then

$$MP_L = \frac{\partial Q}{\partial L} = 7L - 0.036K^3L^2$$

If K takes a positive value less than 4.6 then MP_L will first rise and then fall, eventually becoming negative. Its slope will change from a positive to a negative value as L increases since the slope of MP_L is

$$\frac{\partial MP_L}{\partial L} = 7 - 0.072K^3L$$

Example 10.4

If $q = 20x^{0.6}y^{0.2}z^{0.3}$, find the rate of change of q with respect to x, y and z.

Solution

Although there are now three independent variables instead of two, the same rules still apply, this time with two variables treated as constants. Therefore,

$$\frac{\partial q}{\partial x} = 12x^{-0.4}y^{0.2}z^{0.3}$$

holding y and z constant;

$$\frac{\partial q}{\partial y} = 4x^{0.6}y^{-0.8}z^{0.3}$$

holding x and z constant;

$$\frac{\partial q}{\partial z} = 6x^{0.6}y^{0.2}z^{-0.7}$$

holding x and y constant.

To avoid making mistakes when partially differentiating a function with several variables it may help if you write in the variables that do not change first and then differentiate. In the above example, when differentiating with respect to x for instance, this would mean first writing in $y^{0.2}z^{0.3}$, as y and z are held constant.

Partial differentiation

When a function has a large number of variables, a shorthand notation for the partial derivative may be used. For the function $f = f(x_1, x_2, \ldots, x_n)$ one can write

$$f_1 = \frac{\partial f}{\partial x_1}, \quad f_2 = \frac{\partial f}{\partial x_2} \quad \text{etc.}$$

Example 10.5

For the function

$$f(x_1, x_2, \ldots, x_n) = \sum_{i=1}^{n} 6x_i^{0.5}$$

find f_j where j is any input number.

Solution

This function is a summation of several terms. Only one term, the jth, will contain x_j. If one is differentiating with respect to x_j then all other terms are treated as constants and disappear. Therefore, one only has to differentiate the term $6x_j^{0.5}$ with respect to x_j. Thus $f_j = 3x_j^{-0.5}$.

This notation can also be used to express second-order derivatives. For example, f_{11} denotes the second-order partial derivative $\dfrac{\partial^2 f}{\partial x_1^2}$.

Uses of second-order partial derivatives will be explained in Section 3.

QUESTIONS 10.1

1. Find $\partial y/\partial x$ and $\partial y/\partial z$ when

   (a) $y = 6 + 3x + 16z + 4x^2 + 2z^2$
   (b) $y = 14x^3 z^2$
   (c) $y = 9 + 4xz - 3x^{-2}z^3$

2. Show that the law of diminishing marginal productivity holds for the production function $Q = 12K^{0.4}L^{0.4}$. Will the MP_L schedule take the shape shown in Figure 10.1?

3. Derive formulae for the marginal products of the three inputs in the production function

   $$Q = 40K^{0.3}L^{0.3}R^{0.4}.$$

4. Explain why the production function $Q = 0.4K + 0.7L$ does not obey the law of diminishing marginal productivity.

5. If $Q = 18K^{0.3}L^{0.2}R^{0.5}$, will the marginal products of any of the three inputs K, L and R become negative?

6. For the function

$$Q(x_1, x_2, \ldots, x_n) = \sum_{i=1}^{n} 4x_i^3$$

derive a formula for the partial derivative Q_j, where j is an input number.

10.2 FURTHER APPLICATIONS OF PARTIAL DIFFERENTIATION

Partial differentiation is basically a mathematical application of the assumption of *ceteris paribus* (i.e. other things being held equal) which is frequently used in economic analysis. Because the economy is a complex system to understand, economists often look at the effect of changes in one variable assuming all other influencing factors remain unchanged. When the relationship between the different economic variables can be expressed in a mathematical format, then the analysis of the effect of changes in one variable can be discovered via partial differentiation. We have already seen how partial differentiation can be applied to production functions and in this section we shall examine a few other applications.

Elasticity

The quantity q demanded of a good depends on several factors. These may include the price of the good (p), average consumer income (m), the price of a complement (q_c), the price of a substitute (p_s) and population (n). The relationship can be expressed as

$$q = f(p, m, p_c, p_s, n)$$

In introductory economics courses (and in earlier chapters of this book), price elasticity of demand is usually defined as

$$e = (-1) \frac{\text{percentage change in quantity demanded}}{\text{percentage change in price}}$$

This definition implicitly assumes *ceteris paribus*. This is also assumed in the more precise measure of point (price) elasticity of demand:

$$e = (-1) \frac{p}{q} \frac{1}{dp/dq}$$

Recognizing that quantity demanded depends on factors other than price, then the point elasticity of demand can be more accurately redefined as

$$e = (-1) \frac{p}{q} \frac{1}{\partial p / \partial q} = (-1) \frac{p}{q} \frac{\partial q}{\partial p}$$

(Note that we have employed the *inverse function rule* here. This states that, for any function $y = f(x)$, then

$$\frac{dx}{dy} = \frac{1}{dy/dx}$$

as long as $dy/dx \neq 0$. The rule can also be used for partial derivatives.)

Point elasticity (with respect to own price) can now be determined for given specific demand functions.

Example 10.6

For the demand function

$$q = 35 - 0.4p + 0.15m - 0.25p_c + 0.12p_s + 0.003n$$

where the terms are as defined above, what is the price elasticity of demand when $p = 24$?

Solution

We know the value of p and we can easily derive the partial derivative $\partial q / \partial p = -0.4$. Substituting these into the elasticity formula

$$e = (-1)\frac{p}{q}\frac{\partial q}{\partial p}$$

$$= (-1)\frac{24}{35 - 0.4(24) + 0.15m - 0.25p_c + 0.12p_s + 0.003n}(-0.4)$$

$$= \frac{9.6}{25.4 + 0.15m - 0.25p_c + 0.12p_s + 0.003n}$$

The actual value of elasticity cannot be calculated until specific values for m, p_c, p_s and n are given. Thus this example shows that the value of point elasticity of demand with respect to price will depend on the values of other factors that affect demand and thus determine the position on the demand schedule.

Other measures of elasticity will also depend on the values of the different variables in the demand function. For example, the basic definition of income elasticity of demand is

$$e_m = \frac{\text{percentage increase in quantity demanded}}{\text{percentage increase in income}}$$

If we assume an infinitesimally small change in income and recognize that all other factors are being held constant then

$$e_m = \frac{\Delta q/q}{\Delta m/m} = \frac{m}{q}\frac{\Delta q}{\Delta m} = \frac{m}{q}\frac{\partial q}{\partial m}$$

Thus, for the demand function in Example 10.6 above, income elasticity of demand can be defined as

$$e_m = \frac{m}{q}\frac{\partial q}{\partial m} = \frac{m}{q} \quad (0.15)$$

If the value of m is given as 30, say, then

$$e_m = \frac{30}{35 - 0.4p + 0.15(30) - 0.25p_c + 0.12p_s + 0.003n}\ (0.15)$$

$$= \frac{4.5}{39.5 - 0.4p - 0.25p_c + 0.12p_s + 0.003n}$$

Thus the value of income elasticity of demand will depend on the value of the other factors influencing demand as well as the level of income itself.

Consumer utility functions

The economic theory of consumer behaviour assumes that individuals try to maximize the satisfaction from the goods they consume subject to their budget constraint. The most common approach is to assume that satisfaction is represented by a utility function, and that the amounts of the different goods consumed will determine the level of utility. Thus, a consumer's utility function will be in the form

$$U = U(x_1, x_2, \ldots, x_n)$$

where $x_1, x_2, \ldots, x_n$ represent the amounts of the different goods he or she consumes.

Unlike output in a production function, one cannot actually measure utility and this theoretical concept is only of use in making general predictions about the behaviour of large numbers of consumers, as you should learn in your economics course. Modern economic theory assumes that utility is an 'ordinal concept', meaning that different combinations of goods can be ranked in order of preference but utility itself cannot be quantified in any way. However, economists also work with the concept of 'cardinal' utility where it is assumed that, hypothetically at least, each individual can quantify and compare different levels of utility. It is this cardinal utility concept which is used here.

If we assume that an individual only consumes the two goods A and B, then his/her utility function will take the form

$$U = U(A, B)$$

Marginal utility is defined as the rate of change of total utility with respect to the increase in consumption of one good. Therefore the marginal utility function for good A will be

$$MU_A = \frac{\partial U}{\partial A}$$

and similarly

$$MU_B = \frac{\partial U}{\partial B}$$

Three important principles of utility theory are as follows.

1 The law of diminishing marginal utility says that if, *ceteris paribus*, the quantity consumed of any one good is increased, then eventually its marginal utility will decline.
2 A consumer will consume a good up to the point where its marginal utility is zero if it is a free good or if a fixed payment is made regardless of the quantity consumed, e.g. water rates.
3 A consumer maximizes satisfaction when each good is consumed up to the point where one extra pound spent on one good will derive the same utility as an extra pound spent on any other good.

Some applications of the first two principles are given in the following examples. We shall return to (3) in Chapter 11, when we study constrained optimization.

Example 10.7

Find out whether the law of diminishing marginal utility holds for both goods A and B in the following utility functions:

(i) $U = A^{0.6}B^{0.8}$
(ii) $U = 85AB - 1.6A^2B^2$
(iii) $U = 0.2A^{-1}B^{-1} + 5AB$

Solutions

(i)
$$MU_A = \frac{\partial U}{\partial A} = 0.6A^{-0.4}B^{0.8}$$

Thus MU_A falls as A increases.

$$MU_B = \frac{\partial U}{\partial B} = 0.8A^{0.6}B^{-0.2}$$

Thus MU_B falls as B increases. As both marginal utility functions decline, the law of diminishing marginal utility holds.

(ii)
$$MU_A = \frac{\partial U}{\partial A} = 85B - 3.2AB^2$$

$$MU_B = \frac{\partial U}{\partial B} = 85A - 3.2A^2B$$

Both MU_A and MU_B will be downward-sloping straight lines given that the quantity of the other good is held constant. Therefore the law of diminishing marginal utility holds.

(iii)
$$MU_A = \frac{\partial U}{\partial A} = -0.2A^{-2}B^{-1} + 5B$$

$$MU_B = \frac{\partial U}{\partial B} = -0.2A^{-1}B^{-2} + 5A$$

As A increases, the term $0.2A^{-2}B^{-1}$ gets smaller. As this term is subtracted from $5B$ this means that MU_A rises. Similarly, MU_B will rise as B increases. Therefore the law of diminishing marginal utility does not hold.

Example 10.8

Given the following utility functions, how much of A will be consumed if it is a free good? If necessary give answers in terms of the fixed amount of B.

(i) $U = 96A + 35B - 0.8A^2 - 0.3B^2$
(ii) $U = 72AB - 0.6A^2B^2$
(iii) $U = A^{0.3}B^{0.4}$

Solutions

In each case we need to try to find the value of A where MU_A is zero. (In all cases the law of diminishing marginal utility holds.) Consumers will not consume extra units of A which have negative marginal utility and hence decrease the total utility.

(i)
$$MU_A = \frac{\partial U}{\partial A} = 96 - 0.16A = 0$$

$$96 = 0.16A$$

$$60 = A$$

Thus 60 units of A are consumed if A is free, regardless of the amount of B consumed.

(ii)
$$MU_A = \frac{\partial U}{\partial A} = 72B - 1.2AB^2 = 0$$

$$72B = 1.2AB^2$$

$$60B^{-1} = A$$

Thus the amount of A consumed if it is free will decrease if more B is consumed.

(iii) $$\text{MU}_A = \frac{\partial U}{\partial A} = 0.3A^{-0.7}B^{0.4}$$

This marginal utility function will decline continuously but will not equal zero unless the amount of A consumed becomes infinite. No finite solution can be found.

The Keynesian multiplier

If a government sector and foreign trade are introduced then the basic Keynesian macroeconomic model becomes the accounting identity

$$Y = C + I + G + X - M \tag{1}$$

and the functional relationships of consumption

$$C = cY_d \tag{2}$$

where c is the marginal propensity to consume, and imports

$$M = mY_d \tag{3}$$

where m is the marginal propensity to import.

$$Y_d = (1 - t)Y \tag{4}$$

is the disposable income, c, m and t are given parameters and investment I, government expenditure G and exports X are exogenously determined.

Substituting (2), (3) and (4) into (1) we get

$$Y = c(1 - t)Y + I + G + X - m(1 - t)Y$$

$$Y[1 - c(1 - t) + m(1 - t)] = I + G + X$$

$$Y = \frac{I + G + X}{1 - c(1 - t) + m(1 - t)} = \frac{I + G + X}{1 - (c - m)(1 - t)} \tag{5}$$

In the basic Keynesian model without G and X, the investment multiplier is simply dY/dI, as was explained in Section 8.9.

However, in this extended model one has to assume that G and X are constant to derive the investment multiplier, i.e. the investment multiplier is found by partially differentiating (5) with respect to I which gives

$$\frac{\partial Y}{\partial I} = \frac{1}{1 - (c - m)(1 - t)}$$

You should also be able to see that the government expenditure and export multipliers will also take this format as

$$\frac{\partial Y}{\partial I} = \frac{\partial Y}{\partial G} = \frac{\partial Y}{\partial X} = \frac{1}{1 - (c - m)(1 - t)}$$

Example 10.9

In a Keynesian macroeconomic system, the following relationships hold:

$$Y = C + I + G + X - M \qquad t = 0.2$$
$$C = 0.8Y_d \qquad G = 400$$
$$M = 0.2Y_d \qquad I = 300$$
$$Y_d = (1 - t)Y \qquad X = 288$$

What is the equilibrium level of Y? What increase in G would be necessary to increase Y to 2,500? If this increased expenditure takes place, what will happen to

(i) the government's budget surplus/deficit and
(ii) the balance of payments?

Solution

First we derive the relationship between C and Y. Thus

$$C = 0.8Y_d = 0.8(1 - t)Y = 0.8(1 - 0.2)Y$$

Next we substitute this and the other functional relationships and given values into the accounting identity to find the equilibrium Y. Thus

$$Y = C + I + G + X - M$$
$$= 0.8(1 - 0.2)Y + 300 + 400 + 288 - 0.2(1 - 0.2)Y$$
$$= 0.64Y + 988 - 0.16Y$$
$$(1 - 0.48)Y = 988$$
$$Y = \frac{988}{0.52} = 1,900$$

At this equilibrium level of Y the tax raised will be

$$tY = 0.2(1,900) = 380$$

The imports will be

$$M = 0.2(0.8)1,900 = 304$$

Thus the budget deficit will be

$$G - T = 400 - 380 = 20$$

and the balance of payments will be

$$X - M = 288 - 304 = -16$$

Partial differentiation

The government expenditure multiplier is

$$\frac{\partial Y}{\partial G} = \frac{1}{1 - (c - m)(1 - t)}$$

$$= \frac{1}{1 - (0.8 - 0.2)(1 - 0.2)}$$

$$= \frac{1}{1 - (0.6)(0.8)}$$

$$= \frac{1}{1 - 0.48} = \frac{1}{0.52}$$

The required increase in Y is

$$\Delta Y = 2,500 - 1,900 = 600$$

Given that

$$\Delta G \frac{\partial Y}{\partial G} = \Delta Y$$

where ΔG is the change in government expenditure, then

$$\Delta G \frac{1}{0.52} = 600$$

$$\Delta G = 600(0.52) = 312$$

This is the increase in G required to raise Y to 2,500.
  At the new level of national income, the tax raised will be

$$tY = 0.2(2,500) = 500$$

The new government expenditure will be

$$400 + 312 = 712$$

Therefore, the budget deficit will be

$$G - T = 712 - 500 = 212$$

(an increase of 192 in the deficit). The new level of imports will be

$$M = 0.2(0.8)2,500 = 400$$

and the balance of payments will be

$$X - M = 288 - 400 = -112$$

(increased deficit of 96).

Cost and revenue functions

Some firms produce several different products. When common production facilities are used the costs of the individual products will be related and this will be reflected in the total cost schedules. The marginal cost schedules of the individual products can then be derived by partial differentiation.

Example 10.10

A firm produces two goods q_1 and q_2 and faces the total cost function

$$TC = 45 + 125q_1 + 84q_2 - 6q_1^2q_2^2 + 0.8q_1^3 + 1.2q_2^3$$

What are the two relevant marginal cost functions?

Solution

$$MC_1 = \frac{\partial TC}{\partial q_1} = 125 - 12q_1q_2^2 + 2.4q_1^2$$

$$MC_2 = \frac{\partial TC}{\partial q_2} = 84 - 12q_1^2q_2 + 3.6q_2^2$$

These marginal cost schedules show that the level of marginal cost for one good will depend on the amount of the other good that is produced. In such cases of joint cost functions, profit maximization is not necessarily a case of simply equating the separate marginal cost and marginal revenue schedules. This is explained in Section 10.4.

In other cases, firms may produce different goods which compete with each other in the market place, or are complements. This means that the price of one good will influence the quantity demanded of the other goods sold by the same firm. Marginal revenue for one good will be the partial derivative of the total revenue assuming that the price of the other good is fixed.

Example 10.11

A firm produces goods A and B which are complements. The relevant demand schedules are

$$q_A = 850 - 12.5p_A - 3.8p_B$$
$$q_B = 936 - 4.8p_A - 24p_B$$

What are the marginal revenue schedules for the two goods?

Solution

Marginal revenue is usually expressed as a function of quantity. Therefore, in order to derive total and marginal revenue functions, the demand

Partial differentiation

functions are first rearranged to get price as a function of quantity. For good A,

$$q_A = 850 - 12.5p_A - 3.8p_B$$

Therefore,

$$12.5p_A = 850 - 3.8p_B - q_A$$

$$p_A = \frac{850 - 3.8p_B - q_A}{12.5}$$

$$TR_A = p_A q_A$$

$$= \left(\frac{850 - 3.8p_B - q_A}{12.5}\right) q_A$$

$$= \frac{850q_A - 3.8p_B q_A - q_A^2}{12.5}$$

$$MR_A = \frac{\partial TR}{\partial q_A}$$

$$= \frac{850 - 3.8p_B - 2q_A}{12.5}$$

$$= 68 - 0.304p_B - 0.16q_A \qquad (1)$$

For good B

$$q_B = 936 - 4.8p_A - 24p_B$$

$$24p_B = 936 - 4.8p_A - q_B$$

$$p_B = 39 - 0.2p_A - \frac{q_B}{24}$$

$$TR_B = p_B q_B$$

$$= \left(39 - 0.2p_A - \frac{q_B}{24}\right) q_B$$

$$= 39q_B - 0.2p_A q_B - \frac{q_B^2}{24}$$

$$MR_B = \frac{\partial TR}{\partial q_B}$$

$$= 39 - 0.2p_A - \frac{q_B}{12} \qquad (2)$$

The marginal revenue functions (1) and (2) for MR_A and MR_B confirm that when the demands for two goods are interrelated the marginal revenue function for one good will depend on the price level of the other good.

QUESTIONS 10.2

1. The demand function for a good is

$$q = 56.6 - 0.25p - 0.03m + 0.45p_s + 0.6n$$

where q is the quantity demanded per week, p is the price per unit, m is the average weekly income, p_s is the price of a competing good and n is the population in millions. Given values are $p = 65$, $m = 350$, $p_s = 60$ and $n = 24$.

(a) Calculate the price elasticity of demand.
(b) Find out what would happen to (a) if n rose to 26.
(c) Explain why this is an inferior good.
(d) If producers of the competing product and the manufacturers of this good both increased their prices by the same percentage, what would happen to the quantity demanded (of the original good) assuming that the proportional price change is small and relevant elasticity measures do not alter significantly.

2. Do the following utility functions obey the law of diminishing marginal utility?

(a) $U = 5A + 8B + 2.2A^2B^2 - 0.3A^3B^3$
(b) $U = 24A^{0.8}B^{1.2}$
(c) $U = 6A^{0.7}B^{0.8}$

3. An individual consumes two goods and has the utility function $U = 2A^{0.4}B^{0.4}$, where A and B represent the quantities of the two goods consumed. Will she ever consume either good up to the point where its marginal utility is zero?

4. In a Keynesian macroeconomic model of an economy

$$Y = C + I + G + X - M$$

$C = 0.75Y_d$	$I = 820$	$Y_d = (1 - t)Y$
$M = 0.25Y_d$	$G = 960$	$t = 0.3$
$X = 650$		

What will be the equilibrium value of Y? Use the export multiplier to find out what will happen to the balance of payments if exports exogenously increase by 100.

5. A multiplant firm faces the total cost schedule

$$TC = 850 + 18q_1 + 25q_2 + 0.6q_1^2q_2 + 1.2q_1q_2^2$$

where q_1 and q_2 are output levels in its two plants. What marginal cost schedule does it face if output in plant 2 is expanded while output in plant 1 is kept unchanged?

308

6. In a closed economy (i.e. one with no foreign trade) the following relationships hold:

$$C = 0.6Y_d$$
$$Y_d = (1 - t)Y$$
$$Y = C + I + G$$
$$I = 120$$
$$t = 0.25$$
$$G = 210$$

where C is consumer expenditure, Y_d is disposable income, Y is national income, I is investment, t is the tax rate and G is government expenditure.

What is the marginal propensity to consume out of Y? What is the value of the government expenditure multiplier? How much does government expenditure need to be increased to achieve a national income of 700?

10.3 SECOND-ORDER PARTIAL DERIVATIVES

One of the most important uses of partial differentiation in economics is its application to optimization problems when the objective function has more than one independent variable. Before looking at methods of solution for these optimization problems, it is first necessary to understand how second-order partial derivatives are derived.

In Chapter 9 you learned that, for a single variable function $y = f(x)$,

1 the first-order derivative dy/dx gave a function for its rate of change, i.e. its slope, and
2 the second-order derivative d^2y/dx^2 gave a function for the rate of change of the slope of the original function.

When a function has two independent variables there will be four second-order derivatives. For example, if we consider the production function $Q = 25K^{0.4}L^{0.3}$ the two first-order partial derivatives are

$$\frac{\partial Q}{\partial K} = 10K^{-0.6}L^{0.3} \qquad \frac{\partial Q}{\partial L} = 7.5K^{0.4}L^{-0.7}$$

These represent the marginal product functions for K and L, as was explained in Section 10.1.

Differentiating these functions a second time we get

$$\frac{\partial^2 Q}{\partial K^2} = -6K^{-1.6}L^{0.3} \qquad \frac{\partial^2 Q}{\partial L^2} = -5.25K^{0.4}L^{-1.7}$$

These second-order derivatives represent the rate of change of the marginal product functions. For example, we can see that the slope of MP_L (i.e. $\partial^2 Q/\partial L^2$) will always be negative (assuming positive values of K and L) and as L increases, *ceteris paribus*, the absolute value of its slope diminishes.

We can also find the rate of change of $\partial Q/\partial K$ with respect to changes in L and the rate of change of $\partial Q/\partial L$ with respect to K. These will be

$$\frac{\partial^2 Q}{\partial K \partial L} = 3K^{-0.6}L^{-0.7} \qquad \frac{\partial^2 Q}{\partial L \partial K} = 3K^{-0.6}L^{-0.7}$$

and are known as 'cross partial derivatives'. They show how the rate of change of Q with respect to one input alters when the other input changes. In the example above, the cross partial derivative $\partial^2 Q/\partial L \partial K$ tells us that the rate of change of MP_L with respect to changes in K will be positive and will fall in value as K increases.

You will have noted that $\partial^2 Q/\partial K \partial L = \partial^2 Q/\partial L \partial K$ in this example. In fact, for any continuous function $y = f(x, z)$ it will always be the case that

$$\frac{\partial^2 y}{\partial x \partial z} = \frac{\partial^2 y}{\partial z \partial x}$$

Thus, for a two-variable function $y = f(x, y)$, there will be four second-order partial derivatives:

(i) $\dfrac{\partial^2 y}{\partial x^2}$ (ii) $\dfrac{\partial^2 y}{\partial z^2}$ (iii) $\dfrac{\partial^2 y}{\partial x \partial z}$ (iv) $\dfrac{\partial^2 y}{\partial z \partial x}$

but the cross partial derivatives (iii) and (iv) will be equal.

Example 10.12

Derive the four second-order partial derivatives for the production function

$$Q = 6K + 0.3K^2 L + 1.2L^2$$

and interpret their meaning.

Solution

The two first-order partial derivatives are

$$\frac{\partial Q}{\partial K} = 6 + 0.6KL \qquad \frac{\partial Q}{\partial L} = 0.3K^2 + 2.4L$$

and these represent the marginal product functions for the two inputs.

The four second-order partial derivatives are as follows.

(i) $\dfrac{\partial^2 Q}{\partial K^2} = 0.6L$

This represents the slope of the MP_K function. It tells us that the MP_K function will have a constant slope along its length (i.e. it is linear) but an increase in L will cause an increase in this slope.

(ii) $\dfrac{\partial^2 Q}{\partial L^2} = 2.4$

This represents the slope of the MP_L function and tells us that MP_L is a straight line with slope 2.4.

(iii) $\dfrac{\partial^2 Q}{\partial K \partial L} = 0.6K$

This tells us that MP_K increases if L is increased, for any given value of K. The rate at which MP_K rises as L is increased will depend on the given value of K.

(iv) $\dfrac{\partial^2 Q}{\partial L \partial K} = 0.6K$

This tells us that MP_L will increase if K is increased with L held at a given level and that the rate of this increase will depend on the value of K. Thus, although the slope of the MP_L schedule will always be 2.4, from (ii) above, its actual position will depend on the amount of K used.

Second-order partial derivatives can be derived for any function with more than one independent variable. Some other applications are given below.

Example 10.13

A firm sells two competing products whose demand schedules are

$$q_1 = 120 - 0.8p_1 + 0.5p_2 \qquad q_2 = 160 + 0.4p_1 - 12p_2$$

How will the price of good 2 affect the marginal revenue of good 1?

Solution

To derive the total revenue function for good 1 (TR_1) in terms of q, we first need to rearrange the demand function to get $p_1 = f(q_1)$.
Thus, given

$$q_1 = 120 - 0.8p_1 + 0.5p_2$$
$$0.8p_1 = 120 + 0.5p_2 - q_1$$
$$p_1 = 150 + 0.625p_2 - 1.25q_1$$
$$TR_1 = p_1 q_1$$
$$= (150 + 0.625p_2 - 1.25q_1)q_1$$
$$= 150q_1 + 0.625p_2 q_1 - 1.25q_1^2$$

Thus

$$MR_1 = \frac{\partial TR_1}{\partial q_1}$$

$$= 150 + 0.625p_2 - 2.5q_1$$

This marginal revenue function will have a constant slope of -2.5 regardless of the value of p_2 or the amount of q_1 sold.

The effect of a change in p_2 on MR_1 is shown by the cross partial derivative $\partial^2 TR_1/\partial q_1 \partial p_2 = 0.625$. Thus an increase in p_2 will cause an increase in the marginal revenue from good 1; i.e. although the slope of MR_1 remains constant at -2.5, its position shifts upward.

Note that in order to answer this question, we have formulated the total revenue for good 1 as a function of one price and one quantity, i.e. $TR = f(q_1, p_2)$.

Example 10.14

A firm operates with the production function

$$Q = 820K^{0.3}L^{0.2}$$

and can buy inputs K and L at £65 and £40 respectively per unit. If it can sell its output at a fixed price of £12 per unit, what is the relationship between increases in K and L and total profit? Will a change in K affect the extra profit derived from marginal increases in L?

Solution

The total revenue will be

$$TR = PQ = 12(820K^{0.3}L^{0.2})$$

The total cost will be

$$TC = P_K K + P_L L = 65K + 40L$$

Therefore the profit will be

$$\pi = TR - TC$$
$$= 12(820K^{0.3}L^{0.2}) - (65K + 40L)$$
$$= 9,840K^{0.3}L^{0.2} - 65K - 40L$$

The effects of changes in K and L on profit are shown by the first-order partial derivatives:

$$\frac{\partial \pi}{\partial K} = 2,952K^{-0.7}L^{0.2} - 65 \qquad \frac{\partial \pi}{\partial L} = 1,968K^{0.3}L^{-0.8} - 40$$

312

As long as

$$\frac{2,952L^{0.2}}{K^{0.7}} > 65 \qquad \frac{1,968K^{0.3}}{L^{0.8}} > 40$$

then the effect of increases in K and L on profit will be positive. However, if K is continually increased while L is held constant, then eventually the value of the term $2,952L^{0.2}/K^{0.7}$ will fall below 65 and further increases in K will cause a decrease in profit. Similarly, if L is increased sufficiently, the value of $\partial\pi/\partial L$ will become negative.

To determine the effect of a change in K on the marginal profit function with respect to L, we need to look at the cross partial derivative

$$\frac{\partial^2\pi}{\partial L\partial K} = 0.3(1,968K^{-0.7}L^{-0.8}) = 590.4K^{-0.7}L^{-0.8}$$

As long as K and L are positive, then $590.4K^{-0.7}L^{-0.8}$ will be positive and so, *ceteris paribus*, an increase in K will have a positive effect on the extra profit generated by marginal increases in L.

Second-order and cross partial derivatives can also be derived for functions with three or more independent variables.

If we consider the function with three independent variables $y = f(w, x, z)$, there will be three second-order derivatives with respect to the three independent variables

$$\frac{\partial y}{\partial w} \qquad \frac{\partial y}{\partial x} \qquad \frac{\partial y}{\partial z}$$

plus six cross partial derivatives

$$\frac{\partial^2 y}{\partial w\partial x} = \frac{\partial^2 y}{\partial x\partial w} \qquad \frac{\partial^2 y}{\partial x\partial z} = \frac{\partial^2 y}{\partial z\partial x} \qquad \frac{\partial^2 y}{\partial w\partial z} = \frac{\partial^2 y}{\partial z\partial w}$$

As with the two-variable case, cross partial derivatives will be equal if the two stages of differentiation involve the same two variables, assuming a continuous function. These equalities are shown above.

In this basic mathematics text we shall not be going into the rather complex methods that use these second-order derivatives for functions with more than two independent variables in checking second-order conditions for optimization. However, the example below illustrates how they can be derived.

Example 10.15

Derive the nine second-order derivatives for the production function $Q = 32K^{0.5}L^{0.25}R^{0.4}$ and show that the two cross partial derivatives with respect to each possible pair of independent variables will be equal to each other.

Solution

The three first-order derivatives will be

$$\frac{\partial Q}{\partial K} = 16K^{-0.5}L^{0.25}R^{0.4}$$

$$\frac{\partial Q}{\partial L} = 8K^{0.5}L^{-0.75}R^{0.4}$$

$$\frac{\partial Q}{\partial R} = 12.8K^{0.5}L^{0.25}R^{-0.6}$$

The second-order derivatives will be

$$\frac{\partial^2 Q}{\partial K^2} = -8K^{-1.5}L^{0.25}R^{0.4}$$

$$\frac{\partial^2 Q}{\partial L^2} = -6K^{0.5}L^{-1.75}R^{0.4}$$

$$\frac{\partial^2 Q}{\partial R^2} = -7.68K^{0.5}L^{0.25}R^{-1.6}$$

plus the six cross partial derivatives which can be arranged in three equal pairs:

$$\frac{\partial^2 Q}{\partial K \partial L} = 4K^{-0.5}L^{-0.75}R^{0.4} = \frac{\partial^2 Q}{\partial L \partial K}$$

$$\frac{\partial^2 Q}{\partial L \partial R} = 3.2K^{0.5}L^{-0.75}R^{-0.6} = \frac{\partial^2 Q}{\partial R \partial L}$$

$$\frac{\partial^2 Q}{\partial R \partial K} = 6.4K^{-0.5}L^{0.25}R^{-0.6} = \frac{\partial^2 Q}{\partial K \partial R}$$

QUESTIONS 10.3

1. For the production function $Q = 8K^{0.6}L^{0.5}$ derive a function for the slope of the marginal product of L function. What effect will a marginal increase in K have upon MP_L?
2. Derive all the second-order partial derivatives for the production function $Q = 35KL + 1.4LK^2 + 3.2L^2$ and interpret their meaning.
3. A firm operates three plants with the joint total cost function

$$TC = 58 + 18q_1 + 9q_2q_3 + 0.004q_1^2q_3^2 + 1.2q_1q_2q_3$$

Find the nine second-order derivatives for TC and demonstrate that the cross partial derivatives can be arranged in three equal pairs.

10.4 UNCONSTRAINED OPTIMIZATION: FUNCTIONS WITH TWO VARIABLES

In Chapter 9 it was explained that, for a function to be at a maximum (or minimum) value, not only does the slope of the function have to be zero, but certain second-order conditions also have to be satisfied. The reasons for these second-order conditions were straightforward to explain in the case of a function of one independent variable, but when two or more independent variables are involved the rationale for all the second-order conditions is not quite so easy to understand. We shall therefore just state these second-order conditions and give a brief intuitive explanation before looking at some applications.

For the function $y = f(x, z)$ to be at a maximum or at a minimum, first-order conditions which must be met are

$$\frac{\partial y}{\partial x} = 0 \qquad \frac{\partial y}{\partial z} = 0$$

These are as one would expect. To be at a maximum or minimum, the function must be at a stationary point with respect to changes in both variables.

There are two sets of second-order conditions.

(1) $\dfrac{\partial^2 y}{\partial x^2} < 0$, $\dfrac{\partial^2 y}{\partial z^2} < 0$ for a maximum

$\dfrac{\partial^2 y}{\partial x^2} > 0$, $\dfrac{\partial^2 y}{\partial z^2} > 0$ for a minimum

These follow on from the second-order conditions for the optimization of a single variable function explained in Chapter 9. The rate of change of a function (i.e. its slope) must be decreasing at a stationary point for that point to be a maximum and it must be increasing for a stationary point to be a minimum. The difference here is that these conditions must hold with respect to changes in both independent variables.

(2) The other second-order condition

$$\left(\frac{\partial^2 y}{\partial x^2}\right)\left(\frac{\partial^2 y}{\partial z^2}\right) > \left(\frac{\partial^2 y}{\partial x \partial z}\right)^2$$

must hold at both maximum *and* minimum stationary points.

To get an idea of the reason for this condition, imagine a three-dimensional model with x and z being measured on the two axes of a graph and y being measured by the height above the flat surface on which the x and z axes are drawn. If a point where the y 'hill' has a zero slope is to be a maximum, then one needs to ensure that, whichever direction one moves in, the height will fall and the slope will become steeper. As moves can be made in directions other than parallel to the two axes, it can be mathematically proved that the condition

315

$$\left(\frac{\partial^2 y}{\partial x^2}\right)\left(\frac{\partial^2 y}{\partial z^2}\right) > \left(\frac{\partial^2 y}{\partial x \partial z}\right)^2$$

satisfies this requirement as long as the other second-order conditions for a maximum

$$\frac{\partial^2 y}{\partial x^2} < 0 \qquad \frac{\partial^2 y}{\partial z^2} < 0$$

also hold. On the other hand, when

$$\frac{\partial^2 y}{\partial x^2} > 0 \qquad \frac{\partial^2 y}{\partial z^2} > 0$$

hold at a stationary point, then condition (2) ensures that one is in a 'trough' and will rise no matter which direction one moves in. The function will then be at a minimum.

Note also that all the above conditions refer to the requirements for local maximum or minimum values of a function, which may or may not be global maxima or minima. Refer back to Chapter 9 if you cannot remember the difference between these two concepts.

Let us now look at some applications of these rules for the unconstrained optimization of a function with two independent variables.

Example 10.16

A firm produces two products which are sold in two separate markets with the demand schedules

$$p_1 = 600 - 0.3q_1 \qquad p_2 = 500 - 0.2q_2$$

Production costs are related and the firm faces the total cost schedule

$$TC = 16 + 1.2q_1 + 1.5q_2 + 0.2q_1 q_2$$

If the firm wishes to maximize total profits, how much of each product should it sell? What will the maximum profit level be?

Solution

The total revenue is

$$TR = TR_1 + TR_2$$
$$= p_1 q_1 + p_2 q_2$$
$$= (600 - 0.3q_1)q_1 + (500 - 0.2q_2)q_2$$
$$= 600q_1 - 0.3q_1^2 + 500q_2 - 0.2q_2^2$$

The profit is

$$\pi = \text{TR} - \text{TC}$$

$$= 600q_1 - 0.3q_1^2 + 500q_2 - 0.2q_2^2 - (16 + 1.2q_1 + 1.5q_2 + 0.2q_1q_2)$$

$$= 600q_1 - 0.3q_1^2 + 500q_2 - 0.2q_2^2 - 16 - 1.2q_1 - 1.5q_2 - 0.2q_1q_2$$

$$= -16 + 598.8q_1 - 0.3q_1^2 + 498.5q_2 - 0.2q_2^2 - 0.2q_1q_2$$

First-order conditions for a maximum value of profit are

$$\frac{\partial \pi}{\partial q_1} = 598.8 - 0.6q_1 - 0.2q_2 = 0 \qquad (1)$$

and

$$\frac{\partial \pi}{\partial q_2} = 498.5 - 0.4q_2 - 0.2q_1 = 0 \qquad (2)$$

Multiplying (2) by 3	$1,495.5 - 1.2q_2 - 0.6q_1 = 0$
Rearranging (1)	$598.8 - 0.2q_2 - 0.6q_1 = 0$
Subtracting gives	$896.7 - q_2 = 0$
	$896.7 = q_2$

Substituting this value for q_2 into (1)

$$598.8 - 0.6q_1 - 0.2(896.7) = 0$$

$$598.8 - 179.34 = 0.6q_1$$

$$419.46 = 0.6q_1$$

$$699.1 = q_1$$

Checking second-order conditions:

$$\frac{\partial^2 \pi}{\partial q_1^2} = -0.6 < 0$$

$$\frac{\partial^2 \pi}{\partial q_2^2} = -0.4 < 0$$

$$\frac{\partial^2 \pi}{\partial q_1 \partial q_2} = -0.2$$

Therefore

$$\frac{\partial^2 \pi}{\partial q_1^2} \frac{\partial^2 \pi}{\partial q_2^2} = (-0.6)(-0.4) = 0.24 > 0.04 = (-0.2)^2 = \left(\frac{\partial^2 \pi}{\partial q_1 \partial q_2}\right)^2$$

and so all second-order conditions for a maximum are satisfied when $q_1 = 699.1$ and $q_2 = 896.7$.

The actual profit is found by substituting these values into the profit function

$$\pi = -16 + 598.8q_1 - 0.3q_1^2 + 498.5q_2 - 0.2q_2^2 - 0.2q_1q_2$$

$$= £432,797.02$$

Example 10.17

A firm sells two products which are partial substitutes for each other such that if the price of one product increases then the demand for the other product rises. The demand schedules for the two products are

$$q_1 = 517 - 3.5p_1 + 0.8p_2 \qquad q_2 = 770 - 4.4p_2 + 1.4p_1$$

What prices should the firm charge in each market to maximize total sales revenue?

Solution

For this problem it is more convenient to express total revenue as a function of price rather than quantity. Thus

$$TR = TR_1 + TR_2$$

$$= p_1q_1 + p_2q_2$$

$$= p_1(517 - 3.5p_1 + 0.8p_2) + p_2(770 - 4.4p_2 + 1.4p_1)$$

$$= 517p_1 - 3.5p_1^2 + 0.8p_1p_2 + 770p_2 - 4.4p_2^2 + 1.4p_1p_2$$

$$= 517p_1 - 3.5p_1^2 + 770p_2 - 4.4p_2^2 + 2.2p_1p_2$$

First-order conditions for a maximum are

$$\frac{\partial TR}{\partial p_1} = 517 - 7p_1 + 2.2p_2 = 0 \tag{1}$$

and

$$\frac{\partial TR}{\partial p_2} = 770 - 8.8p_2 + 2.2p_1 = 0 \tag{2}$$

Multiplying (1) by 4 $2,068 - 28p_1 + 8.8p_2 = 0$

Rearranging and adding (2) $\underline{770 + 2.2p_1 - 8.8p_2 = 0}$

gives $2,838 - 25.8p_1 = 0$

                                  $2,838 = 25.8p_1$

                                     $110 = p_1$

Substituting this value of p_1 into (1)

$$517 - 7(110) + 2.2p_2 = 0$$

Partial differentiation

$$2.2p_2 = 770 - 517 = 253$$

$$p_2 = 115$$

Checking second-order conditions:

$$\frac{\partial^2 \text{TR}}{\partial p_1^2} = -7 < 0$$

$$\frac{\partial^2 \text{TR}}{\partial p_2^2} = -8.8 < 0$$

$$\frac{\partial^2 \text{TR}}{\partial p_1 \partial p_2} = 2.2$$

$$(-7)(-8.8) = 61.6 > 4.84 = (2.2)^2$$

Therefore all second-order conditions for a maximum value of total revenue are satisfied when $p_1 = £110$ and $p_2 = £253$.

Example 10.18

A multiplant monopoly operates two plants whose total cost schedules are

$$\text{TC}_1 = 8.5 + 0.03q_1^2 \qquad \text{TC}_2 = 5.2 + 0.04q_2^2$$

If it faces the demand schedule

$$p = 60 - 0.04q$$

where $q = q_1 + q_2$, how much should it produce in each plant in order to maximize profits?

Solution

The total revenue is

$$\text{TR} = pq = (60 - 0.04q)q = 60q - 0.04q^2$$

Substituting $(q_1 + q_2)$ for q gives

$$\text{TR} = 60(q_1 + q_2) - 0.04(q_1 + q_2)^2$$

$$= 60q_1 + 60q_2 - 0.04q_1^2 - 0.08q_1q_2 - 0.04q_2^2$$

The profit is

$$\pi = \text{TR} - \text{TC}_1 - \text{TC}_2$$

$$= 60q_1 + 60q_2 - 0.04q_1^2 - 0.08q_1q_2 - 0.04q_2^2 - 8.5 - 0.03q_1^2 - 5.2 - 0.04q_2^2$$

$$= -13.7 + 60q_1 + 60q_2 - 0.07q_1^2 - 0.08q_2^2 - 0.08q_1q_2$$

319

First-order conditions for a maximum value of π require

$$\frac{\partial \pi}{\partial q_1} = 60 - 0.14q_1 - 0.08q_2 = 0 \qquad (1)$$

and

$$\frac{\partial \pi}{\partial q_2} = 60 - 0.16q_2 - 0.08q_1 = 0 \qquad (2)$$

Multiplying (1) by 2 $120 - 0.28q_1 - 0.16q_2 = 0$

Rearranging and subtracting (2) $\underline{60 - 0.08q_1 - 0.16q_2 = 0}$

$$60 - 0.2q_1 = 0$$
$$60 = 0.2q_1$$
$$300 = q_1$$

Substituting this value of q_1 into (1)

$$60 - 0.14(300) - 0.08q_2 = 0$$
$$18 = 0.08q_2$$
$$225 = q_2$$

Checking second-order conditions:

$$\frac{\partial^2 \pi}{\partial q_1^2} = -0.14 < 0$$

$$\frac{\partial^2 \pi}{\partial q_2^2} = -0.16 < 0$$

$$\frac{\partial^2 \pi}{\partial q_1 \partial q_2} = -0.08$$

$$(-0.14)(-0.16) = 0.0224 > 0.0064 = (-0.08)^2$$

Therefore second-order conditions confirm that profit is maximized when $q_1 = 300$ and $q_2 = 225$.

We can check that the total profit is positive for these output levels. The total output is

$$q = q_1 + q_2 = 300 + 225 = 525$$

Therefore

$$p = 60 - 0.04q = 60 - 0.04(525) = 39$$
$$TR = pq = 39(525) = 20{,}475$$
$$TC = TC_1 + TC_2$$
$$= [8.5 + 0.03(300)^2] + [5.2 + 0.04(225)^2]$$
$$= 2{,}708.5 + 2{,}030.2 = 4{,}738.7$$

$$\pi = \text{TR} - \text{TC} = 20{,}475 - 4{,}738.7 = £15{,}736.30$$

Note that this method could also be used to solve the multiplant monopoly problems in Chapter 5 that only involved linear functions. The unconstrained optimization method used here is, of course, a more general method that can be used for both linear and non-linear functions.

Example 10.19

A firm sells its output in a perfectly competitive market at a fixed price of £200 per unit. It buys the two inputs K and L at prices of £42 per unit and £5 per unit respectively, and faces the production function

$$q = 3.1K^{0.3}L^{0.25}$$

What combination of K and L should it use to maximize profit?

Solution

$$\text{TR} = pq = 200(3.1K^{0.3}L^{0.25}) = 620K^{0.3}L^{0.25}$$
$$\text{TC} = 42K + 5L$$
$$\pi = \text{TR} - \text{TC} = 620K^{0.3}L^{0.25} - 42K - 5L$$

First-order conditions for a maximum require

$$\frac{\partial \pi}{\partial K} = 186K^{-0.7}L^{0.25} - 42 = 0 \qquad \frac{\partial \pi}{\partial L} = 155K^{0.3}L^{-0.75} - 5 = 0$$

We thus have

$$186L^{0.25} = 42K^{0.7}$$
$$L^{0.25} = \frac{42}{186}K^{0.7} \tag{1}$$

and

$$155K^{0.3} = 5L^{0.75}$$
$$31K^{0.3} = L^{0.75} \tag{2}$$

Taking (1) to the power of 3

$$L^{0.75} = \left(\frac{42}{186}K^{0.7}\right)^3 = \frac{42^3}{186^3}K^{2.1} \tag{3}$$

Setting (3) equal to (2)

$$\frac{42^3}{186^3}K^{2.1} = 31K^{0.3}$$

321

$$K^{1.8} = \frac{31(186)^3}{42^3} = 2,692.481$$

$$K = 80.471179 \tag{4}$$

Substituting (4) into (1)

$$L^{0.25} = \frac{42}{186}(80.471179)^{0.7} = 4.8717455$$

$$L = (L^{0.25})^4 = (4.8717455)^4 = 563.29822$$

Therefore, first-order conditions suggest optimum values are at the stationary point where $K = 80.47$ and $L = 563.3$ (to 2 decimal places).
  Checking second-order conditions:

$$\frac{\partial^2 \pi}{\partial K^2} = (-0.7)186K^{-1.7}L^{0.25}$$

$$= -130.2(80.47)^{-1.7}(563.3)^{0.25}$$

$$= -0.3653576 < 0$$

$$\frac{\partial^2 \pi}{\partial L^2} = (-0.75)155K^{0.3}L^{-1.75}$$

$$= -116.25(80.47)^{0.3}(563.3)^{-1.75}$$

$$= -0.0066572 < 0$$

$$\frac{\partial^2 \pi}{\partial K \partial L} = (0.25)186K^{-0.7}L^{-0.75}$$

$$= 46.5(80.47)^{-0.7}(563.3)^{-0.75}$$

$$= 0.0186404$$

$$\frac{\partial^2 \pi}{\partial K^2} \frac{\partial^2 \pi}{\partial L^2} = (-0.3653576)(-0.0066572) = 0.0024323$$

$$\left(\frac{\partial^2 \pi}{\partial K \partial L}\right)^2 = (0.0186404)^2 = 0.0003475$$

Therefore

$$\frac{\partial^2 \pi}{\partial K^2} \frac{\partial^2 \pi}{\partial L^2} > \left(\frac{\partial^2 \pi}{\partial K \partial L}\right)^2$$

and so second-order conditions confirm that when $K = 80.47$ and $L = 563.3$ profit is at its maximum.

The actual profit will be

$$\pi = 620K^{0.3}L^{0.25} - 42K - 5L$$
$$= 620(80.47)^{0.3}(563.3)^{0.25} - 42(80.47) - 5(563.3)$$
$$= 11,265.924 - 3,379.74 - 2,816.5$$
$$= £5,069.68$$

Note that in this problem, and other similar ones in this section, the indices in the Cobb–Douglas production function add up to less than unity, giving decreasing returns to scale and hence rising average and marginal (long-run) cost schedules. If there were increasing returns to scale and the average and marginal cost schedules continued to fall, a firm facing a fixed price would wish to expand output indefinitely and so no profit-maximizing solution would be found by this method.

Example 10.20

A multiplant monopoly operates two plants whose total cost schedules are

$$TC_1 = 36 + 0.003q_1^3 \qquad TC_2 = 45 + 0.005q_2^3$$

If its total output is sold in a market where the demand schedule is $p = 320 - 0.1q$, where $q = q_1 + q_2$, how much should it produce in each plant to maximize total profits?

Solution

The total revenue is

$$TR = pq = (320 - 0.1q)q = 320q - 0.1q^2$$

Substituting $q_1 + q_2 = q$ gives

$$TR = 320(q_1 + q_2) - 0.1(q_1 + q_2)^2$$
$$= 320q_1 + 320q_2 - 0.1(q_1^2 + 2q_1q_2 + q_2^2)$$
$$= 320q_1 + 320q_2 - 0.1q_1^2 - 0.2q_1q_2 - 0.1q_2^2$$

Thus profit will be

$$\pi = TR - TC$$
$$= TR - TC_1 - TC_2$$
$$= (320q_1 + 320q_2 - 0.1q_1^2 - 0.2q_1q_2 - 0.1q_2^2) - (36 + 0.003q_1^3) - (45 + 0.005q_2^3)$$
$$= 320q_1 + 320q_2 - 0.1q_1^2 - 0.2q_1q_2 - 0.1q_2^2 - 36 - 0.003q_1^3 - 45 - 0.005q_2^3$$

First-order conditions for a maximum require

$$\frac{\partial \pi}{\partial q_1} = 320 - 0.2q_1 - 0.2q_2 - 0.009q_1^2 = 0 \tag{1}$$

and

$$\frac{\partial \pi}{\partial q_2} = 320 - 0.2q_1 - 0.2q_2 - 0.015q_2^2 = 0 \tag{2}$$

Subtracting (2) from (1)

$$-0.009q_1^2 + 0.015q_2^2 = 0$$

$$q_2^2 = \frac{0.009}{0.015} q_1^2 = 0.6q_1^2$$

$$q_2 = (0.6q_1^2)^{0.5} = 0.7746q_1 \tag{3}$$

Substituting (3) into (1)

$$320 - 0.2q_1 - 0.2(0.7746q_1) - 0.009q_1^2 = 0$$
$$320 - 0.2q_1 - 0.15492q_1 - 0.009q_1^2 = 0$$
$$0 = 0.009q_1^2 + 0.35492q_1 - 320 \tag{4}$$

Using the quadratic formula to solve (4)

$$q_1 = \frac{-b \pm \sqrt{b^2 - 4ac}}{2a} = \frac{-0.35492 \pm \sqrt{(0.35492)^2 - 4(0.009)(-320)}}{0.018}$$

$$= \frac{-0.35492 \pm \sqrt{11.645968}}{0.018}$$

$$= \frac{-0.35492 \pm 3.412619}{0.018}$$

Disregarding the negative solution we have

$$q_1 = \frac{3.057699}{0.018}$$

$$= 169.87216 = 169.87 \quad \text{(to 2 dp)}$$

Substituting into (3)

$$q_2 = 0.7746(169.87216) = 131.58 \quad \text{(to 2 dp)}$$

Checking second-order conditions:

$$\frac{\partial^2 \pi}{\partial q_1^2} = -0.2 - 0.018q_1 = -0.2 - 0.018(169.87) = -3.25766 < 0$$

$$\frac{\partial^2 \pi}{\partial q_2^2} = -0.2 - 0.03q_2 = -0.2 - 0.03(131.58) = -4.1474 < 0$$

$$\frac{\partial^2 \pi}{\partial q_1 \partial q_2} = -0.2$$

$$(-3.25766)(-4.1474) = 13.51 > 0.04 = (-0.2)^2$$

Therefore all second-order conditions for a maximum value of profit are satisfied when $q_1 = 169.87$ and $q_2 = 131.58$.

When a function involves more than two independent variables, the second-order conditions for a maximum or minimum become even more complex. In Section 10.3 we learned that if a function has three independent variables there will be nine second derivatives. When the number of variables exceeds three, the number of second derivatives which have to be taken into account expands rapidly and one really needs to use matrix algebra to analyse their properties.

Those of you who go on to study mathematical economics will encounter the second-order tests for a maximum or minimum of a multivariable function. In this basic mathematics text, however, we shall just consider how the first-order conditions for a maximum or minimum can be used to determine optimum values for economic problems involving three or more independent variables. From the way these problems are constructed it will be obvious whether or not a maximum or a minimum value is being sought, and it will be assumed that second-order conditions are satisfied for the values of the function that meet the first-order conditions.

Example 10.21

A firm operates with the production function

$$Q = 95K^{0.3}L^{0.2}R^{0.25}$$

and buys the three inputs K, L and R at prices of £30, £16 and £12 respectively per unit. If it can sell its output at a fixed price of £4 a unit, what is the maximum profit it can make? (Assume that second-order conditions are met at stationary points.)

Solution

$$\pi = \text{TR} - \text{TC}$$
$$= PQ - (P_K K + P_L L + P_R R)$$
$$= 400(95K^{0.3}L^{0.2}R^{0.25}) - (30K + 16L + 12R)$$
$$= 3{,}80K^{0.3}L^{0.2}R^{0.25} - 30K - 16L - 12R$$

Basic mathematics for economists

First-order conditions for a maximum are

$$\frac{\partial \pi}{\partial K} = 114K^{-0.7}L^{0.2}R^{0.25} - 30 = 0 \qquad (1)$$

$$\frac{\partial \pi}{\partial L} = 76K^{0.3}L^{-0.8}R^{0.25} - 16 = 0 \qquad (2)$$

$$\frac{\partial \pi}{\partial R} = 95K^{0.3}L^{0.2}R^{-0.75} - 12 = 0 \qquad (3)$$

From (1)

$$114L^{0.2}R^{0.25} = 30K^{0.7}$$

$$R^{0.25} = \frac{30K^{0.7}}{114L^{0.2}} \qquad (4)$$

Substituting (4) into (2)

$$76K^{0.3}L^{-0.8}\left[\frac{30K^{0.7}}{114L^{0.2}}\right] = 16$$

$$76K^{0.3}(30K^{0.7}) = 16L^{0.8}(114L^{0.2})$$

$$2{,}280K = 1{,}824L$$

$$K = 0.8L \qquad (5)$$

Substituting (5) into (4)

$$R^{0.25} = \frac{30(0.8L)^{0.7}}{114L^{0.2}} = \frac{30(0.8)^{0.7}L^{0.5}}{114} \qquad (6)$$

Cubing each side

$$R^{0.75} = \frac{27{,}000(0.8)^{2.1}L^{1.5}}{114^3}$$

Inverting

$$R^{-0.75} = \frac{114^3}{27{,}000(0.8)^{2.1}L^{1.5}} \qquad (7)$$

Substituting (7) and (5) into (3)

$$\frac{95(0.8L)^{0.3}L^{0.2}(114)^3}{27{,}000(0.8)^{2.1}L^{1.5}} = 12$$

$$\frac{95(0.8)^{0.3}L^{0.3}L^{0.2}(114)^3}{(0.8)^{2.1}L^{1.5}} = 324{,}000$$

$$\frac{95(114)^3}{(0.8)^{1.8}L} = 324{,}000$$

$$\frac{95(114)^3}{324{,}000(0.8)^{1.8}} = L$$

$$649.12924 = L \tag{8}$$

Substituting (8) into (5)

$$K = 0.8(649.12924) = 519.3034 \tag{9}$$

Substituting (8) into (6)

$$R^{0.25} = \frac{30(0.8)^{0.7}(649.12924)^{0.5}}{114}$$

$$R = \frac{30^4(0.8)^{2.8}(649.12924)^2}{114^4}$$

$$= 1{,}081.882$$

It is assumed that the second-order conditions for a maximum are met when K, L and R take these values.

The actual profit will be (taking quantities to 1dp)

$$\pi = 3800(519.3)^{0.3}(649.1)^{0.2}(1{,}081.9)^{0.25}$$

$$- 30(519.3) - 16(649.1) - 12(1{,}081.9)$$

$$= 51{,}929.98 - 15{,}579 - 10{,}385.6 - 12{,}982.8$$

$$= \pounds 12{,}982.58$$

QUESTIONS 10.4

1. A firm produces two products which are sold in separate markets with the demand schedules

$$p_1 = 210 - 0.4q_1^2 \qquad p_2 = 491 - 6q_2$$

Production costs are related and the firm's total cost schedule is

$$TC = 32 + 0.8q_1^2 + 0.7q_2^2 + 0.1q_1q_2$$

How much should the firm sell in each market in order to maximize total profits?

2. The same firm produces two competing products whose demand schedules are

$$q_1 = 219 - 1.8p_1 + 0.5p_2 \qquad q_2 = 303 - 2.1p_2 + 0.8p_1$$

What price should it charge in the two markets to maximize total sales revenue?

3. A price-discriminating monopoly sells in two separable markets with demand schedules

$$p_1 = 215 - 0.012q_1 \qquad p_2 = 324 - 0.023q_2$$

and faces the total cost schedule

$$TC = 4{,}200 + 0.3q^2$$

where $q = q_1 + q_2$. What should it sell in each market to maximize total profit? (Note that negative quantities are not allowed. Remember the price discrimination rules explained in Chapter 5.)

4. A monopoly sells its output in two separable markets with the demand schedules

$$p_1 = 20 - \frac{q_1}{6} \qquad p_2 = 13.75 - \frac{q_2}{8}$$

If it faces the total cost schedule

$$TC = 74 + 2.26q + 0.01q^2$$

where $q = q_1 + q_2$, what is the maximum profit it can make?

5. A multiplant monopoly operates two plants whose cost schedules are

$$TC_1 = 2.4 + 0.015q_1^2 \qquad TC_2 = 3.5 + 0.012q_2^2$$

and sells its total output in a market where

$$p = 32 - 0.02q$$

How much should it produce in each plant to maximize total profits?

6. A firm operates two plants with the total cost schedules

$$TC_1 = 62 + 0.00018q_1^3 \qquad TC_2 = 48 + 0.00014q_2^3$$

and faces the demand schedule

$$p = 2{,}360 - 0.15q$$

How much should it produce in each plant to maximize total profits?

7. A firm faces the production function $Q = 0.8K^{0.4}L^{0.3}$. It sells its output at a fixed price of £450 a unit and can buy the inputs K and L at £15 per unit and £8 per unit respectively. What input mix will maximize profit?

8. If a firm selling in a perfectly competitive market where the ruling price is £40 can buy inputs K and L at prices per unit of £20 and £6 respectively and operates with the production function $Q = 21K^{0.4}L^{0.2}$, what is the maximum profit it can make?

9. Explain why a firm facing the production function $Q = 2.4K^{0.6}L^{0.2}$, where K costs £25 per unit and L costs £9 per unit, and selling Q at a fixed price of £82 per unit cannot make a profit of more than £20,000, no matter how efficiently it plans its input mix.

10. A firm can buy inputs K and L at £32 per unit and £20 per unit respectively and sell its output at a fixed price of £5 per unit. How should it organize production to ensure maximum profit if it faces the production function $Q = 82K^{0.5}L^{0.3}$?

10.5 TOTAL DIFFERENTIALS AND TOTAL DERIVATIVES

(Note that the mathematical methods developed in this section are mainly used for the proofs of different economic theories rather than for direct numerical applications. These proofs may be omitted if your course does not include these areas of economics as they are not essential in order to understand the following chapters.)

In Chapter 8, when the concept of differentiation was introduced, you learned that the derivative dy/dx measured the rate of change of y with respect to x for infinitesimally small changes in x and y. For any non-linear function $y = f(x)$, the value of dy/dx will alter if x and y alter. It is therefore not possible to predict the effect of a given increase in x on y with complete accuracy. However, for a very small change (Δx) in x, we can say that it will be approximately true that the resulting change in y will be

$$\Delta y = \frac{dy}{dx} \Delta x$$

The closer the function $y = f(x)$ is to a straight line, the more accurate will be the prediction, as the following example demonstrates.

Example 10.22

For the functions below assume that the value of x increases from 10 to 11. Predict the effect on y using the derivative dy/dx evaluated at the first value of x and check the answer against the new value of the function.

(i) $y = 2x$ (ii) $y = 2x^2$ (iii) $y = 2x^3$

Solution

In all cases $\Delta x = 11 - 10 = 1$.

(i) $\frac{dy}{dx} = 2$

Therefore, predicted $\Delta y = \Delta x(dy/dx) = 2(1) = 2$. The actual values are

$$y = 2(10) = 20 \text{ when } x = 10$$
$$y = 2(11) = 22 \text{ when } x = 11$$

The actual change is

$$22 - 20 = 2 \text{ (accuracy of prediction 100\%)}$$

(ii) $\dfrac{dy}{dx} = 4x = 4(10) = 40$

Therefore, predicted $\Delta y = \Delta x (dy/dx) = 40(1) = 40$. The actual values are

$$y = 2(10)^2 = 200 \text{ when } x = 10$$
$$y = 2(11)^2 = 242 \text{ when } x = 11$$

The actual change is

$$242 - 200 = 42 \text{ (accuracy of prediction 95\%)}$$

(iii) $\dfrac{dy}{dx} = 6x^2 = 6(10)^2 = 600$

Therefore, predicted $\Delta y = \Delta x (dy/dx) = 600(1) = 600$. The actual values are

$$y = 2(10)^3 = 2{,}000 \text{ when } x = 10$$
$$y = 2(11)^3 = 2{,}662 \text{ when } x = 11$$

The actual change is

$$2{,}662 - 2{,}000 = 662 \text{ (accuracy of prediction 89.7\%)}$$

The above method of predicting approximate actual changes in a variable can itself be useful for practical purposes. However, in economic theory this mathematical method is taken a stage further and helps yield some important results.

If the changes in x and y become infinitesimally small then even for non-linear functions

$$\Delta y = \frac{dy}{dx} \Delta x$$

These infinitesimally small changes in x and y are known as 'differentials'. When y is a function of more than one independent variable, e.g. $y = f(x, z)$, and there are infinitesimally small changes in all variables, then the total effect will be

$$\Delta y = \frac{\partial y}{\partial x} \Delta x + \frac{\partial y}{\partial z} \Delta z$$

This is known as the 'total differential' as it shows the total effect on y of changes in all independent variables. It is usual to write dy, dx, dz etc. to represent infinitesimally small changes instead of Δy, Δx, Δz, which usually represent small, but finite, changes. Thus

$$dy = \frac{\partial y}{\partial x} dx + \frac{\partial y}{\partial z} dz$$

330

Partial differentiation

Example 10.23

What is the total differential of $y = 6x^2 + 8z^2 - 0.3xz$?

Solution

The total differential is

$$dy = \frac{\partial y}{\partial x} dx + \frac{\partial y}{\partial z} dz$$

$$= (12x - 0.3z) dx + (16z - 0.3x) dz$$

We can now demonstrate some examples of how the concept of a total differential can be used in economic theory.

In production theory, the slope of an isoquant represents the marginal rate of technical substitution (MRTS) between two inputs. The use of the total differential can help demonstrate that the MRTS will equal the ratio of the marginal products of the two inputs.

An approximate measure of the MRTS of K for L (usually written as $MRTS_{KL}$) is the amount of K that would be needed to compensate for the loss of one unit of L so that the production level remains unchanged. For infinitesimally small changes in K and L the $MRTS_{KL}$ measures the rate at which K needs to be substituted for L to keep output unchanged, i.e. it is equal to the slope of the isoquant through the given values of K and L when K is measured on the vertical axis and L on the horizontal axis. For any given output level, K is effectively a function of L (and vice versa) and so

$$MRTS_{KL} = \frac{dK}{dL} \tag{1}$$

For the production function $Q = f(K, L)$, the total differential is

$$dQ = \frac{\partial Q}{\partial K} dK + \frac{\partial Q}{\partial L} dL$$

If we are looking at changes along an isoquant then output is unchanged and so dQ is zero and thus

$$\frac{\partial Q}{\partial K} dK + \frac{\partial Q}{\partial L} dL = 0$$

$$\frac{\partial Q}{\partial K} dK = -\frac{\partial Q}{\partial L} dL$$

$$\frac{dK}{dL} = -\frac{\partial Q/\partial L}{\partial Q/\partial K} \tag{2}$$

We already know that $\partial Q/\partial L$ and $\partial Q/\partial K$ represent the marginal products of K and L. Therefore, from (1) and (2) above,

331

$$\text{MRTS} = -\frac{\text{MP}_\text{L}}{\text{MP}_\text{K}}$$

Another use of the total differential is to prove *Euler's theorem* and demonstrate the conditions for the 'exhaustion of the total product'. This relates to the marginal productivity theory of factor pricing and the normative idea of what might be considered a 'fair wage', which was debated for many years by political economists.

Consider a firm that uses several different inputs. Each will contribute a different amount to total production. One suggestion for what might be considered a 'fair wage' was that each input, including labour, should be paid the 'value of its marginal product' (VMP). This is defined as the marginal product of the last unit of the input (MP_i) multiplied by the price that the finished good is sold at (P_x), i.e.

$$\text{VMP}_i = P_x \, \text{MP}_i$$

for each input i. Any such suggestion is, of course, a normative concept and the value judgements on which it is based can be questioned. However, what we are concerned with here is whether it is even *possible* to pay each input the value of its marginal product. If it is not possible, then it would not be a practical idea to set this as an objective even if it seemed a 'fair' principle.

Before looking at Euler's theorem we can illustrate how the conditions for product exhaustion can be derived for a Cobb–Douglas production function with two inputs. This example also shows how the product price is irrelevant and it is the properties of the production function that matter.

Assume that a firm sells its output Q at a given price P and that $Q = AK^\alpha L^\beta$ where A, α and β are constants. If each input is paid the value of its marginal product then the prices of the two inputs K and L would be

$$P_\text{K} = P\,\text{MP}_\text{K} = P\,\frac{\partial Q}{\partial K} \qquad P_\text{L} = P\,\text{MP}_\text{L} = P\,\frac{\partial Q}{\partial L}$$

The total expenditure on inputs would therefore be

$$\text{TC} = K P_\text{K} + L P_\text{L}$$

$$= K\left(P\,\frac{\partial Q}{\partial K}\right) + L\left(P\,\frac{\partial Q}{\partial L}\right)$$

$$= P\left(K\,\frac{\partial Q}{\partial K} + L\,\frac{\partial Q}{\partial L}\right) \tag{1}$$

On the other hand, total revenue from the sale of the firm's output will be

$$\text{TR} = PQ$$

Total expenditure on inputs (which are paid the value of their marginal product) will equal the total revenue when $\text{TR} = \text{TC}$. Therefore, from (1),

$$PQ = P\left(K\frac{\partial Q}{\partial K} + L\frac{\partial Q}{\partial L}\right)$$

Cancelling P, this gives

$$Q = K\frac{\partial Q}{\partial K} + L\frac{\partial Q}{\partial L} \tag{2}$$

Thus the conditions of product exhaustion are based on the physical properties of the production function. If (2) holds then the product is exhausted. If it does not hold then there will be either not enough revenue or a surplus.

For the Cobb–Douglas production function $Q = AK^{\alpha}L^{\beta}$ we know that

$$\frac{\partial Q}{\partial K} = \alpha AK^{\alpha-1}L^{\beta} \qquad \frac{\partial Q}{\partial L} = \beta AK^{\alpha}L^{\beta-1}$$

Substituting into (2), this gives

$$\begin{aligned}
Q &= K(\alpha AK^{\alpha-1}L^{\beta}) + L(\beta AK^{\alpha}L^{\beta-1}) \\
&= \alpha AK^{\alpha}L^{\beta} + \beta AK^{\alpha}L^{\beta} \\
&= \alpha Q + \beta Q \\
&= Q(\alpha + \beta)
\end{aligned} \tag{3}$$

The condition required for (3) to hold is that $\alpha + \beta = 1$. This means that product exhaustion occurs for a Cobb–Douglas production function when there are constant returns to scale. We can also see from (3) and (1) that

(i) when there are decreasing returns to scale and $\alpha + \beta < 1$, then $TC = P(\alpha + \beta)Q < PQ = TR$ and so there will be a surplus, and
(ii) when there are increasing returns to scale and $\alpha + \beta > 1$, then $TC = P(\alpha + \beta)Q > PQ = TR$ and so there will not be enough revenue to pay each input its VMP.

Euler's theorem looks at the case of a general production function

$$Q = f(x_1, x_2, \ldots, x_n)$$

The previous example showed that the price will always cancel in the TR and TC formulae and what we are interested in is whether or not

$$Q = x_1\frac{\partial Q}{\partial x_1} + x_2\frac{\partial Q}{\partial x_2} + \ldots + x_n\frac{\partial Q}{\partial x_n} \tag{1}$$

Using the notation

$$f_1 = \frac{\partial Q}{\partial x_1}, \ f_2 = \frac{\partial Q}{\partial x_2}, \ \ldots \ \text{etc.}$$

the total differential of this production function will be

$$dQ = f_1\,dx_1 + f_2\,dx_2 + \ldots + f_n\,dx_n \tag{2}$$

Assume that all inputs are increased by the same proportion λ. Thus

$$\frac{dx_1}{x_i} = \lambda \text{ for all } i$$

and so

$$dx_i = \lambda x_i \tag{3}$$

Substituting (3) into (2) this gives

$$dQ = f_1 \lambda x_1 + f_2 \lambda x_2 + \ldots + f_n \lambda x_n$$
$$= \lambda(f_1 x_1 + f_2 x_2 + \ldots + f_n x_n)$$

$$\frac{dQ}{\lambda} = f_1 x_1 + f_2 x_2 + \ldots + f_n x_n \tag{4}$$

Multiplying top and bottom of the left-hand side of (4) by Q gives

$$\left(\frac{1}{\lambda}\frac{dQ}{Q}\right)Q = f_1 x_1 + f_2 x_2 + \ldots + f_n x_n$$

Thus, product exhaustion will only hold if

$$\frac{1}{\lambda}\frac{dQ}{Q} = 1$$

$$\frac{dQ}{Q} = \lambda$$

If this does hold it means that output increases by the same proportion as the inputs, i.e. there are constant returns to scale. If there are decreasing returns to scale, output increases by a smaller proportion than the inputs. Therefore,

$$\frac{dQ}{Q} < \lambda$$

and so

$$\frac{1}{\lambda}\frac{dQ}{Q} < 1$$

This means that

$$f_1 x_1 + f_2 x_2 + \ldots + f_n x_n < Q$$

so that if each input is paid the value of its marginal product there will be some surplus left over.

Similarly, if there are increasing returns to scale then

$$\frac{dQ}{Q} > \lambda$$

334

Partial differentiation

Therefore,

$$\frac{1}{\lambda}\frac{dQ}{Q} > 1$$

and

$$f_1 x_1 + f_2 x_2 + \ldots + f_n x_n > Q$$

which means that the total cost of paying each input the value of its marginal product will sum to more than the total revenue earned, i.e. it will not be possible.

To sum up, Euler's theorem proves that if each input is paid the value of its marginal product the total cost of the inputs will

1 equal total revenue if there are constant returns to scale;
2 be less than total revenue if there are decreasing returns to scale; and
3 be greater than total revenue if there are increasing returns to scale.

Example 10.24

Is it possible for a firm to pay each input the value of its marginal product if it operates with the production function $Q = 14K^{0.6}L^{0.8}$?

Solution

If each input is paid its VMP then the price of input K will be

$$P_K = \text{VMP}_K = P\,\text{MP}_K = P\frac{\partial Q}{\partial K} = P(8.4K^{-0.4}L^{0.8})$$

where P is the price of the final product. For input L,

$$P_L = \text{VMP}_L = P\,\text{MP}_L = P\frac{\partial Q}{\partial L} = P(11.2K^{0.6}L^{-0.2})$$

The total cost of inputs will therefore be

$$\begin{aligned}
TC &= P_K K + P_L L \\
&= P(8.4K^{-0.4}L^{0.8})K + P(11.2K^{0.6}L^{-0.2})L \\
&= P(8.4K^{0.6}L^{0.8}) + P(11.2K^{0.6}L^{0.8}) \\
&= P(8.4K^{0.6}L^{0.8} + 11.2K^{0.6}L^{0.8}) \\
TC &= P(19.6K^{0.6}L^{0.8})
\end{aligned}$$

The total revenue will be

$$TR = PQ = P(14K^{0.6}L^{0.8})$$

335

Therefore,

$$\frac{TC}{TR} = \frac{P(19.6K^{0.6}L^{0.8})}{P(14K^{0.6}L^{0.8})} = \frac{19.6}{14} = 1.4$$

Thus the total revenue is not enough to pay each input the value of its marginal product. This checks out with the predictions of Euler's theorem, given that there are increasing returns to scale for this production function.

Total Derivatives

In partial differentiation it is assumed that one variable changes while all other independent variables are held constant. However, in some instances there may be a connection between the independent variables and so this *ceteris paribus* assumption will not apply. For example, in a production function the amount of one input used may affect the amount of another input that can be used with it.

From the total differential of a function we can derive a total derivative which can cope with this additional effect. Assume $y = f(x, z)$ and also that $x = g(z)$. Thus any change in z will affect y directly via the function $f(x, z)$ and indirectly by changing x via the function $g(z)$ which in turn will affect y via the function $f(x, z)$.

The total differential of $y = f(x, z)$ is

$$dy = \frac{\partial y}{\partial x} dx + \frac{\partial y}{\partial z} dz$$

Dividing through by dz gives

$$\frac{dy}{dz} = \frac{\partial y}{\partial x} \frac{dx}{dz} + \frac{\partial y}{\partial z}$$

The first term shows the indirect effect of z, via x, and the second term shows the direct effect.

Example 10.25

If $Q = 25K^{0.4}L^{0.5}$ and $K = 0.8L^2$ what is the total effect of a change in L on Q? Identify the direct and indirect effects.

Solution

The total differential is

$$dQ = \frac{\partial Q}{\partial K} dK + \frac{\partial Q}{\partial L} dL$$

336

Partial differentiation

The total derivative with respect to L will be

$$\frac{dQ}{dL} = \frac{\partial Q}{\partial K}\frac{dK}{dL} + \frac{\partial Q}{\partial L} \tag{1}$$

From the functions given in the question we can derive

$$\frac{\partial Q}{\partial K} = 10K^{-0.6}L^{0.5} \qquad \frac{\partial Q}{\partial L} = 12.5K^{0.4}L^{-0.5} \qquad \frac{dK}{dL} = 1.6L$$

Substituting these derivatives into (1), we get

$$\frac{dQ}{dL} = (10K^{-0.6}L^{0.5})1.6L + 12.5K^{0.4}L^{-0.5}$$

$$= 16K^{-0.6}L^{1.5} + 12.5^{0.4}L^{-0.5}$$

The first term shows the indirect effect of changes in L on Q and the second term shows the direct effect.

QUESTIONS 10.5

1. Derive the total differentials of the following production functions:

   (a) $Q = 20K^{0.6}L^{0.4}$
   (b) $Q = 48K^{0.3}L^{0.2}R^{0.4}$
   (c) $Q = 6K^{0.8} + 5L^{0.7} + 0.8K^2L^2$

2. If each input is paid the value of its marginal product, will this exhaust a firm's total revenue if the relevant production function is

   (a) $Q = 4K + 1.5L$?
   (b) $Q = 8K^{0.4}L^{0.3}$?
   (c) $Q = 3.5K^{0.25}L^{0.35}R^{0.3}$?

3. If $y = 40x^{0.4}z^{0.3}$ and $x = 5z^{0.25}$, find the total effect of a change in z on y.

4. A consumer spends all her income on the two goods A and B. The quantity of good A bought is determined by the demand function $Q_A = f(P_A, P_B, M)$ where P_A and P_B are the prices of the two goods and M is real income. A change in the price of A will also affect real income M via the function $M = g(P_A, P_B, £M)$ where $£M$ is money income. Derive an expression for the total effect of a change in P_A on Q_A.

11

Constrained optimization

11.1 CONSTRAINED OPTIMIZATION AND RESOURCE ALLOCATION

Chapters 9 and 10 dealt with the optimization of functions without any constraints imposed. However, in economics we often come across resource allocation problems that involve the optimization of some variable subject to certain limitations. For example, a firm may try to maximize output subject to the constraint of a given budget for expenditure on inputs, or it may wish to minimize costs subject to a certain minimum output being produced. We have already seen in Chapter 5 how constrained optimization problems with linear constraints and objective functions can be tackled using linear programming. This chapter explains how problems involving the constrained optimization of non-linear functions can be tackled.

We shall consider two methods: (i) constrained optimization by substitution and (ii) the Lagrange multiplier.

The Lagrange multiplier method can be used for most types of constrained optimization problems. The substitution method is mainly suitable for problems where a function with only two variables is maximized or minimized subject to one constraint. We shall consider this simpler substitution method first.

11.2 CONSTRAINED OPTIMIZATION BY SUBSTITUTION

Consider the example of a firm that wishes to maximize output given the production function $Q = f(K, L)$ and facing set prices P_K and P_L for K and L and a fixed budget M. This problem is illustrated in Figure 11.1. The firm needs to find the combination of K and L that will allow it to reach the optimum point X which is on the highest possible isoquant within the budget constraint with intercepts M/P_K and M/P_L.

To determine a solution for the above type of problem mathematically one first needs to learn how to set up an economic resource allocation problem as a mathematical constrained optimization problem. The following examples suggest ways in which this can be done.

Constrained optimization

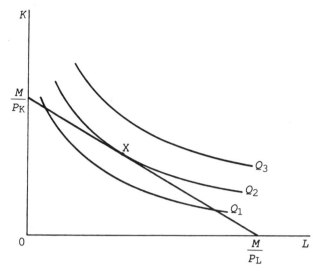

Figure 11.1

Example 11.1

A firm faces the production function $Q = 12K^{0.4}L^{0.4}$ and can buy the inputs K and L at prices per unit of £40 and £5 respectively. If it has a budget of £800 what combination of K and L should it use in order to produce the maximum possible output?

Solution

The problem is to maximize the function $Q = 12K^{0.4}L^{0.4}$ subject to the budget constraint

$$40K + 5L = 800 \tag{1}$$

(In all problems in this chapter, it is assumed that each constraint 'bites'; e.g. all the budget is used in this example.)

The theory of the firm tells us that a firm is optimally allocating a fixed budget if the last £1 spent on each input adds the same amount to output, i.e. marginal product over price should be equal for all inputs. This optimization condition can be written as

$$\frac{MP_K}{P_K} = \frac{MP_L}{P_L} \tag{2}$$

The marginal products can be determined by partial differentiation:

$$MP_K = \frac{\partial Q}{\partial K} = 4.8K^{-0.6}L^{0.4} \tag{3}$$

339

$$MP_L = \frac{\partial Q}{\partial L} = 4.8K^{0.4}L^{-0.6} \tag{4}$$

Substituting (3) and (4) and the given prices for P_K and P_L into (2)

$$\frac{4.8K^{-0.6}L^{0.4}}{40} = \frac{4.8K^{0.4}L^{-0.6}}{5}$$

Dividing both sides by 4.8 and multiplying by 40 gives

$$K^{-0.6}L^{0.4} = 8K^{0.4}L^{-0.6}$$

Multiplying both sides by $K^{0.6}L^{0.6}$ gives

$$L = 8K \tag{5}$$

Substituting (5) into the budget constraint (1) gives

$$40K + 5(8K) = 800$$
$$40K + 40K = 800$$
$$80K = 800$$
$$K = 10$$

Therefore, from (5), $L = 80$.

Note that although this method allows us to derive optimum values of K and L that satisfy condition (2) above, it does not provide a check on whether there is one unique solution to (2) corresponding to a maximum value of Q, i.e. there is no second-order condition check.

However, it may be assumed that in all the problems in this section the objective function is maximized (or minimized depending on the question) when the basic economic rules for an optimum are satisfied.

The above is not the only way of tackling this problem by substitution. An alternative approach might be as follows.

Example 11.1 (reworked)

Solution

From the budget constraint

$$40K + 5L = 800$$
$$5L = 800 - 40K \tag{1}$$

Therefore

$$L = 160 - 8K \tag{2}$$

Substituting (2) into the objective function $Q = 12K^{0.4}L^{0.4}$ gives

$$Q = 12K^{0.4}(160 - 8K)^{0.4} \tag{3}$$

Constrained optimization

We are now faced with the unconstrained optimization problem of finding the value of K that maximizes the function (3) which has the budget constraint (1) 'built in' it by substitution, which entails setting $dQ/dK = 0$.

It is not straightforward to differentiate the function in (3), however, and we must wait until further topics in calculus have been covered before proceeding with this solution (see Chapter 12, Example 12.9).

Example 11.2

A firm faces the production function $Q = 20K^{0.4} L^{0.6}$. It can buy inputs K and L for £400 a unit and £200 a unit respectively. What combination of L and K should be used to achieve the maximum output that can be produced with a budget of £6,000?

Solution

$$MP_L = \frac{\partial Q}{\partial L} = 12K^{0.4} L^{-0.4} \qquad MP_K = \frac{\partial Q}{\partial K} = 8K^{-0.6} L^{0.6}$$

Optimal input mix requires

$$\frac{MP_L}{P_L} = \frac{MP_K}{P_K}$$

Therefore

$$\frac{12K^{0.4} L^{-0.4}}{200} = \frac{8K^{-0.6} L^{0.6}}{400}$$

Cross multiplying gives

$$4,800K = 1,600L$$

$$3K = L$$

Substituting this result into the budget constraint

$$200L + 400K = 6,000$$

gives

$$200(3K) + 400K = 6,000$$
$$600K + 400K = 6,000$$
$$1,000K = 6,000$$
$$K = 6$$

Therefore

$$L = 3K = 18$$

The examples of constrained optimization considered so far have only involved output maximization when a firm faces a Cobb–Douglas production function, but the same technique can also be applied to other forms of production functions.

Example 11.3

A firm faces the production function

$$Q = 120L + 200K - L^2 - 2K^2$$

for positive values of Q. It can buy L at £5 a unit and K at £8 a unit and has a budget of £70. What is the maximum output it can produce?

Solution

$$MP_L = \frac{\partial Q}{\partial L} = 120 - 2L \qquad MP_K = \frac{\partial Q}{\partial L} = 200 - 4K$$

For optimal input combination

$$\frac{MP_L}{P_L} = \frac{MP_K}{P_K}$$

Therefore, substituting MP_K and MP_L from above and the given prices of £5 and £8,

$$\frac{120 - 2L}{5} = \frac{200 - 4K}{8}$$

$$8(120 - 2L) = 5(200 - 4K)$$

$$960 - 16L = 1,000 - 20K$$

$$20K = 40 + 16L$$

$$K = 2 + 0.8L \qquad (1)$$

Substituting (1) into the budget constraint

$$5L + 8K = 70$$

we get

$$5L + 8(2 + 0.8L) = 70$$

$$5L + 16 + 6.4L = 70$$

$$11.4L = 54$$

$$L = 4.74 \qquad \text{(to 2 dp)}$$

Substituting this result into (1)

$$K = 2 + 0.8(4.74) = 5.79$$

Therefore maximum output is

$$Q = 120L + 200K - L^2 - 2K^2$$
$$= 120(4.74) + 200(5.79) - (4.74)^2 - 2(5.79)^2$$
$$= 1,637.28$$

We can also apply this technique to the theoretical model of consumer behaviour, where a typical consumer is assumed to maximize the utility derived from consuming different goods subject to a budget constraint.

Example 11.4

The utility a consumer derives from consuming the two goods A and B can be assumed to be determined by the utility function

$$U = 40A^{0.25} B^{0.5}$$

If A costs £4 a unit and B costs £10 a unit and the consumer's income is £600, what combination of A and B will maximize utility?

Solution

The marginal utility of A is

$$MU_A = \frac{\partial U}{\partial A} = 10A^{-0.75} B^{0.5}$$

The marginal utility of B is

$$MU_B = \frac{\partial U}{\partial B} = 20A^{0.25} B^{-0.5}$$

Consumer theory tells us that total utility will be maximized when the utility derived from the last pound spent on each good is equal to the utility derived from the last pound spent on any other good. This optimization rule can be expressed as

$$\frac{MU_A}{P_A} = \frac{MU_B}{P_B}$$

Therefore, substituting the above MU functions and the given prices of £4 and £10, this condition becomes

$$\frac{10A^{-0.75} B^{0.5}}{4} = \frac{20A^{0.25} B^{-0.5}}{10}$$

$$100B = 80A$$

$$B = 0.8A \qquad (1)$$

Substituting this value of B into the budget constraint

$$4A + 10B = 600$$

gives

$$4A + 10(0.8A) = 600$$
$$4A + 8A = 600$$
$$12A = 600$$
$$A = 50$$

Thus from (1)

$$B = 0.8(50) = 40$$

The substitution method can also be used for constrained minimization problems. If output is given and a firm is required to minimize the cost of this output, then one variable can be eliminated from the production function before it is substituted into the cost function which is to be minimized.

Example 11.5

A firm operates with the production function $Q = 4K^{0.6}L^{0.4}$ and buys inputs K and L at prices per unit of £40 and £15 respectively. What is the cheapest way of producing 600 units of output?

Solution

The output constraint is

$$600 = 4K^{0.6}L^{0.4}$$

Therefore

$$\frac{150}{K^{0.6}} = L^{0.4}$$

$$\left(\frac{150}{K^{0.6}}\right)^{2.5} = L$$

$$\frac{275{,}567.6}{K^{1.5}} = L \tag{1}$$

The total cost of inputs, which is to be minimized, is

$$TC = 40K + 15L \tag{2}$$

Substituting (1) into (2) gives

$$TC = 40K + 15(275{,}567.6)\,K^{-1.5}$$

Differentiating to find a stationary point

Constrained optimization

$$\frac{dTC}{dK} = 40 - 22.5(275,567.6)\,K^{-2.5} = 0$$

$$40 = \frac{22.5(275,567.6)}{K^{2.5}}$$

$$K^{2.5} = \frac{22.5(275,567.6)}{40} = 155,006.78$$

$$K = 119.16268$$

Substituting this value into (1) gives

$$L = \frac{275,567.6}{(119.16268)^{1.5}} = 211.84478$$

Checking the second-order condition:

$$\frac{d^2 TC}{dK^2} = (2.5)22.5(275,567.6)\,K^{-3.5} > 0 \text{ for any } K > 0$$

This confirms that these values minimize TC. We can also check that these values give 600 when substituted into the production function.

$$Q = 4K^{0.6}L^{0.4} = 4(119.16268)^{0.6}(211.84478)^{0.4} = 600$$

Thus cost minimization is achieved when $K = 119.16$ and $L = 212.84$ (to 2 decimal places) and total production costs will be

$$TC = 40(119.16) + 15(211.84) = £7,944$$

QUESTIONS 11.1

1. If a firm has a budget of £378 what combination of K and L will maximize output given the production function $Q = 40K^{0.6}L^{0.3}$ and prices for K and L of £20 per unit and £6 per unit respectively?
2. A firm faces the production function $Q = 6K^{0.4}L^{0.5}$. If it can buy input K at £32 a unit and input L at £8 a unit, what combination of L and K should it use to maximize production if it is constrained by a fixed budget of £36,000?
3. A consumer spends all her income of £120 on the two goods A and B. Good A costs £10 a unit and good B costs £15. What combination of A and B will she purchase if her utility function is $U = 4A^{0.5}B^{0.5}$?
4. If a firm faces the production function $Q = 4K^{0.5}L^{0.5}$ what is the maximum output it can produce for a budget of £200? The prices of K and L are given as £4 per unit and £2 per unit respectively.
5. Make up your own constrained optimization problem for an objective function with two independent variables and solve it using the substitution method.

6. A firm faces the production function $Q = 2K^{0.2}L^{0.6}$ and can buy L at £240 a unit and K at £4 a unit.

   (a) If it has a budget of £16,000 what combination of K and L should it use to maximize output?
   (b) If it is given a target output of 40 units of Q what combination of K and L should it use to minimize the cost of this output?

7. A firm has a budget of £1,140 and can buy inputs K and L at £3 and £8 respectively a unit. Its output is determined by the production function

$$Q = 6K + 20L - 0.025K^2 - 0.05L^2$$

   for positive values of Q. What is the maximum output it can produce?

8. A firm operates with the production function $Q = 30K^{0.4}L^{0.2}$ and buys inputs K and L at £12 per unit and £5 per unit respectively. What is the cheapest way of producing 750 units of output? (Work to nearest whole units of K and L.)

11.3 THE LAGRANGE MULTIPLIER: CONSTRAINED MAXIMIZATION WITH TWO VARIABLES

The best way to explain how to use the Lagrange multiplier is with an example and so we shall work through the problem in Example 11.1 from the last section using the Lagrange multiplier method.

The firm is trying to maximize $Q = 12K^{0.4}L^{0.4}$ subject to the budget constraint $40K + 5L = 800$. The first step is to rearrange the budget constraint so that zero appears on one side of the equality sign.
Therefore

$$0 = 800 - 40K - 5L \tag{1}$$

We then write the 'Lagrange equation' or 'Lagrangian' in the form

$$\mathcal{L} = \text{(function to be optimized)} + \lambda \text{(constraint)}$$

where $\mathcal{L}$ is just the value of the Lagrangian function and λ is known as the 'Lagrangian multiplier'. (Do not worry about where these terms come from or what their actual values are. They are just introduced to help the analysis.)

In this problem the Lagrange function is thus

$$\mathcal{L} = 12K^{0.4}L^{0.4} + \lambda(800 - 40K - 5L) \tag{2}$$

Next, derive the partial differentials of $\mathcal{L}$ with respect to K, L and λ and set them equal to zero, i.e. find the stationary points of $\mathcal{L}$ that satisfy the first-order conditions for a maximum.

$$\frac{\partial \mathcal{L}}{\partial K} = 4.8K^{-0.6}L^{0.4} - 40\lambda = 0 \tag{3}$$

Constrained optimization

$$\frac{\partial \mathcal{L}}{\partial L} = 4.8K^{0.4}L^{-0.6} - 5\,\lambda = 0 \tag{4}$$

$$\frac{\partial \mathcal{L}}{\partial \lambda} = 800 - 40K - 5L = 0 \tag{5}$$

You will note that (5) is the same as the budget constraint (1). We now have a set of three linear simultaneous equations in three unknowns to solve for K and L. The Lagrange multiplier λ can be eliminated as, from (3),

$$0.12K^{-0.6}L^{0.4} = \lambda$$

and from (4)

$$0.96K^{0.4}L^{-0.6} = \lambda$$

Therefore

$$0.12K^{-0.6}L^{0.4} = 0.96K^{0.4}L^{-0.6}$$

Multiplying both sides by $K^{0.6}L^{0.6}$,

$$0.12L = 0.96K$$

$$L = 8K \tag{6}$$

Substituting (6) into (5),

$$800 - 40K - 5(8K) = 0$$

$$800 = 80K$$

$$10 = K$$

Substituting back into (5),

$$800 - 40(10) - 5L = 0$$

$$400 = 5L$$

$$80 = L$$

These are the same values of K and L as those obtained by the substitution method. Thus, the values of K and L that satisfy the first-order conditions for a maximum value of the Lagrangian function $\mathcal{L}$ are the values that will maximize output subject to the given budget constraint. We shall just accept this result without going into the proof of why this is so.

Strictly speaking we should now check the second-order conditions in the above problem to be sure that we actually have a maximum rather than a minimum. These, however, are rather complex, involving an examination of the function at and near the stationary points found, and we shall discuss these in the next section. For the time being you can assume that once the stationary points of a Lagrangian function have been found the second-order conditions for a maximum will automatically be met. Some more examples are worked through so that you can become familiar with this method.

Example 11.6

A firm can buy two inputs K and L at £18 per unit and £8 per unit respectively and it faces the production function $Q = 24K^{0.6}L^{0.3}$. What is the maximum output it can produce for a budget of £50,000? (Assume fractions of Q can be produced but work to nearest whole units of K and L.)

Solution

The budget constraint is $50,000 - 18K - 8L = 0$ and the function to be maximized is $Q = 24K^{0.6}L^{0.3}$. The Lagrangian for this problem is therefore

$$\mathcal{L} = 24K^{0.6}L^{0.3} + \lambda(50,000 - 18K - 8L)$$

We partially differentiate to find the stationary points of $\mathcal{L}$.

$$\frac{\partial \mathcal{L}}{\partial K} = 14.4K^{-0.4}L^{0.3} - 18\lambda = 0$$

Therefore

$$\frac{14.4L^{0.3}}{18K^{0.4}} = \lambda \tag{1}$$

$$\frac{\partial \mathcal{L}}{\partial L} = 7.2K^{0.6}L^{-0.7} - 8\lambda = 0$$

Therefore

$$\frac{7.2K^{0.6}}{8L^{0.7}} = \lambda \tag{2}$$

$$\frac{\partial \mathcal{L}}{\partial \lambda} = 50,000 - 18K - 8L = 0 \tag{3}$$

Setting (1) equal to (2) to eliminate λ

$$\frac{14.4L^{0.3}}{18K^{0.4}} = \frac{7.2K^{0.6}}{8L^{0.7}}$$

$$115.2L = 129.6K$$

$$L = 1.125K \tag{4}$$

Substituting (4) into (3)

$$50,000 - 18K - 8(1.125K) = 0$$

$$50,000 - 18K - 9K = 0$$

$$50,000 = 27K$$

$$1,851.8519 = K \tag{5}$$

Substituting (5) into (4)

$$L = 1.125(1,851.8519) = 2,083.3334$$

Thus, to the nearest whole unit, optimum values of K and L are 1,852 and 2,083 respectively.

We can check that when these whole values of K and L are used the total cost will be

$$TC = 18K + 8L = 18(1,852) + 8(2,083) = 33,336 + 16,664 = £50,000$$

and so the budget constraint is satisfied. The actual maximum output level will be

$$Q = 24K^{0.6}L^{0.3} = 24(1,852)^{0.6}(2,083)^{0.3} = 21,697.146 \text{ units}$$

The same mathematical method is used whatever the economic application. One must use one's knowledge of economics to set up the mathematical problem, however, as the example below demonstrates.

Example 11.7

A consumer has the utility function $U = 40A^{0.5}B^{0.5}$. The prices of A and B, the two goods consumed, are initially £20 per unit and £5 per unit respectively, and the consumer's income is £600. The price of A then falls to £10. Work out the income and substitution effects of this price change on the amount of A consumed using Hicks' method and say whether A and B are normal or inferior goods.

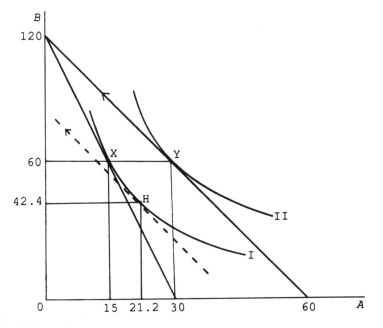

Figure 11.2

Solution

To help solve this problem the relevant budget schedules and indifference curves are drawn in Figure 11.2, although the indifference curves are not accurately drawn to scale. The original optimum is at X. The price fall for A causes the budget line to become flatter and swing round, giving a new equilibrium at Y.

Hicks' method for splitting the total change in A into its income and substitution effects requires one to draw a 'ghost' budget line parallel to the new budget line (reflecting the new relative prices) but tangential to the original indifference curve. This is shown by the broken line tangential to indifference curve I at H. From X to H is the substitution effect and from H to Y is the income effect of the price change. This problem requires us to find the corresponding values of A and B for the three tangency points X, Y and H and then to comment on the direction of these changes.

The original equilibrium is the combination of A and B that maximizes the utility function $U = 40A^{0.5} B^{0.5}$ subject to the budget constraint $600 = 20A + 5B$. These values of A and B can be found by deriving the stationary points of the Lagrange function

$$\mathcal{L} = 40A^{0.5} B^{0.5} + \lambda(600 - 20A - 5B)$$

Thus

$$\frac{\partial \mathcal{L}}{\partial A} = 20A^{-0.5} B^{0.5} - 20\,\lambda = 0 \quad \text{giving } A^{-0.5} B^{0.5} = \lambda \tag{1}$$

$$\frac{\partial \mathcal{L}}{\partial B} = 20A^{0.5} B^{-0.5} - 5B = 0 \quad \text{giving } 4A^{0.5} B^{-0.5} = \lambda \tag{2}$$

$$\frac{\partial \mathcal{L}}{\partial \lambda} = 600 - 20A - 5B = 0 \tag{3}$$

Setting (1) equal to (2)

$$A^{-0.5} B^{0.5} = 4A^{0.5} B^{-0.5}$$

$$B = 4A \tag{4}$$

Substituting (4) into (3)

$$600 - 20A - 5(4A) = 0$$
$$600 = 40A$$
$$15 = A$$

Substituting this value into (4)

$$B = 4(15) = 60$$

Thus, $A = 15$ and $B = 60$ at the original equilibrium at X.

When the price of A falls to 10, the budget constraint becomes

$$600 = 10A + 5B$$

and so the new Lagrange function is

$$\mathcal{L} = 40A^{0.5}B^{0.5} + \lambda(600 - 10A - 5B)$$

New stationary points will be where

$$\frac{\partial \mathcal{L}}{\partial A} = 20A^{-0.5}B^{0.5} - 10\lambda = 0 \quad \text{giving } 2A^{-0.5}B^{0.5} = \lambda \qquad (5)$$

$$\frac{\partial \mathcal{L}}{\partial B} = 20A^{0.5}B^{-0.5} - 5\lambda = 0 \quad \text{giving } 4A^{0.5}B^{-0.5} = \lambda \qquad (6)$$

$$\frac{\partial \mathcal{L}}{\partial \lambda} = 600 - 10A - 5B = 0 \qquad (7)$$

Setting (5) equal to (6)

$$2A^{-0.5}B^{0.5} = 4A^{0.5}B^{-0.5}$$

$$B = 2A \qquad (8)$$

Substituting (8) into (7)

$$600 - 10A - 5(2A) = 0$$
$$600 - 10A - 10A = 0$$
$$600 = 20A$$
$$30 = A$$

Substituting this value into (8) gives $B = 60$. Thus, the total effect of the price change is to increase consumption of A from 15 to 30 units and leave consumption of B unchanged at 60.

There are several ways of finding the values of A and B that correspond to point H. We know that H is on the same indifference curve as point X, and therefore the utility function will take the same value at both points. At X, $A = 15$ and $B = 60$ and so

$$U = 40A^{0.5}B^{0.5} = 40(15)^{0.5}(60)^{0.5} = 40(900)^{0.5} = 40(30) = 1,200$$

Thus, at any point on the indifference curve I

$$40A^{0.5}B^{0.5} = 1,200$$

and so

$$B^{0.5} = 30A^{-0.5}$$
$$B = 900A^{-1} \qquad (9)$$

The slope of the indifference curve I will therefore be

$$\frac{dB}{dA} = -900A^{-2} \tag{10}$$

At point X, the indifference curve I is tangential to the new budget line whose slope will be

$$\frac{-P_A}{P_B} = \frac{-10}{5} = -2 \tag{11}$$

Therefore, from (10) and (11)

$$-900A^{-2} = -2$$

$$450 = A^2$$

$$21.2132 = A$$

Substituting this value into (9)

$$B = 900(21.2132)^{-1} = 42.4264$$

Thus the substitution effect of A's price fall, from X to H, increases consumption of A from 15 to 21.2132 units and decreases consumption of B from 60 to 42.4264 units.

The income effect, from H to Y, increases consumption of A from 21.2132 to 30 units and also increases consumption of B from 42.4264 back to its original 60 unit level. As both income effects are positive, both A and B must be normal goods.

QUESTIONS 11.2

Use the Lagrange method to answer questions 1, 2, 3, 4, 6(a) and 7 from Questions 11.1 above.

11.4 THE LAGRANGE MULTIPLIER: SECOND-ORDER CONDITIONS

Inasmuch as it involves setting the first derivatives of the objective function equal to zero, the Lagrange method of solving constrained optimization problems is similar to the method of solving unconstrained optimization problems involving functions of several variables that was explained in Chapter 10. However, one cannot simply apply the same set of second-order conditions to check for a maximum or minimum because of the special role that the Lagrange multiplier takes. The mathematics required to prove why this is so, and to explain what additional second-order conditions are necessary for a Lagrangian function to be a maximum or minimum, becomes rather complex. Therefore, in this basic mathematics text, we shall just look at an intuitive explanation of what these conditions involve. First, we shall consider the conditions for a maximum.

If we assume that a function has two independent variables then both the function to be maximized, f(A, B), and the constraint could take on several possible forms, as illustrated in Figure 11.3. These diagrams are all constructed on the same basis as isoquant maps. Thus the lines I, II and III represent different levels of the objective function with its value increasing as one moves away from the origin.

In Figure 11.3(a), the objective function is convex to the origin and the constraint CD is linear. Maximization of the objective function occurs at the tangency point T.

In Figure 11.3(b), the objective function is concave to the origin and so it is maximized subject to the linear constraint CD at the corner point C. Thus the tangency point T does *not* determine the maximum value.

In Figure 11.3(c), the objective function is convex to the origin but the constraint CD is non-linear and more sharply curved than the objective function. Thus the tangency point T is *not* the maximum value. Higher values of the objective function can be found at points R and S, for example.

From the above examples we can see that, for the two-variable case, a linear constraint and an objective function that is convex to the origin will ensure maximization at the point of tangency. In other cases tangency may not ensure maximization.

If a problem involves the maximization of production subject to a linear budget constraint this means that output is maximized where the slope of the isoquant is equal to the slope of the budget constraint.

We have already seen in Chapter 10 how a Cobb–Douglas production function in the form $Q = AK^\alpha L^\beta$ will correspond to a set of isoquants which continually decline and become flatter as L is increased, i.e. are convex to the origin. Thus in any constrained optimization problem where one is attempting to maximize a production function in the Cobb–Douglas format subject to a linear budget constraint, the input combination that satisfies the first-order conditions will be a maximum.

If the function represents another concept, such as utility, the same conditions apply. In all such cases one assumes that the independent variables in the objective function must take positive values and so any negative mathematical solutions can be disregarded.

Although we cannot illustrate functions with more than two variables diagrammatically, the same basic principles apply when one is attempting to maximize a function with three or more variables subject to a linear constraint. Thus any Cobb–Douglas production function with more than two inputs will be at a maximum, subject to a specified linear budget constraint, when the first-order conditions for optimization of the relevant Lagrange equation are met. For the purpose of answering the problems in this text, and for most constrained maximization problems that you will encounter in a first-year economics course, it can be assumed that the stationary points of the Lagrange function will satisfy the second-order conditions for a maximum.

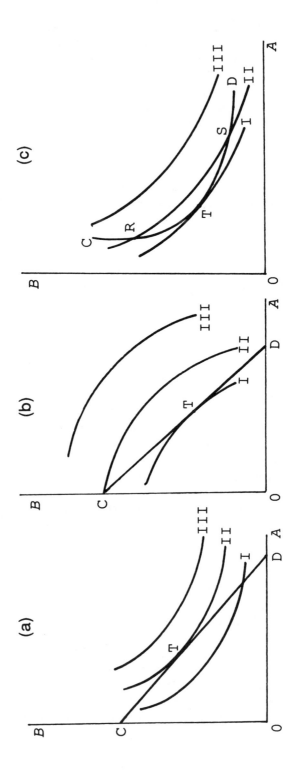

Figure 11.3

The properties of the objective function and constraint that guarantee that a Lagrange function is minimized when first-order conditions are met are the reverse of the properties required for a maximum, i.e. the objective function must be linear and the constraint must be convex to the origin. Thus if one is required to find values of K and L that minimize a budget function in the form

$$TC = P_K K + P_L L$$

subject to a given output $\overline{Q}$ being produced via the production function $Q = AK^\alpha L^\beta$, then the corresponding Lagrange function is

$$\mathscr{L} = P_K K + P_L L + \lambda(\overline{Q} - AK^\alpha L^\beta)$$

and the values of K and L which satisfy the first-order conditions

$$\frac{\partial \mathscr{L}}{\partial K} = 0 \qquad \frac{\partial \mathscr{L}}{\partial L} = 0$$

will also satisfy second-order conditions for a minimum.

If you refer back to Figure 11.3(a) you can see the rationale for this rule in the two-input case. If input prices are given and one is trying to minimize the cost of the output represented by isoquant II, then one needs to find the budget constraint with slope equal to the negative of the price ratio which is nearest the origin and still goes through this isoquant. The linear objective function and the constraint convex to the origin guarantee that this will be at the tangency point T.

Some examples of minimization problems that use this rule are given in the next section.

To conclude this section, let us reiterate what we have learned about second-order conditions and Lagrangians.

When the objective function is in the Cobb–Douglas format and the constraint is linear, then the second-order conditions for a maximum are met at the stationary points of the Lagrange function. This rule is reversed for minimization.

It is important to note that other forms of functions are not covered by these rules. However, they still allow you to use the Lagrange multiplier method to tackle most of the constrained optimization problems that you will encounter unless you go on to study mathematical economics.

11.5 CONSTRAINED MINIMIZATION USING THE LAGRANGE MULTIPLIER

As was explained in Section 11.4, the same principles used to construct a Lagrange function for a constrained maximization problem are used to construct a Lagrange function for a constrained minimization problem. The difference is that the components of the function are reversed, as is shown in the following examples. In all these cases the constraints and objective function take formats which guarantee that second-order conditions for a minimum are met.

Example 11.8

A firm operates with the production function $Q = 4K^{0.6}L^{0.5}$ and can buy K at £15 a unit and L at £8 a unit. What input combination will minimize the cost of producing 200 units of output?

Solution

The output constraint is $200 = 4K^{0.6}L^{0.5}$ and the objective function to be minimized is the total cost function $TC = 15K + 8L$. The corresponding Lagrangian function is therefore

$$\mathcal{L} = 15K + 8L + \lambda(200 - 4K^{0.6}L^{0.5})$$

Partially differentiating $\mathcal{L}$ and setting equal to zero to get the first-order conditions we have

$$\frac{\partial \mathcal{L}}{\partial K} = 15 - \lambda 2.4K^{-0.4}L^{0.5} = 0 \quad \text{giving} \quad \frac{15K^{0.4}}{2.4L^{0.5}} = \lambda \tag{1}$$

$$\frac{\partial \mathcal{L}}{\partial L} = 8 - \lambda 2K^{0.6}L^{-0.5} = 0 \quad \text{giving} \quad \frac{4L^{0.5}}{K^{0.6}} = \lambda \tag{2}$$

$$\frac{\partial \mathcal{L}}{\partial \lambda} = 200 - 4K^{0.6}L^{0.5} = 0 \tag{3}$$

Setting (1) equal to (2) to eliminate λ

$$\frac{15K^{0.4}}{2.4L^{0.5}} = \frac{4L^{0.5}}{K^{0.6}}$$

$$15K = 9.6L$$

$$1.5625K = L \tag{4}$$

Substituting (4) into (3)

$$200 - 4K^{0.6}(1.5625K)^{0.5} = 0$$

$$200 = 4K^{0.6}(1.5625)^{0.5}K^{0.5}$$

$$\frac{200}{4(1.5625)^{0.5}} = K^{1.1}$$

$$40 = K^{1.1}$$

$$28.603434 = K$$

Substituting this value into (4)

$$1.5625(28.603434) = L$$

$$44.692866 = L$$

Thus the optimal input combination is 28.6 units of K plus 44.7 units of L (to 1 decimal place).

We can check that these input values correspond to the given output level by substituting them back into the production function. Thus

$$Q = 4K^{0.6}L^{0.5} = 4(28.6)^{0.6}(44.7)^{0.5} = 200$$

which is correct. The actual cost entailed will be

$$TC = 15K + 8L = 15(28.6) + 8(44.7) = £786.60$$

Example 11.9

The prices of inputs K and L are given as £12 per unit and £3 per unit respectively, and a firm operates with the production function $Q = 25K^{0.5}L^{0.5}$.

(i) What is the minimum cost of producing 1,250 units of output?
(ii) Demonstrate that the maximum output that can be produced for this budget will be the 1,250 units specified in (i) above.

Solution

This question essentially asks us to demonstrate that the constrained maximization and minimization methods give consistent answers.

(i) The output constraint is that

$$1,250 = 25K^{0.5}L^{0.5}$$

The objective function to be minimized is the cost function

$$TC = 12K + 3L$$

The corresponding Lagrange function is therefore

$$\mathscr{L} = 12K + 3L + \lambda(1,250 - 25K^{0.5}L^{0.5})$$

First-order conditions require

$$\frac{\partial \mathscr{L}}{\partial K} = 12 - \lambda 12.5K^{-0.5}L^{0.5} = 0 \quad \text{giving} \quad \frac{12K^{0.5}}{12.5L^{0.5}} = \lambda \quad \text{(1)}$$

$$\frac{\partial \mathscr{L}}{\partial L} = 3 - \lambda 12.5K^{0.5}L^{0.5} = 0 \quad \text{giving} \quad \frac{3L^{0.5}}{12.5K^{0.5}} = \lambda \quad \text{(2)}$$

$$\frac{\partial \mathscr{L}}{\partial \lambda} = 1,250 - 25K^{0.5}L^{0.5} = 0 \quad \text{(3)}$$

Setting (1) equal to (2)

$$\frac{12K^{0.5}}{12.5L^{0.5}} = \frac{3L^{0.5}}{12.5K^{0.5}}$$

$$4K = L \quad \text{(4)}$$

Substituting (4) into (3)

$$1,250 - 25K^{0.5}(4K)^{0.5} = 0$$
$$1,250 = 25K^{0.5}(4)^{0.5}K^{0.5}$$
$$1,250 = 50K$$
$$25 = K$$

Substituting this value into (4)

$$4(25) = L$$
$$100 = L$$

When these optimum values of K and L are used the actual minimum cost will be

$$TC = 12K + 3L = 12(25) + 3(100) = 300 + 300 = £600$$

(ii) This part of the question requires us to find the values of K and L that will maximize output subject to a budget of £600, i.e. the answer to (i) above.

   The constraint is therefore $12K + 3L = 600$ and the objective function to be maximized is

$$Q = 25K^{0.5}L^{0.5}$$

The corresponding Lagrange equation is therefore

$$\mathcal{L} = 25K^{0.5}L^{0.5} + \lambda(600 - 12K - 3L)$$

First-order conditions require

$$\frac{\partial\mathcal{L}}{\partial K} = 12.5K^{-0.5}L^{0.5} - 12\lambda = 0 \quad \text{giving} \quad \frac{12.5L^{0.5}}{12K^{0.5}} = \lambda \tag{5}$$

$$\frac{\partial\mathcal{L}}{\partial L} = 12.5K^{0.5}L^{-0.5} - 3\lambda = 0 \quad \text{giving} \quad \frac{12.5K^{0.5}}{3L^{0.5}} = \lambda \tag{6}$$

$$\frac{\partial\mathcal{L}}{\partial\lambda} = 600 - 12K - 3L = 0 \tag{7}$$

Setting (5) equal to (6)

$$\frac{12.5L^{0.5}}{12K^{0.5}} = \frac{12.5K^{0.5}}{3L^{0.5}}$$
$$3L = 12K$$
$$L = 4K \tag{8}$$

Substituting (8) into (7)

$$600 - 12K - 3(4K) = 0$$

$$600 - 12K - 12K = 0$$

$$600 = 24K$$

$$25 = K$$

Substituting this value into (8)

$$L = 4(25) = 100$$

These are the same optimum values of K and L that were found in part (i) above. The actual output produced by 25 of K plus 100 of L will be $Q = 25K^{0.5}L^{0.5} = 25(25)^{0.5}(100)^{0.5} = 1,250$ which is the amount specified in the question.

Although most of the examples of constrained optimization presented in this chapter are concerned with a firm's output and costs, or a consumer's utility level and income, the Lagrange method can be applied to other areas of economics. For instance, in environmental economics one may wish to find the cheapest way of securing a given level of environmental cleanliness.

Example 11.10

Assume that there are two uncontrollable sources of pollution into a lake. The local water authority can clean up the discharges and reduce pollution levels from these sources but there are, of course, costs involved. The damage effects of each pollution source are measured on a 'pollution scale'. The lower the pollution level the greater the cost of achieving it, as is shown by the cost schedules for cleaning up the two pollution sources:

$$Z_1 = 478 - 2C_1^{0.5}$$

$$Z_2 = 600 - 3C_2^{0.5}$$

where Z_1 and Z_2 are pollution levels and C_1 and C_2 are expenditure levels (in £000s) on reducing pollution.

To secure an acceptable level of water purity in the lake the water authority's objective is to reduce the total pollution level to 1,000 by the cheapest method. How can it do this?

Solution

This can be formulated as a constrained optimization problem where the constraint is the total amount of pollution $Z_1 + Z_2 = 1000$ and the objective function to be minimized is the cost of pollution control $TC = C_1 + C_2$. Thus the Lagrange function is

$$\mathcal{L} = C_1 + C_2 + \lambda(1,000 - Z_1 - Z_2)$$

Basic mathematics for economists

Substituting in the cost functions for Z_1 and Z_2, this becomes

$$\mathcal{L} = C_1 + C_2 + \lambda[1{,}000 - (478 - 2C_1^{0.5}) - (600 - 3C_2^{0.5})]$$

$$\mathcal{L} = C_1 + C_2 + \lambda(-78 + 2C_1^{0.5} + 3C_2^{0.5})$$

$$\frac{\partial \mathcal{L}}{\partial C_1} = 1 + \lambda C_1^{-0.5} = 0 \quad \text{giving } \lambda = -C_1^{0.5} \tag{1}$$

$$\frac{\partial \mathcal{L}}{\partial C_2} = 1 + \lambda 1.5 C_2^{-0.5} = 0 \quad \text{giving } \lambda = \frac{-C_2^{0.5}}{1.5} \tag{2}$$

$$\frac{\partial \mathcal{L}}{\partial \lambda} = -78 + 2C_1^{0.5} + 3C_2^{0.5} = 0 \tag{3}$$

Equating (1) and (2)

$$-C_1^{0.5} = \frac{-C_2^{0.5}}{1.5}$$

$$1.5 C_1^{0.5} = C_2^{0.5} \tag{4}$$

Substituting (4) into (3)

$$-78 + 2C_1^{0.5} + 3(1.5C_1^{0.5}) = 0$$

$$2C_1^{0.5} + 4.5C_1^{0.5} = 78$$

$$6.5C_1^{0.5} = 78$$

$$C_1^{0.5} = 12 \tag{5}$$

$$C_1 = 144$$

Substituting (5) into (4)

$$C_2^{0.5} = 1.5(12) = 18$$

$$C_2 = 324$$

We can check the total pollution level:

$$Z_1 + Z_2 = (478 - 2C_1^{0.5}) + (600 - 3C_2^{0.5})$$

$$= 478 - 2(12) + 600 - 3(18)$$

$$= 1{,}000$$

which is the required level.

Thus the water authority should spend £144,000 on reducing the first pollution source and £324,000 on reducing the second source.

QUESTIONS 11.3

1. Use the Lagrange multiplier to answer Questions 6(b) and 8 from Questions 11.1 above.

2. What is the cheapest way of producing 850 units of output if a firm operates with the production function $Q = 30K^{0.5}L^{0.5}$ and can buy input K at £75 a unit and L at £40 a unit?

3. Two pollution sources can be cleaned up if money is spent on them according to the functions $Z_1 = 780 - 12C_1^{0.5}$ and $Z_2 = 600 - 8C_2^{0.5}$ where Z_1 and Z_2 are the pollution levels from the two sources and C_1 and C_2 are expenditure levels (in £000s) on pollution reduction. What is the cheapest way of reducing the total pollution level from 1,380, which it would be without any controls, to 1,000?

4. A firm buys inputs K and L at £70 a unit and £30 a unit respectively and faces the production function $Q = 40K^{0.5}L^{0.5}$. What is the cheapest way it can produce an output of 500 units?

11.6 CONSTRAINED OPTIMIZATION WITH MORE THAN TWO VARIABLES

The same procedures that were used for two-variable problems are also used for applying the Lagrange method to constrained optimization problems with three or more variables. The only difference is that one has a more complex set of simultaneous equations to solve for the optimum values that satisfy the first-order conditions. Although some of these sets of equations may initially look rather awkward to work with, they can usually be greatly simplified and solutions can be found by basic algebra, as the following examples show. As with the two-variable problems, it is assumed that second-order conditions for a maximum (or minimum) are satisfied at stationary points of the Lagrange function given the formats of the objective functions and the constraints in the problems set out here.

Example 11.11

A firm has a budget of £300 to spend on the three inputs x, y and z whose prices per unit are £4, £1 and £6 respectively. What combination of *x*, *y* and z should it employ to maximize output if it faces the production function $Q = 24x^{0.3}y^{0.2}z^{0.3}$?

Solution

The budget constraint is

$$300 - 4x - y - 6z = 0$$

and the objective function to be maximized is

$$Q = 24x^{0.3}y^{0.2}z^{0.3}$$

Thus the Lagrange function is

$$\mathscr{L} = 24x^{0.3}y^{0.2}z^{0.3} + \lambda(300 - 4x - y - 6z)$$

Differentiating

$$\frac{\partial \mathscr{L}}{\partial x} = 7.2x^{-0.7}y^{0.2}z^{0.3} - 4\lambda = 0 \qquad \lambda = 1.8x^{-0.7}y^{0.2}z^{0.3} \qquad (1)$$

$$\frac{\partial \mathscr{L}}{\partial y} = 4.8x^{0.3}y^{-0.8}z^{0.3} - \lambda = 0 \qquad \lambda = 4.8x^{0.3}y^{-0.8}z^{0.3} \qquad (2)$$

$$\frac{\partial \mathscr{L}}{\partial z} = 7.2x^{0.3}y^{0.2}z^{-0.7} - 6\lambda = 0 \qquad \lambda = 1.2x^{0.3}y^{0.2}z^{-0.7} \qquad (3)$$

$$\frac{\partial \mathscr{L}}{\partial \lambda} = 300 - 4x - y - 6z = 0 \qquad (4)$$

A simultaneous three-linear-equation system in the three unknowns x, y and z can now be set up if λ is eliminated. There are several ways in which this can be done. In the method below we set (1) and then (3) equal to (2) to eliminate x and z and then substitute into (4) to solve for y.

Whichever method is used, the objective is to arrive at functions for any two of the unknown variables in terms of the remaining third variable.

Setting (1) equal to (2)

$$1.8x^{-0.7}y^{0.2}z^{0.3} = 4.8x^{0.3}y^{-0.8}z^{0.3}$$

Multiplying both sides by $x^{0.7}y^{0.8}$ and dividing by $z^{0.3}$ gives

$$1.8y = 4.8x$$

$$0.375y = x \qquad (5)$$

We have now eliminated z and obtained a function for x in terms of y. Next we need to eliminate x and obtain a function for z in terms of y. Setting (2) equal to (3)

$$4.8x^{0.3}y^{-0.8}z^{0.3} = 1.2x^{0.3}y^{0.2}z^{-0.7}$$

Multiplying through by $z^{0.7}y^{0.8}$ and dividing by $x^{0.3}$ gives

$$4.8z = 1.2y$$

$$z = 0.25y \qquad (6)$$

Substituting (5) and (6) into (4)

$$300 - 4(0.375y) - y - 6(0.25y) = 0$$

$$300 - 1.5y - y - 1.5y = 0$$

$$300 = 4y$$

$$75 = y$$

Therefore from (5)

$$x = 28.12$$

and from (6)

$$z = 18.75$$

If these optimal values of x, y and z are used, output will be $Q = 24x^{0.3}y^{0.2}z^{0.3} = 24(28.12)^{0.3}(75)^{0.2}(18.75)^{0.3} = 373.1$ units.

Example 11.12

A firm uses the three inputs K, L and R to manufacture good Q and faces the production function

$$Q = 50K^{0.4}L^{0.2}R^{0.2}$$

It has a budget of £24,000 and can buy K, L and R at £80, £12 and £10 respectively per unit. What combination of inputs will maximize its output?

Solution

The objective function to be maximized is $Q = 50K^{0.4}L^{0.2}R^{0.2}$ and the budget constraint is

$$24{,}000 - 80K - 12L - 10R = 0$$

Thus the Lagrange equation is

$$\mathcal{L} = 50K^{0.4}L^{0.2}R^{0.2} + \lambda(24{,}000 - 80K - 12L - 10R)$$

Differentiating

$$\frac{\partial \mathcal{L}}{\partial K} = 20K^{-0.6}L^{0.2}R^{0.2} - 80\lambda = 0 \qquad \lambda = 0.25K^{-0.6}L^{0.2}R^{0.2} \tag{1}$$

$$\frac{\partial \mathcal{L}}{\partial L} = 10K^{0.4}L^{-0.8}R^{0.2} - 12\lambda = 0 \qquad \lambda = \frac{10}{12}K^{0.4}L^{-0.8}R^{0.2} \tag{2}$$

$$\frac{\partial \mathcal{L}}{\partial R} = 10K^{0.4}L^{0.2}R^{-0.8} - 10\lambda = 0 \qquad \lambda = K^{0.4}L^{0.2}R^{-0.8} \tag{3}$$

$$\frac{\partial \mathcal{L}}{\partial \lambda} = 24{,}000 - 80K - 12L - 10R = 0 \tag{4}$$

Equating (1) and (2) to eliminate R

$$0.25K^{-0.6}L^{0.2}R^{0.2} = \frac{10}{12}K^{0.4}L^{-0.8}R^{0.2}$$

$$3L = 10K$$

$$0.3L = K \tag{5}$$

363

Equating (2) and (3) to eliminate K and get R in terms of L

$$\frac{10}{12}K^{0.4}L^{-0.8}R^{0.2} = K^{0.4}L^{0.2}R^{-0.8}$$

$$10R = 12L$$

$$R = 1.2L \qquad (6)$$

Substituting (5) and (6) into (4)

$$24,000 - 80(0.3L) - 12L - 10(1.2L) = 0$$

$$24,000 = 24L + 12L + 12L$$

$$24,000 = 48L$$

$$500 = L$$

Substituting this value into (5) and (6)

$$K = 0.3(500) = 150 \qquad R = 1.2(500) = 600$$

If the firm uses these optimal values for K, L and R, its output will be

$$Q = 50K^{0.4}L^{0.2}R^{0.2} = 50(150)^{0.4}(500)^{0.2}(600)^{0.2} = 4,622 \text{ units}$$

Example 11.13

A firm buys the four inputs K, L, R and M at per-unit prices of £50, £30, £25 and £20 respectively and operates with the production function

$$Q = 160K^{0.3}L^{0.25}R^{0.2}M^{0.25}$$

What is the maximum output it can make for a total cost of £30,000?

Solution

The relevant Lagrange function is

$$\mathscr{L} = 160K^{0.3}L^{0.25}R^{0.2}M^{0.25} + \lambda(30,000 - 50K - 30L - 25R - 20M)$$

Differentiating to find stationary points

$$\frac{\partial \mathscr{L}}{\partial K} = 48K^{-0.7}L^{0.25}R^{0.2}M^{0.25} - 50\lambda = 0 \qquad \lambda = \frac{48L^{0.25}R^{0.2}M^{0.25}}{50K^{0.7}} \qquad (1)$$

$$\frac{\partial \mathscr{L}}{\partial L} = 40K^{0.3}L^{-0.75}R^{0.2}M^{0.25} - 30\lambda = 0 \qquad \lambda = \frac{4K^{0.3}R^{0.2}M^{0.25}}{3L^{0.75}} \qquad (2)$$

$$\frac{\partial \mathscr{L}}{\partial R} = 32K^{0.3}L^{0.25}R^{-0.8}M^{0.25} - 25\lambda = 0 \qquad \lambda = \frac{32K^{0.3}L^{0.25}M^{0.25}}{25R^{0.8}} \qquad (3)$$

$$\frac{\partial \mathscr{L}}{\partial M} = 40K^{0.3}L^{0.25}R^{0.2}M^{-0.75} - 20\lambda = 0 \qquad \lambda = \frac{2K^{0.3}L^{0.25}R^{0.2}}{M^{0.75}} \qquad (4)$$

$$\frac{\partial \mathcal{L}}{\partial \lambda} = 30{,}000 - 50K - 30L - 25R - 20M = 0 \tag{5}$$

Equating (1) and (2)

$$\frac{48L^{0.25}R^{0.2}M^{0.25}}{50K^{0.7}} = \frac{4K^{0.3}R^{0.2}M^{0.25}}{3L^{0.75}}$$

Dividing through by $R^{0.2}M^{0.25}$ and cross multiplying

$$144L = 200K$$
$$0.72L = K \tag{6}$$

In the above equality it is neater to express K as a fraction of L rather than vice versa. We must now find R and M in terms of L and so (2) must be equated with (3) and (4) to ensure L is not cancelled out in each set of equalities. Thus, equating (2) and (3)

$$\frac{4K^{0.3}R^{0.2}M^{0.25}}{3L^{0.75}} = \frac{32K^{0.3}L^{0.25}M^{0.25}}{25R^{0.8}}$$

Cancelling out $K^{0.3}M^{0.25}$ and cross multiplying

$$100R = 96L$$
$$R = 0.96L \tag{7}$$

Equating (2) and (4)

$$\frac{4K^{0.3}R^{0.2}M^{0.25}}{3L^{0.75}} = \frac{2K^{0.3}L^{0.25}R^{0.2}}{M^{0.75}}$$

Cancelling $K^{0.3}R^{0.2}$ and cross multiplying

$$4M = 6L$$
$$M = 1.5L \tag{8}$$

Substituting (6), (7) and (8) into (5)

$$30{,}000 - 50(0.72L) - 30L - 25(0.96L) - 20(1.5L) = 0$$
$$30{,}000 - 36L - 30L - 24L - 30L = 0$$
$$30{,}000 = 120L$$
$$250 = L$$

Substituting this value into (6)

$$K = 0.72(250) = 180$$

Substituting this value into (7)

$$R = 0.96(250) = 240$$

Substituting this value into (8)

$$M = 1.5(250) = 375$$

When these optimal values of K, L, M and R are used, output will be

$$Q = 160K^{0.3}L^{0.25}R^{0.2}M^{0.25}$$
$$= 160(180^{0.3})(250^{0.25})(240^{0.2})(375^{0.25}) = 39{,}786.6 \text{ units}$$

Example 11.14

A firm operates with the production function $Q = 20K^{0.5}L^{0.25}R^{0.4}$. The input prices per unit are £20 for K, £10 for L and £5 for R. What is the cheapest way of producing 1,200 units of output?

Solution

The constraint is $Q = 20K^{0.5}L^{0.25}R^{0.4} = 1{,}200$ and the objective function to be minimized is $TC = 20K + 10L + 5R$. The corresponding Lagrange function is therefore

$$\mathscr{L} = 20K + 10L + 5R + \lambda(1{,}200 - 20K^{0.5}L^{0.25}R^{0.4})$$

Differentiating to get stationary points

$$\frac{\partial \mathscr{L}}{\partial K} = 20 - \lambda 10K^{-0.25}L^{0.25}R^{0.4} = 0 \qquad \lambda = \frac{2K^{0.5}}{L^{0.25}R^{0.4}} \tag{1}$$

$$\frac{\partial \mathscr{L}}{\partial L} = 10 - \lambda 5K^{0.5}L^{-0.75}R^{0.4} = 0 \qquad \lambda = \frac{2L^{0.75}}{K^{0.5}R^{0.4}} \tag{2}$$

$$\frac{\partial \mathscr{L}}{\partial R} = 5 - \lambda 8K^{0.5}L^{0.25}R^{-0.6} = 0 \qquad \lambda = \frac{5R^{0.6}}{8K^{0.5}L^{0.25}} \tag{3}$$

$$\frac{\partial \mathscr{L}}{\partial \lambda} = 1{,}200 - 20K^{0.5}L^{0.25}R^{0.4} = 0 \tag{4}$$

Equating (1) and (2)

$$\frac{2K^{0.5}}{L^{0.25}R^{0.4}} = \frac{2L^{0.75}}{K^{0.5}R^{0.4}}$$

$$K = L \tag{5}$$

Equating (2) and (3)

$$\frac{2L^{0.75}}{K^{0.5}R^{0.4}} = \frac{5R^{0.6}}{8K^{0.5}L^{0.25}}$$

$$16L = 5R$$

$$3.2L = R \tag{6}$$

Constrained optimization

Substituting (5) and (6) into (4) to eliminate R and K

$$1{,}200 - 20(L)^{0.5}L^{0.25}(3.2L)^{0.4} = 0$$
$$1{,}200 - 20(3.2)^{0.4}L^{1.15} = 0$$
$$1{,}200 = 31.848574L^{1.15}$$
$$37.678296 = L^{1.15}$$
$$23.47 = L$$

Substituting this value into (5)

$$K = 23.47$$

and into (6)

$$R = 3.2(23.47) = 75.1$$

Checking that these values do give the required output:

$$Q = 20(23.47)^{0.5}(23.47)^{0.25}(75.1)^{0.4} = 1{,}200$$

The actual cost will be

$$20K + 10L + 5R = 20(23.4) + 10(23.47) + 5(75.1) = £1{,}079.60$$

Example 11.15

A firm operates with the production function

$$Q = 45K^{0.4}L^{0.3}R^{0.3}$$

and can buy input K at £80 a unit, L at £35 and R at £50. What is the cheapest way it can produce an output of 70,000 units?

Solution

The constraint is $Q = 45K^{0.4}L^{0.3}R^{0.3} = 70{,}000$ and the objective function to be minimized is $TC = 80K + 35L + 50R$. The corresponding Lagrange function is thus

$$\mathscr{L} = 80K + 35L + 50R + \lambda(70{,}000 - 45K^{0.4}L^{0.3}R^{0.3})$$

Differentiating to get first-order conditions for a minimum

$$\frac{\partial \mathscr{L}}{\partial K} = 80 - \lambda 18K^{-0.6}L^{0.3}R^{0.3} = 0 \qquad \lambda = \frac{80K^{0.6}}{18L^{0.3}R^{0.3}} \qquad (1)$$

$$\frac{\partial \mathscr{L}}{\partial L} = 35 - \lambda 13.5K^{0.4}L^{-0.7}R^{0.3} = 0 \qquad \lambda = \frac{35L^{0.7}}{13.5K^{0.4}R^{0.3}} \qquad (2)$$

$$\frac{\partial \mathscr{L}}{\partial R} = 50 - \lambda\, 13.5 K^{0.4} L^{0.3} R^{-0.7} = 0 \qquad\qquad \lambda = \frac{50 R^{0.7}}{13.5 K^{0.4} L^{0.3}} \tag{3}$$

$$\frac{\partial \mathscr{L}}{\partial \lambda} = 75{,}000 - 45 K^{0.4} L^{0.3} R^{0.3} = 0 \tag{4}$$

Equating (1) and (2)

$$\frac{80 K^{0.6}}{18 L^{0.3} R^{0.3}} = \frac{35 L^{0.7}}{13.5 K^{0.4} R^{0.3}}$$

$$1{,}080 K = 630 L$$

$$\frac{12 K}{7} = L \tag{5}$$

As we have L in terms of K we now need to use (1) and (3) to get R in terms of K. Thus equating (1) and (3)

$$\frac{80 K^{0.6}}{18 L^{0.3} R^{0.3}} = \frac{50 R^{0.7}}{13.5 K^{0.4} L^{0.3}}$$

$$1{,}080 K = 900 R$$

$$1.2 K = R \tag{6}$$

Substituting (5) and (6) into (4)

$$75{,}000 - 45 K^{0.4} \left(\frac{12 K}{7}\right)^{0.3} (1.2 K)^{0.3} = 0$$

$$75{,}000 - 45 \left(\frac{12}{7}\right)^{0.3} (1.2)^{0.3} K^{0.4} K^{0.3} K^{0.3} = 0$$

$$75{,}000 = 55.871697 K$$

$$1{,}342.3612 = K$$

Substituting this value into (5)

$$L = \frac{12}{7}(1{,}342.3612) = 2{,}301.1907$$

Substituting into (6)

$$R = 1.2(1{,}342.3612) = 1{,}610.8334$$

Thus, optimum values are

$$K = 1{,}342.4 \qquad L = 2{,}301.2 \qquad R = 1{,}610.8 \qquad \text{(to 1 dp)}$$

Total expenditure on inputs will then be

$$80 K + 35 L + 50 R = 80(1{,}342.4) + 35(2{,}301.2) + 50(1{,}610.8) = £268{,}474$$

QUESTIONS 11.4

1. A firm has a budget of £570 to spend on the three inputs x, y and z whose prices per unit are respectively £4, £6 and £3. What combination of x, y and z will maximize output given the production function $Q = 2x^{0.2}y^{0.3}z^{0.45}$?

2. A firm uses inputs K, L and R to manufacture good Q. It has a budget of £828 and its production function (for positive values of Q) is

$$Q = 20K + 16L + 12R - 0.2K^2 - 0.1L^2 - 0.3R^2$$

If $P_L = £20$, $P_K = £10$ and $P_R = £6$ what is the maximum output it can produce? Assume that second-order conditions for a maximum are satisfied for the relevant Lagrangian.

3. What amounts of the inputs x, y and z should a firm use to maximize output if it faces the production function $Q = 2x^{0.4}y^{0.2}z^{0.6}$ and it has a budget of £600, given that the prices of x, y and z are respectively £4, £1 and £2 per unit?

4. A firm buys the inputs x, y and z for £5, £10 and £2 respectively per unit. If its production function is $Q = 60x^{0.2}y^{0.4}z^{0.5}$ how much can it produce for an outlay of £8,250?

5. Inputs K, L, R and M cost £10, £6, £15 and £3 respectively per unit. What is the cheapest way of producing an output of 900 units if a firm operates with the production function $Q = 20K^{0.4}L^{0.3}R^{0.2}M^{0.25}$?

6. Make up your own constrained optimization problem for an objective function with three variables and solve it.

7. A firm faces the production function $Q = 50K^{0.5}L^{0.2}R^{0\,25}$ and is required to produce an output level of 1,913 units. What is the cheapest way of doing this if the per-unit costs of inputs K, L and R are £80, £24 and £45 respectively?

369

12

Further topics in calculus

12.1 OVERVIEW

In this chapter some techniques are introduced that can be used to differentiate functions that are rather more complex than those encountered in Chapters 8, 9, 10 and 11. These are the chain rule, the product rule and the quotient rule. As you will see in the examples that are worked through, it is often necessary to use several of these methods to differentiate some functions. We shall also look at the concept of integration, which is essentially 'differentiation in reverse', and how it can be used in economics.

12.2 THE CHAIN RULE

The chain rule is used to differentiate 'functions within functions'. For example, if we have the function

$$y = f(z)$$

and we also know that there is a second functional relationship

$$z = g(x)$$

then we can write y as a function of x in the form

$$y = f[g(x)]$$

To differentiate y with respect to x in this type of function we use the chain rule which states that

$$\frac{dy}{dx} = \frac{dy}{dz}\frac{dz}{dx}$$

One economics example of a function within a function occurs in the marginal revenue productivity theory of the demand for labour, which is explained later. However, we shall first look at what is perhaps the most frequent use of the chain rule, which is to break down an awkward function artificially into two components in order to allow differentiation via the chain rule.

370

Further topics in calculus

Assume, for example, that you wish to find an expression for the slope of the non-linear demand function

$$p = (150 - 0.2q)^{0.5} \tag{1}$$

The basic rules for differentiation explained in Chapter 8 cannot cope with this sort of function. However, if we let

$$z = 150 - 0.2q \tag{2}$$

then (1) above can be rewritten as

$$p = z^{0.5} \tag{3}$$

(Note that in both (1) and (3) the functions are assumed to hold for $p \geq 0$ only, i.e. negative roots are ignored.)
Differentiating (2) and (3) we get

$$\frac{dz}{dq} = -0.2 \qquad \frac{dp}{dz} = 0.5z^{-0.5}$$

Thus, using the chain rule,

$$\frac{dp}{dq} = \frac{dp}{dz}\frac{dz}{dq} = 0.5z^{-0.5}(-0.2) = -0.1z^{-0.5} = \frac{-0.1}{(150 - 0.2q)^{0.5}}$$

Some more examples of the use of the chain rule are set out below.

Example 12.1

The present value of a payment of £1 due in 8 years' time is given by the formula

$$PV = \frac{1}{(1 + i)^8}$$

where i is the given interest rate. What is the rate of change of PV with respect to i?

Solution

If we let

$$z = 1 + i \tag{1}$$

then we can write

$$PV = \frac{1}{z^8} = z^{-8} \tag{2}$$

Differentiating (1) and (2) gives

$$\frac{dz}{di} = 1 \qquad \frac{dPV}{dz} = -8z^{-9}$$

371

Therefore, using the chain rule, the rate of change of PV with respect to i will be

$$\frac{dPV}{di} = \frac{dPV}{dz}\frac{dz}{di} = -8z^{-9} = \frac{-8}{(1+i)^9}$$

Example 12.2

If $y = (48 + 20x^{-1} + 4x + 0.3x^2)^4$ what is dy/dx?

Solution

Let

$$z = 48 + 20x^{-1} + 4x + 0.3x^2 \tag{1}$$

and so

$$\frac{dz}{dx} = -20x^{-2} + 4 + 0.6x \tag{2}$$

Substituting (1) into the function given in the question

$$y = z^4$$

and so

$$\frac{dy}{dz} = 4z^3 \tag{3}$$

Therefore, using the chain rule and substituting (2) and (3)

$$\frac{dy}{dx} = \frac{dy}{dz}\frac{dz}{dx}$$

$$= 4z^3(-20x^{-2} + 4 + 0.6x)$$

$$= 4(48 + 20x^{-1} + 4x + 0.3x^2)^3 (-20x^{-2} + 4 + 0.6x)$$

In the marginal revenue productivity theory of the demand for labour, the rule for profit maximization is to employ additional units of labour as long as the extra revenue generated by selling the extra output produced by an additional unit of labour exceeds the marginal cost of employing this additional unit of labour. This decision is assumed to be made in the short run when inputs other than labour are assumed fixed.

The optimal amount of labour is employed when

$$MRP_L = MC_L$$

where MRP_L is the marginal revenue product of labour, defined as the additional revenue generated by an additional unit of labour, and MC_L is the marginal cost of an additional unit of labour, which is equal to the wage rate unless the firm is a monopsonist (sole buyer) in the labour market.

If all relevant functions are assumed to be continuous then the above definitions can be rewritten as

$$MRP_L = \frac{dTR}{dL} \qquad MC_L = \frac{dTC_L}{dL}$$

where TR is total sales revenue (i.e. pq) and TC_L is the total cost of labour.

If a firm is a monopoly seller of a good, then we effectively have to deal with two functions in order to derive its MRP_L function since total revenue will depend on output, i.e. $TR = f(q)$, and output will depend on labour input, i.e. $q = f(L)$. Therefore, using the chain rule,

$$MRP_L = \frac{dTR}{dL} = \frac{dTR}{dq}\frac{dq}{dL}$$

We already know that

$$\frac{dTR}{dq} = MR \qquad \frac{dq}{dL} = MP_L$$

Therefore,

$$MRP_L = (MR)(MP_L)$$

This is the rule for the profit-maximizing amount of labour to employ which you should encounter in your microeconomics course.

Example 12.3

A firm is a monopoly seller of good q and faces the demand schedule $p = 200 - 2q$, where p is the price in pounds, and the short-run production function $q = 4L^{0.5}$. If it can buy labour at a fixed wage of £8 then how much L should be employed to maximize profit?

Solution

Using the chain rule we need to derive a formula for MRP_L in terms of L and then set it equal to £8, given that MC_L is fixed at this wage rate. As

$$MRP_L = \frac{dTR}{dL} = \frac{dTR}{dq}\frac{dq}{dL}$$

we need to find dTR/dq and dq/dL.

Given $p = 200 - 2q$, then

$$TR = pq = (200 - 2q)q = 200q - 2q^2$$

373

Therefore

$$\frac{d\text{TR}}{dq} = 200 - 4q \tag{1}$$

Given $q = 4L^{0.5}$, then

$$\frac{dq}{dL} = 2L^{-0.5} \tag{2}$$

Thus from (1) and (2)

$$\text{MRP}_L = (200 - 4q)2L^{-0.5}$$

Setting MRP_L equal to the wage rate we get

$$\frac{400 - 8q}{L^{0.5}} = 8$$

$$400 - 8q = 8L^{0.5} \tag{3}$$

Substituting $q = 4L^{0.5}$ into (3), as we are trying to derive a formula in terms of L, we get

$$400 - 8(4L^{0.5}) = 8L^{0.5}$$
$$400 - 32L^{0.5} = 8L^{0.5}$$
$$400 = 40L^{0.5}$$
$$10 = L^{0.5}$$
$$100 = L$$

which is the optimal employment level.

In the example above the idea of a 'short-run production function' was used to simplify the analysis. Now that you understand how an MRP_L function can be derived we can work with full production functions where $\text{MP}_L = \partial Q/\partial L$.

The chain rule is used in exactly the same way in partial differentiation as in the differentiation of single-variable functions.

Example 12.4

A firm operates with the production function

$$q = 45K^{0.7}L^{0.4}$$

and faces the demand function

$$p = 6,980 - 6q$$

Derive its MRP_L function.

Solution

We wish to find $\text{MRP}_L = \partial \text{TR}/\partial L$ assuming K is fixed. We know that

$$\text{TR} = pq = (6{,}980 - 6q)q = 6{,}980q - 6q^2$$

Therefore

$$\frac{d\text{TR}}{dq} = 6{,}980 - 12q \tag{1}$$

From the production function $q = 45K^{0.7}L^{0.4}$ we can derive

$$\text{MP}_L = \frac{\partial q}{\partial L} = 18\,K^{0.7}L^{-0.6} \tag{2}$$

Using the chain rule and substituting (1) and (2)

$$\text{MRP}_L = \frac{\partial \text{TR}}{\partial L} = \frac{d\text{TR}}{dq}\frac{\partial q}{\partial L} = (6{,}980 - 12q)18K^{0.7}L^{-0.6} \tag{3}$$

As we wish to derive MRP_L as a function of L, we substitute the production function given in the question into (3). Thus

$$\text{MRP}_L = [6{,}980 - 12(45K^{0.7}L^{0.4})]18K^{0.7}L^{-0.6}$$

$$= 125{,}640K^{0.7}L^{-0.6} - 9{,}720K^{1.4}L^{-0.2}$$

The chain rule can also be used to calculate point elasticity of demand for some non-linear demand functions.

Example 12.5

Find the point elasticity of demand when $q = 10$ for the demand function $p = (120 - 2q)^{0.5}$.

Solution

Point elasticity is

$$(-1)\frac{p}{q}\frac{1}{dp/dq} \tag{1}$$

Create a new variable z such that $z = 120 - 2q$ and thus $p = z^{0.5}$. Differentiating:

$$\frac{dz}{dq} = -2 \qquad \frac{dp}{dz} = 0.5z^{-0.5}$$

Therefore

$$\frac{dp}{dq} = \frac{dp}{dz}\frac{dz}{dp}$$

$$= 0.5\, z^{-0.5}(-2)$$

$$= 0.5(120 - 2q)^{-0.5}(-2)$$

$$= \frac{-1}{(120 - 2q)^{0.5}}$$

and so

$$\frac{1}{dp/dq} = -(120 - 2q)^{0.5}$$

When $q = 10$, then from the original demand function
$p = (120 - 20)^{0.5} = 100^{0.5} = 10$. Thus, substituting into formula (1), the point
elasticity will be

$$(-1)\frac{10}{10}(-1)(120 - 2q)^{0.5} = (120 - 20)^{0.5} = 100^{0.5} = 10.$$

Sometimes it may be possible to simplify an expression in order to be
able to differentiate it, but one may instead use the chain rule if it is more
convenient. The same result will be obtained by both methods, of course.

Example 12.6

Differentiate the function $y = (6 + 4x)^2$.

Solution

(i) By multiplying out

$$y = (6 + 4x)^2 = 36 + 48x + 16x^2$$

Therefore

$$\frac{dy}{dx} = 48 + 32x$$

(ii) Using the chain rule, let $z = 6 + 4x$ so that $y = z^2$. Thus

$$\frac{dy}{dx} = \frac{dy}{dz}\frac{dz}{dx} = 2z \times 4 = 2(6 + 4x)4 = 48 + 32x$$

QUESTIONS 12.1

1. A firm operates in the short run with the production function $q = 2L^{0.5}$
   and faces the demand schedule $p = 60 - 4q$ where p is price in pounds.
   If it can employ labour at a wage rate of £4 per hour, how much should
   it employ to maximize profit?
2. If a supply schedule is given by $p = (2 + 0.05q)^2$ show (a) by multiplying
   out and (b) by using the chain rule that its slope is 2.2 when q is 400.

3. The return R on a sum M invested at $i\%$ for 3 years is given by the formula

$$R = M(1 + i)^3$$

What is the rate of change of R with respect to i?

4. If $y = (3 + 0.6x^2)^{0.5}$ what is dy/dx?

5. If a firm faces the total cost function $TC = (6 + x)^{0.5}$, what is its marginal cost function?

6. A firm operates with the production function $q = 0.4K^{0.5}L^{0.5}$ and sells its output in a market where it is a monopoly with the demand schedule $p = 60 - 2q$. If K is fixed at 25 units and the wage rate is £7 per unit of L, derive the MRP_L function and work out how much L the firm should employ to maximize profit.

7. A firm faces the demand schedule $p = 650 - 3q$ and the production function $q = 4K^{0.5}L^{0.5}$ and has to pay £8 per unit to buy L. If K is fixed at 4 units how much L should the firm use if it wishes to maximize profits?

8. If a firm operates with the total cost function $TC = 4 + 10(9 + q^2)^{0.5}$, what is its marginal cost when q is 4?

9. Given the production function $q = (6K^{0.5} + 0.5L^{0.5})^{0.5}$, find MP_L when K is 16 and L is 576.

12.3 THE PRODUCT RULE

The product rule allows one to differentiate two functions which are multiplied together.

If $y = uv$ where u and v are functions of x, then according to the product rule

$$\frac{dy}{dx} = u\frac{dv}{dx} + v\frac{du}{dx}$$

As with the chain rule, one may find it convenient to split a single awkward function into two artificial functions even if these functions do not necessarily have any particular economic meaning. The following examples show how this rule can be used.

Example 12.7

If $y = (7.5 + 0.2x^2)(4 + 8x^{-1})$ what is dy/dx?

Solution

This function could in fact be multiplied out and differentiated without using the product rule. However, let us first use the product rule and then we can compare the answers obtained by the two methods. They should, of course, be the same.

We are given the function

$$y = (7.5 + 0.2x^2)(4 + 8x^{-1})$$

so let

$$u = 7.5 + 0.2x^2 \qquad v = 4 + 8x^{-1}$$

Therefore

$$\frac{du}{dx} = 0.4x \qquad \frac{dv}{dx} = -8x^{-2}$$

We can now use the product rule and substitute these functions. Thus

$$\frac{dy}{dx} = u\frac{dv}{dx} + v\frac{du}{dx}$$

$$= (7.5 + 0.2x^2)(-8x^{-2}) + (4 + 8x^{-1})0.4x$$

$$= -60x^{-2} - 1.6 + 1.6x + 3.2$$

$$= 1.6 + 1.6x - 60x^{-2} \tag{1}$$

If the original function is multiplied out we get

$$y = (7.5 + 0.2x^2)(4 + 8x^{-1}) = 30 + 60x^{-1} + 0.8x^2 + 1.6x$$

and so

$$\frac{dy}{dx} = -60x^{-2} + 1.6x + 1.6 \tag{2}$$

The answers (1) and (2) are the same, as we expected.

In many cases it is not possible to multiply out the different components of a function and then one must use the product rule to differentiate. One may also need to use the chain rule to help differentiate the different subfunctions.

Example 12.8

A firm faces the non-linear demand function $p = (650 - 0.25q)^{1.5}$. What output should it sell to maximize total revenue?

Solution

$TR = pq = (650 - 0.25q)^{1.5}q$ when the demand function in the question is substituted for p. To differentiate TR using the product rule let

$$u = (650 - 0.25q)^{1.5} \qquad v = q$$

giving

$$\frac{du}{dq} = 1.5(650 - 0.25q)^{0.5}(-0.25) = -0.375(650 - 0.25q)^{0.5}$$

using the chain rule and

$$\frac{dv}{dq} = 1$$

Therefore,

$$\frac{dTR}{dq} = u\frac{dv}{dq} + v\frac{du}{dq}$$

$$= (650 - 0.25q)^{1.5} + q(-0.375)(650 - 0.25q)^{0.5}$$

$$= (650 - 0.25q)^{0.5}(650 - 0.25q - 0.375q)$$

$$= (650 - 0.25q)^{0.5}(650 - 0.625q) \qquad (1)$$

For a stationary point

$$\frac{dTR}{dq} = (650 - 0.25q)^{0.5}(650 - 0.625q) = 0$$

Therefore, either $650 - 0.25q = 0$ or $650 - 0.625q = 0$,
i.e. $\qquad\qquad\qquad 2{,}600 = q \qquad$ or $\qquad 1{,}040 = q$.
We now need to check which of these values of q satisfies the second-order condition for a maximum. (You should immediately be able to see why it will not be 2,600 by observing what happens when this quantity is substituted into the demand function.)

To derive d^2TR/dq^2 we need to use the product rule again to differentiate dTR/dq. From (1) above

$$\frac{dTR}{dq} = (650 - 0.25q)^{0.5}(650 - 0.625q)$$

Let $u = (650 - 0.25q)^{0.5}$ and $v = 650 - 0.625q$ giving

$$\frac{du}{dq} = 0.5(650 - 0.25q)^{-0.5}(-0.25) = -0.125(650 - 0.25q)^{-0.5}$$

using the chain rule and

$$\frac{dv}{dq} = -0.625$$

Therefore,

$$\frac{d^2TR}{dq^2} = u\frac{dv}{dq} + v\frac{du}{dq}$$

$$= (650 - 0.25q)^{0.5}(-0.625) + (650 - 0.625q)(-0.125)(650 \cdots 0.25q)^{-0.5}$$

$$= \frac{(650 - 0.25q)(-0.625) + (650 - 0.625q)(-0.125)}{(650 - 0.25q)^{0.5}} \qquad (2)$$

Substituting $q = 1,040$ into (2) gives

$$\frac{d^2 TR}{dq^2} = \frac{(390)(-0.625) + 0}{390^{0.5}} = -12.34 < 0$$

Therefore, TR is maximized when $q = 1,040$.

We can double check that the other stationary point will not maximize TR by substituting $q = 2,600$ into (2) giving

$$\frac{d^2 TR}{dq^2} = \frac{0 + (-975)(-0.125)}{0} \rightarrow +\infty$$

This second value for q obviously does not satisfy second-order conditions for a maximum.

Example 12.9

When is the function $Q = 12 K^{0.4}(160 - 8K)^{0.4}$ at a maximum? (This is Example 11.1 (reworked) which was not completed in the last chapter.)

Solution

The first-order condition for a maximum is $dQ/dK = 0$. To differentiate $Q = 12K^{0.4}(160 - 8K)^{0.4}$ using the product rule let

$$u = 12K^{0.4} \text{ and } v = (160 - 8K)^{0.4} \text{ giving}$$

$$du/dK = 4.8K^{-0.6} \text{ and}$$

$$\frac{dv}{dK} = 0.4(160 - 8K)^{-0.6}(-8)$$

$$= -3.2(160 - 8K)^{-0.6}$$

Therefore,

$$\frac{dQ}{dK} = 12K^{0.4}(-3.2)(160 - 8K)^{-0.6} + (160 - 8K)^{0.4}4.8K^{-0.6}$$

$$= \frac{-38.4K + (160 - 8K)4.8}{(160 - 8K)^{0.6}K^{0.6}}$$

$$= \frac{768 - 76.8K}{(160 - 8K)^{0.6}K^{0.6}} \tag{1}$$

Setting (1) equal to zero for a stationary point must mean $768 - 76.8K = 0$, giving $K = 10$.

As we have already left this example in mid-solution once already, it will not do any harm to leave it once again. Although the second-order condition could be worked out using the product rule it is more convenient to use the quotient rule in this case and so we shall continue this problem in Example 12.13.

Example 12.10

In a perfectly competitive market the demand schedule is $p = 120 - 0.5q^2$ and the supply schedule is $p = 20 + 2q^2$. If the government imposes a per-unit tax t on the good sold in this market, what level of t will maximize the government's tax yield?

Solution

With the tax the supply schedule shifts upwards and becomes $p = 20 + 2q^2 + t$. In equilibrium demand price equals supply price. Therefore

$$120 - 0.5q^2 = 20 + 2q^2 + t$$

$$100 - t = 2.5q^2$$

$$40 - 0.4t = q^2$$

$$(40 - 0.4t)^{0.5} = q \qquad (1)$$

The government's tax yield (TY) is tq. Substituting (1) gives

$$TY = t(40 - 0.4t)^{0.5} \qquad (2)$$

We need to set $dTY/dt = 0$ for the first-order condition for a maximum.
From (2) let $u = t$ and $v = (40 - 0.4t)^{0.5}$ giving

$$\frac{du}{dt} = 1$$

$$\frac{dv}{dt} = 0.5(40 - 0.4t)^{-0.5}(-0.4)$$

$$= -0.2(40 - 0.4)^{-0.5}$$

Therefore, using the product rule

$$\frac{dTY}{dt} = t(-0.2)(40 - 0.4t)^{-0.5} + (40 - 0.4t)^{0.5}$$

$$= \frac{-0.2t + 40 - 0.4t}{(40 - 0.4t)^{0.5}}$$

$$= \frac{40 - 0.6t}{(40 - 0.4t)^{0.5}} = 0 \qquad (3)$$

From (3) we can see that, for finite values of t, $dTY/dt = 0$ when $40 - 0.6t = 0$, i.e. $66.67 = t$.

To check second-order conditions for this stationary point we need to find d^2TY/dt^2. From (3) we can write

$$\frac{dTY}{dt} = (40 - 0.6t)(40 - 0.4t)^{-0.5}$$

Let $u = 40 - 0.6t$ and $v = (40 - 0.4t)^{-0.5}$ giving

$$\frac{du}{dt} = -0.6$$

$$\frac{dv}{dt} = -0.5(40 - 0.4t)^{-1.5}(-0.4)$$

Therefore, using the product rule,

$$\frac{d^2 TY}{dt^2} = (40 - 0.6t)[0.2(40 - 0.4t)^{-1.5}] + (40 - 0.4t)^{-0.5}(-0.6) \qquad (4)$$

When $t = 66.67$ then $40 - 0.6t = 0$ and so the first term in (4) disappears giving

$$\frac{d^2 TY}{dt^2} = [40 - 0.4(66.67)]^{-0.5}(-0.6) = -0.1644 < 0$$

Therefore, the second-order condition for a maximum is satisfied when $t = 66.67$. Maximum tax revenue is raised when the per-unit tax is £66.67.

QUESTIONS 12.2

1. If $y = (6x + 7)^{0.5}(2.6x^2 - 1.9)$, what is dy/dx?
2. What output will maximize total revenue given the non-linear demand schedule $p = (60 - 2q)^{1.5}$?
3. Derive a function for the marginal product of L given the production function

$$Q = 85(0.5K^{0.8} + 3L^{0.5})^{0.6}$$

4. If $Q = 120K^{0.5}(250 - 0.5K)^{0.3}$ at what value of K will $dQ/dK = 0$? (That is, find the first-order condition for maximization of Q.)
5. In a perfectly competitive market the demand schedule is $p = 600 - 4q^{0.5}$ and the supply schedule is $p = 30 + 6q^{0.5}$. What level of a per-unit tax levied on the good sold in this market will maximize the government's tax yield?
6. Make up your own function involving the product of two subfunctions and then differentiate it using the product rule.
7. For the demand function $p = (60 - 0.1q)^{0.5}$:

   (a) derive an expression for the slope of the demand schedule;
   (b) demonstrate that this slope gets flatter as q increases from 0 to 600;
   (c) find the output at which total revenue is a maximum.

12.4 THE QUOTIENT RULE

The quotient rule allows one to differentiate two functions where one function is divided by the other function.

Further topics in calculus

If $y = u/v$ where u and v are functions of x, then according to the quotient rule

$$\frac{dy}{dx} = \frac{v\frac{du}{dx} - u\frac{dv}{dx}}{v^2}$$

Example 12.11

What is dy/dx if

$$y = \frac{4x^2}{8 + 0.2x}?$$

Solution

Let $u = 4x^2$ and $v = 8 + 0.2x$ giving

$$\frac{du}{dx} = 8x \qquad \frac{dv}{dx} = 0.2$$

Therefore, according to the quotient rule,

$$\frac{dy}{dx} = \frac{v\,du/dx - u\,dv/dx}{v^2}$$

$$= \frac{(8 + 0.2x)8x - 4x^2(0.2)}{(8 + 0.2x)^2}$$

$$= \frac{64x + 1.6x^2 - 0.8x^2}{(8 + 0.2x)^2}$$

$$= \frac{64x + 0.8x^2}{(8 + 0.2x)^2} \qquad (1)$$

This solution could actually have been found using the product rule, since any function in the form $y = u/v$ can be written as $y = uv^{-1}$. We can check this by reworking Example 12.11 and differentiating the function $y = 4x^2(8 + 0.2x)^{-1}$.

Let $u = 4x^2$ and $v = (8 + 0.2x)^{-1}$ giving

$$\frac{du}{dx} = 8x \qquad \frac{dv}{dx} = -0.2(8 + 0.2x)^{-2}$$

Using the product rule

$$\frac{dy}{dx} = u\frac{dv}{dx} + v\frac{du}{dx}$$

$$= 4x^2[-0.2(8 + 0.2x)^{-2}] + (8 + 0.2x)^{-1}8x$$

383

$$= \frac{-0.8x^2 + (8 + 0.2x)8x}{(8 + 0.2x)^2}$$

$$= \frac{-0.8x^2 + 64x + 1.6x^2}{(8 + 0.2x)^2}$$

$$= \frac{64x + 0.8x^2}{(8 + 0.2x)^2} \qquad (2)$$

The answers (1) and (2) are identical, as expected. Whether one uses the quotient rule or the product rule depends on the functions to be differentiated. Only practice will give you an idea of which will be the easier to use for specific examples.

Example 12.12

If a monopoly faces the non-linear demand function

$$p = \frac{252}{(4 + q)^{0.5}}$$

derive a function for marginal revenue.

Solution

$$TR = pq = \frac{252q}{(4 + q)^{0.5}}$$

Let $u = 252q$ and $v = (4 + q)^{0.5}$ and so

$$\frac{du}{dq} = 252 \qquad \frac{dv}{dq} = 0.5(4 + q)^{-0.5}$$

We need to find $MR = dTR/dq$. Using the quotient rule

$$\frac{dTR}{dq} = \frac{v\,du/dq - u\,dv/dq}{v^2}$$

$$= \frac{(4 + q)^{0.5}252 - 252q(0.5)(4 + q)^{-0.5}}{4 + q}$$

$$= \frac{(4 + q)252 - 126q}{(4 + q)^{1.5}}$$

$$= \frac{1{,}008 + 252q - 126q}{(4 + q)^{1.5}}$$

Thus

$$MR = \frac{1{,}008 + 126q}{(4 + q)^{1.5}}$$

Further topics in calculus

Note that, in this example, MR only becomes zero when q becomes infinitely large, as negative quantities are not possible. TR will therefore rise continually as q increases.

All three rules may be used in some problems. In particular, one may find it convenient to use the chain rule and the product rule to derive the first-order condition in an optimization problem and then use the quotient rule to check the second-order condition. If we return to the unfinished Example 12.9 we can now see how the quotient rule can be used to check the second-order condition.

Example 12.13

In Example 12.9 we were trying to find the value of K which maximized the function $Q = 12K^{0.4}(160 - 8K)^{0.4}$. First-order conditions were satisfied when

$$\frac{dQ}{dK} = \frac{768 - 76.8K}{(160 - 8K)^{0.6}K^{0.6}} = 0$$

giving $K = 10$. To derive d^2Q/dK^2 let $u = 768 - 76.8K$ and $v = (160 - 8K)^{0.6}K^{0.6}$. Therefore,

$$\frac{du}{dK} = -76.8 \tag{1}$$

and, using the product rule,

$$\frac{dv}{dK} = (160 - 8K)^{0.6}0.6K^{-0.4} + K^{0.6}0.6(160 - 8K)^{-0.4}(-8)$$

$$= \frac{(160 - 8K)0.6 - 4.8K}{K^{0.4}(160 - 8K)^{0.4}}$$

$$= \frac{96 - 9.6K}{K^{0.4}(160 - 8K)^{0.4}} \tag{2}$$

Using the quotient rule and (1) and (2)

$$\frac{d^2Q}{dK^2} = \frac{(160 - 8K)^{0.6}K^{0.6}(-76.8) - (768 - 76.8K)\dfrac{96 - 9.6K}{K^{0.4}(160 - 8K)^{0.4}}}{(160 - 8K)^{1.2}K^{1.2}}$$

$$= \frac{(160 - 8K)K(-76.8) - 76.8(10 - K)9.6(10 - K)}{(160 - 8K)^{1.2}K^{1.2}}$$

When $K = 10$ then

$$\frac{d^2Q}{dK^2} = \frac{-76.8(800)}{(800)^{1.6}} < 0$$

Therefore, the second-order condition for a maximum is satisfied when $K = 10$.

Basic mathematics for economists

In your introductory economics course you were probably given an intuitive geometrical explanation of why a marginal cost schedule cuts a U-shaped average cost curve at its minimum point. The quotient rule can now be used to prove this rule.

In the short run, with only one variable input, assume that total cost (TC) is a function of q.

Thus, $MC = dTC/dq$ (as explained in Chapter 8) and, by definition, $AC = TC/q$. To differentiate AC using the quotient rule let $u = TC$ and $v = q$ giving

$$\frac{du}{dq} = \frac{dTC}{dq} = MC \qquad \frac{dv}{dq} = 1$$

Therefore

$$\frac{dAC}{dq} = \frac{qMC - TC}{q^2} \qquad (1)$$

Setting (1) equal to zero to find the value of q that satisfies first-order conditions for a maximum we get $qMC - TC = 0$ (or $q \to \infty$, which we disregard), giving

$$MC = \frac{TC}{q} \qquad (2)$$

But $TC/q = AC$ by definition. Therefore, $MC = AC$ when AC is at a stationary point.

To check second-order conditions we need to find d^2AC/dq^2. From (1) above we know that

$$\frac{dAC}{dq} = \frac{qMC - TC}{q^2}$$

Again use the quotient rule and let $u = qMC - TC$ and $v = q^2$ giving

$$\frac{du}{dq} = \left(q\frac{dMC}{dq} + MC\right) - MC = q\frac{dMC}{dq}$$

and $dv/dq = 2q$. Therefore

$$\frac{d^2AC}{dq^2} = \frac{q^2\left(q\frac{dMC}{dq}\right) - (qMC - TC)2q}{q^4}$$

$$= \frac{q^3(2q\,dMC/dq - 2MC + 2q^{-1}TC)}{q^4}$$

$$= \frac{q\,dMC/dq - 2MC + 2q^{-1}TC}{q^2}$$

$$= \frac{q\,dMC/dq + 2(q^{-1}TC - MC)}{q^2} \qquad (3)$$

386

Further topics in calculus

The first-order condition for a minimum is satisfied when $MC = TC/q$, from (2) above. Substituting this result into (3) we get

$$\frac{d^2 AC}{dq^2} = \frac{1}{q^2}\left[q\frac{dMC}{dq} + 2\left(\frac{TC}{q} - \frac{TC}{q}\right)\right]$$

$$= \frac{1}{q}\frac{dMC}{dq} > 0 \qquad \text{when} \qquad \frac{dMC}{dq} > 0$$

i.e. when MC is rising.

Therefore, the second-order condition for a minimum is satisfied when $MC = AC$ and MC is rising. Thus, although MC may cut AC at another point when MC is falling, when MC is rising it cuts AC at its minimum point.

Individual labour supply

Not all of you will have encountered the theory of individual labour supply. Nevertheless you should now be able to understand the following example which shows how the utility-maximizing combination of work and leisure hours can be found when an individual's utility function, wage rate and maximum working day are specified.

Example 12.14

In the theory of individual labour supply it is assumed that an individual derives utility from both leisure (L) and income (I). Income is determined by hours of work (H) multiplied by the hourly wage rate (w), i.e. $I = wH$.

Assume that each day a total of 12 hours is available for an individual to split between leisure and work, the wage rate is given as £4 an hour and that the individual's utility function is $U = L^{0.5}I^{0.75}$. How will this individual balance leisure and income so as to maximize utility?

Solution

$H = 12 - L$ given a maximum working day of 12 hours. Therefore, given an hourly wage of £4,

$$I = wH = w(12 - L) = 4(12 - L) = 48 - 4L \tag{1}$$

Substituting (1) into the utility function

$$U = L^{0.5}I^{0.75} = L^{0.5}(48 - 4L)^{0.75} \tag{2}$$

To differentiate U using the product rule let $u = L^{0.5}$ and $v = (48 - 4L)^{0.75}$, giving $du/dL = 0.5L^{-0.5}$ and

$$\frac{dv}{dL} = 0.75(48 - 4L)^{-0.25}(-4)$$

$$= -3(48 - 4L)^{-0.25}$$

Therefore

$$\frac{dU}{dL} = L^{0.5}[-3(48 - 4L)^{-0.25}] + (48 - 4L)^{0.75}(0.5L^{-0.5})$$

$$= \frac{-3L + (48 - 4L)0.5}{(48 - 4L)^{0.25}L^{0.5}}$$

$$= \frac{24 - 5L}{(48 - 4L)^{0.25}L^{0.5}} = 0$$

for a stationary point. Therefore

$$24 - 5L = 0$$
$$24 = 5L$$
$$4.8 = L$$

and

$$H = 12 - 4.8 = 7.2 \text{ hours}$$

To check the second-order condition we need to differentiate

$$\frac{dU}{dL} = \frac{24 - 5L}{(48 - 4L)^{0.25}L^{0.5}}$$

to get d^2U/dL^2. Let $u = 24 - 5L$ and $v = (48 - 4L)^{0.25}L^{0.5}$, giving $du/dL = -5$ and

$$\frac{dv}{dL} = (48 - 4L)^{0.25}0.5L^{-0.5} + L^{0.5}0.25(48 - 4L)^{-0.75}(-4)$$

$$= \frac{(48 - 4L)0.5 - L}{L^{0.5}(48 - 4L)^{0.75}}$$

$$= \frac{24 - 3L - L}{L^{0.5}(48 - 4L)^{0.75}}$$

Therefore, using the quotient rule,

$$\frac{d^2U}{dL^2} = \frac{(48 - 4L)^{0.25}L^{0.5}(-5) - (24 - 5L)[(24 - 3L)/L^{0.5}(48 - 4L)^{0.75}]}{(48 - 4L)^{0.5}L}$$

When $L = 4.8$ then $24 - 5L = 0$ and so the second part of the numerator disappears. Dividing through by $(48 - 4L)^{0.25}L^{0.5}$ we therefore get

$$\frac{d^2U}{dL^2} = \frac{-5}{(48 - 4L)^{0.25}L^{0.5}} = -0.985 < 0$$

Further topics in calculus

and so the second-order condition for a maximum is satisfied.

Thus, this individual will maximize utility when 7.2 hours are worked and 4.8 hours are taken as leisure.

QUESTIONS 12.3

1. If

$$y = \frac{(3x + 0.4x^2)}{(8 - 6x^{1.5})^{0.5}}$$

what is dy/dx?

2. Derive a function for marginal revenue for the demand schedule

$$p = \frac{720}{(25 + q)^{0.5}}$$

3. Using your answer from Questions 12.2, no. 4, show that the second-order condition for a maximum value of the function $Q = 120K^{0.5}(250 - 0.5K)^{0.3}$ is satisfied when K is 312.5, and evaluate d^2Q/dK^2.

4. For the demand schedule $p = (800 - 0.4q)^{0.5}$ find which value of q will maximize total revenue, using the quotient rule to check the second-order condition.

5. Assume that an individual can choose the number of hours per day worked up to a maximum of 12 hours. This individual attempts to maximize utility which is derived from income (I) and leisure (L) according to the utility function $U = L^{0.4}I^{0.6}$. L is defined as hours not worked out of the 12 hour maximum working day, and I is defined as hours worked (H) multiplied by the hourly wage rate of £15. What mix of leisure and work will be chosen?

6. Show that when a firm faces a U-shaped short-run average variable cost (AVC) schedule, its marginal cost schedule will always cut the AVC schedule at its minimum point when MC is rising.

12.5 INTEGRATION

Integrating a function means finding another function which, when it is differentiated, gives the first function. It is basically differentiation in reverse, and the rules for integration are the reverse of those for differentiation. Unlike differentiation, which we have seen to be very useful in optimization problems, the mathematical technique of integration is not as widely used in economics and so we shall only look at some of the basic ideas involved.

Assume that you wish to integrate the function

$$f'(x) = 12x + 24x^2$$

This means that you wish to find a function $y = f(x)$ such that

$$\frac{dy}{dx} = f'(x) = 12x + 24x^2$$

From your knowledge of differentiation you should be able to work out that if $y = 6x^2 + 8x^3$ then

$$\frac{dy}{dx} = 12x + 24x^2$$

However, the same derivative can be obtained from other functions. For example, if $y = 35 + 6x^2 + 8x^3$ then

$$\frac{dy}{dx} = 12x + 24x^2$$

and if $y = -7.5 + 6x^2 + 8x^3$ then

$$\frac{dy}{dx} = 12x + 24x^2$$

Because constant numbers disappear when a function is differentiated, we cannot know what constant should appear in an integrated function unless further information is available. We therefore simply include a 'constant of integration', denoted here by C, in the 'integral', which is the integrated version of a function. Thus the integral of $12x + 24x^2$ will be $6x^2 + 8x^3 + C$.

The usual sign for integration is

$$y = \int f'(x)\, dx$$

This means that y is the integral of the function $f'(x)$. The sign $\int$ is known as the integration sign. The 'dx' signifies that y will equal $f'(x)$ when it is differentiated with respect to x. We can therefore write

$$y = \int (12x + 24x^2)\, dx = 6x^2 + 8x^3 + C$$

The general rule for the integration of individual terms in an expression is

$$\int ax^n\, dx = \frac{ax^{n+1}}{n+1} + C$$

where a and n are parameters and $n \neq -1$. As this is the reverse of the rule for differentiation you should have no problems in seeing how the answers in the example below are derived.

Example 12.15

Find the following integrals: Solutions:

(i) $\displaystyle\int 30x^4\,dx$ $y = 6x^5 + C$

(ii) $\displaystyle\int (24 + 7.2x)\,dx$ $y = 24x + 3.6x^2 + C$

(iii) $\displaystyle\int 0.5x^{-0.5}\,dx$ $y = x^{0.5} + C$

(iv) $\displaystyle\int (48x - 0.4x^{-1.4})\,dx$ $y = 24x^2 + x^{-0.4} + C$

(v) $\displaystyle\int (65 + 1.5x^{-2.5} + 1.5x^2)\,dx$ $y = 65x - x^{-1.5} + 0.5x^3 + C$

There are rules for integrating more complex functions which are based on the chain, product and quotient rules for differentiation. However, they are awkward to use and will not be much use to you at present and so they are not covered in this text. The special case when $n = -1$, i.e. the integral $\int (1/x)\,dx$, will be dealt with in Chapter 14 when we look at exponential functions.

In earlier chapters we have seen how differentiation of total cost, total revenue and other functions gives the corresponding marginal function. For example,

$$\frac{dTC}{dq} = MC \qquad \frac{dTR}{dq} = MR$$

using the usual terminology. Therefore the integration of the marginal function will give the corresponding total function, apart from the unknown constant.

Total cost functions can usually be split into fixed and variable components. The integral of marginal cost without a zero constant of integration will give total variable costs. For example, if we are given the information that total variable cost is

$$TVC = 25q - 6q^2 + 0.8q^3$$

and total fixed cost is

$$TFC = 10$$

then, by definition,

$$TC = TVC + TFC$$

$$= 10 + 25q - 6q^2 + 0.8q^3$$

Thus, marginal cost will be

$$MC = \frac{dTC}{dq} = 25 - 12q + 2.4q^2$$

On the other hand, if we were given the information that total fixed cost was 10 and that

$$MC = 25 - 12q + 2.4q^2$$

then we could find total variable cost by integration as

$$TVC = \int MC \, dq = 25q - 6q^2 + 0.8q^3$$

Thus

$$TC = TFC + TVC = 10 + 25q - 6q^2 + 0.8q^3$$

Example 12.16

If a firm spends £650 on fixed costs and its marginal cost function is $MC = 82 - 16q + 1.8q^2$, what is its total cost function?

Solution

We know that for any cost function

$$\int MC \, dq = TVC$$

Therefore

$$\int (82 - 16q + 1.8q^2) dq = TVC$$

$$82q - 8q^2 + 0.6q^3 = TVC$$

We are told that $TFC = 650$ and so

$$TC = TFC + TVC = 650 + 82q - 8q^2 + 0.6q^3$$

If one is given a firm's marginal revenue function then one can integrate this to find the total revenue function. For example, if $MR = 360 - 2.5q$ then

$$\int MR \, dq = 360q - 1.25q^2 + C$$

When q is zero, TR must also be zero. Thus there is no constant term in a TR function and so in the above case we can say $C = 0$ and

$$\int MR \, dq = 360q - 1.25q^2 = TR \qquad (1)$$

Further topics in calculus

This result, that the constant of integration is zero, holds for all MR schedules. We can easily check the answer in the above example as we know that, for a linear demand schedule, the corresponding MR schedule will have the same intercept on the price axis but twice the slope.

Thus, given that

$$MR = 360 - 2.5q$$

the demand schedule is

$$p = 360 - 1.25q$$

and

$$TR = pq = (360 - 1.25q)q = 360q - 1.25q^2$$

This checks out with the solution (1) above.

With non-linear demand and MR functions this short-cut method cannot be used and one has to use integration to derive TR.

Example 12.17

If $MR = 520 - 3q^{0.5}$ what is the corresponding TR function?

Solution

$$TR = \int MR\, dq = \int (520 - 3q^{0.5})dq = 520q - 2q^{1.5}$$

Once the TR function corresponding to a given MR function has been derived then one simply has to divide this through by q to arrive at the demand function.

Example 12.18

If a firm's marginal revenue function is

$$MR = 960 - 0.15q^2$$

what total revenue will it earn if it charges a price of £715?

Solution

$$TR = \int MR\, dq = \int (960 - 0.15q^2)dq = 960q - 0.05q^3$$

In this example we need to use the TR function to find the price. Since $TR = pq$ then $p = TR/q$ and so

393

$$p = \frac{1}{q}(960q - 0.05q^3) = 960 - 0.05q^2$$

$$0.05q^2 = 960 - p$$

$$q^2 = 19{,}200 - 20p$$

$$q = (19{,}200 - 20p)^{0.5}$$

When $p = 715$ then

$$q = (19{,}200 - 14{,}300)^{0.5} = 4{,}900^{0.5} = 70$$

Thus, maximum total revenue will be

$$pq = 715(70) = £50{,}050$$

If both MC and MR functions are specified then one can use integration to work out what the actual profit is at any given output, provided that TFC is specified.

Example 12.19

If a firm faces the marginal cost schedule

$$MC = 180 + 0.3q^2$$

and the marginal revenue schedule

$$MR = 540 - 0.6q^{1.5}$$

and total fixed costs are £65, what is the maximum profit it can make?

Solution

Profit is maximized when MC = MR. Therefore,

$$180 + 0.3q^2 = 540 - 0.6q^2$$

$$0.9q^2 = 360$$

$$q^2 = 400$$

$$q = 20$$

To find the actual profit (π), we now integrate to get TR and TC and then subtract TC from TR:

$$TR = \int MR\,dq = \int (540 - 0.6q^2)dq = 540q - 0.2q^3$$

$$TC = \int MC\,dq + TFC = \int (180 + 0.3q^2)dq + 65 = 180q + 0.1q^3 + 65$$

$$\pi = TR - TC$$

$$= 540q - 0.2q^3 - (180q + 0.1q^3 + 65)$$
$$= 540q - 0.2q^3 - 180q - 0.1q^3 - 65$$
$$= 360q - 0.3q^3 - 65 \tag{1}$$

Thus when $q = 20$ the maximum profit is

$$\pi = 360(20) - 0.3(20)^3 - 65 = £4{,}735$$

QUESTIONS 12.4

1. Find the integrals for the following functions:

(a) $25x$

(b) $5 + 1.2x + 0.15x^2$

(c) $120x^4 - 60x^3$

(d) $42 - 18x^{-2}$

(e) $90x^{0.5} - 44x^{-1.2}$

2. Find the total variable cost functions corresponding to the following marginal cost functions:

(a) $MC = 4 + 0.1q$

(b) $MC = 42 - 18q + 6q^2$

(c) $MC = 35 + 0.9q^2$

(d) $MC = 62 - 16q + 1.5q^2$

(e) $MC = 185 - 24q + 1.2q^3$

12.6 DEFINITE INTEGRALS

The integrals we have looked at so far are called 'indefinite integrals'. Another form of integral is the 'definite integral'. This is specified with two values of the independent variable and is defined as the value of the integral at one value minus its value at the other value.

For example, the definite integral $\int_3^8 6x^2\,dx$ is the value of this integral when x is 8 minus its value when x is 3. Given that

$$\int 6x^2\,dx = 2x^3 + C$$

then

$$\int_3^8 = [2(8)^3 + C] - [2(3)^3 + C]$$
$$= 1024 + C - 54 - C = 970$$

In any definite integral the two constants of integration will always cancel out, as in the example above. The usual notation used to evaluate a definite integral is

$$\int_3^8 6x^2\,dx = [2x^3]_3^8 = 2(8)^3 - 2(3)^3 = 1{,}024 - 54 = 970$$

The same procedure is used for more complex functions.

Example 12.20

Evaluate the definite integral

$$\int_{5}^{6} (6x^{0.5} - 3x^{-2} + 85x^{4})\, dx$$

Solution

$$\int_{5}^{6} (6x^{0.5} - 3x^{-2} + 85x^{4})\, dx = [4x^{1.5} + 3x^{-1} + 17x^{5}]_{5}^{6}$$

$$= (58.787752 + 0.5 + 132,192) - (44.72136 + 0.6 + 53,125)$$

$$= 132,251.29 - 53,170.32$$

$$= 79,080.97$$

An important feature of definite integrals is that they are equal to the area between a function and the horizontal axis and between the two specified values of the independent variable. For example, assume that you wished to find the area between $x = 1$ and $x = 3$ under the function $y = 20 + 4x$ which is illustrated in Figure 12.1, i.e. the area BFDA. This would be equal to the definite integral

$$\int_{1}^{3} (20 + 4x)\, dx = [20x + 2x^{2}]_{1}^{3}$$

$$= (60 + 18) - (20 + 2)$$

$$= 78 - 22 = 56$$

(Although the scale on the x axis in Figure 12.1 is elongated for clarity, if the same unit of measurement is used for both x and y then this area is measured in abstract 'square units'.)

For this example of a linear function this answer can be checked using basic geometry. The area of the rectangle ABCD is

$$2 \times 24 = 48$$

The area of the triangle BFC is

$$\tfrac{1}{2}\text{ area EBCF} = \tfrac{1}{2}(2 \times 8) = 8$$

$$\text{total area} = 56$$

This concept of the definite integral has several applications in economics.

To evaluate TVC from an MC function for a given value of output one simply substitutes the given quantity into the TVC function. This is the same as evaluating the definite integral between zero and the given quantity. For example, assume that you wished to find the value of TVC when $q = 8$ and you are given the function $MC = 7.5 + 0.3q^{2}$. This value would be

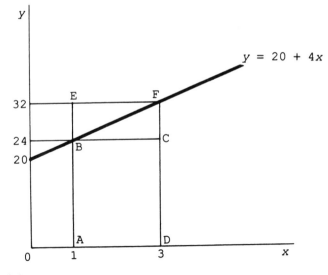

Figure 12.1

$$\int_0^8 (7.5 + 0.3q^2)\,dq = [7.5q + 0.1q^3]_0^8 = 60 + 51.2 = 111.2$$

Therefore, TVC is equal to the area under the MC schedule between zero and the given quantity.

We can also see that the increase in TVC between two quantities will be equal to the area under the corresponding MC schedule between the given quantities.

Assume that marginal cost is the function $MC = 20 + 4x$, where x is output and cost is in pounds (Figure 12.1), and you wish to determine the increase in TVC when output is increased from 1 to 3 units. This will be the area BFDA which we have already found to be 56 'square units', of £56 when the function represents MC.

This must be so because this area represents the definite integral $[20x + 2x^2]_1^3$ which is the value of TVC when quantity is 3 minus its value when quantity is 1.

The definite integral of a function between two given quantities has been shown to be equal to the area under the function between the two quantities but above the horizontal axis. If a function takes negative values, i.e. it goes below the horizontal axis, then areas below the axis but above the function are treated as 'negative' areas.

This phenomenon is illustrated in Figure 12.2 which shows the linear demand schedule $p = 60 - 2q$ and the linear marginal revenue schedule $MR = 60 - 4q$. The corresponding total revenue schedule $TR = 60q - 2q^2$ is shown in the lower part of the diagram.

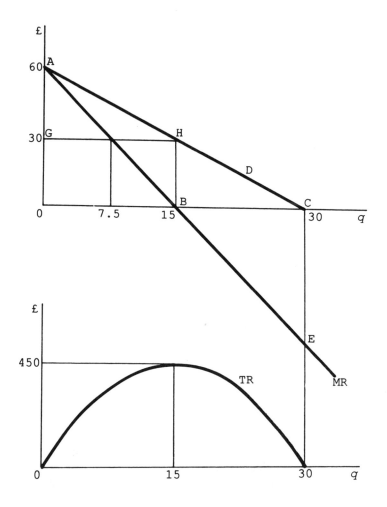

Figure 12.2

Total revenue is at its maximum when

$$MR = 60 - 4q = 0$$

$$q = 15$$

Therefore,

$$p = 60 - 2(15) = 30$$

and so the maximum value of TR is $pq = 450$. Given the linear demand and marginal revenue schedules we can see that TR rises from 0 to 450 when q

398

increases from 0 to 15 and then falls back again to zero when q increases from 15 to 30.

These changes in TR correspond to the values of the definite integrals over these quantity ranges and are represented by the area between the MR schedule and the quantity axis.

When q is 15, TR will be equal to the area 0AB which is

$$\int_0^{15} \text{MR} \, dq = \int_0^{15} (60 - 4q)dq = [60q - 2q^2]_0^{15} = 900 - 450 = 450$$

The change in TR when q increases from 15 to 30 will be the 'negative' area BCE which lies above the MR schedule and below the quantity axis. This will be equal to

$$\int_{15}^{30} \text{MR} \, dq = \int_{15}^{30} (60 - 4q)dq$$
$$= [60q - 2q^2]_{15}^{30}$$
$$= (1,800 - 1,800) - (900 - 450) = -450$$

This checks with our initial assessment. Total revenue rises by 450 and then falls by the same amount.

Finally, let us see what happens when we look at the definite integral of the MR function over the entire output range 0–30. This will be

$$\int_0^{30} \text{MR} \, dq = \int_0^{30} (60 - 4q)dq = [60q - 2q^2]_0^{30} = 1,800 - 1,800 = 0$$

The negative area BCE has exactly cancelled out the positive area 0AB, giving zero TR when q is 30, which is correct.

We can also use the demand schedule in Figure 12.2 to determine *consumer surplus*. This is defined as the area below a demand schedule but above the ruling price. This difference between what consumers are willing to pay for a good and what they actually have to pay is often used as a measure of welfare and both its uses and its shortcomings should be explained in your economics course.

Returning to Figure 12.2, if price were zero then consumer surplus would be the entire area under the demand schedule, the triangle 0AC. Geometrically this area can be calculated as

$$\tfrac{1}{2} (\text{height} \times \text{base}) = \tfrac{1}{2} (60 \times 30) = 900$$

Using the definite integral of the demand function the area will be

$$\int_0^{30} (60 - 2q)dq = [60q - q^2]_0^{30} = 1,800 - 900 = 900$$

Both answers are, of course, the same.

If price is 30 then consumer surplus is the area AHG. The corresponding quantity is 15 and so the area AHB0 is equal to the definite integral

$$\int_0^{15} (60-2q)dq = [60q-q^2]_0^{15} = 900-225 = 675$$

Thus $AHG = AHB0 - GHB0 = 675 - 450 = 225$

The above examples of linear functions were used so that the integration method of finding areas under functions could be easily compared with the geometric solutions. The same principles can also be applied to a non-linear function.

Example 12.21

For the non-linear demand function $p = 1,800 - 0.6q^2$ and the corresponding marginal revenue function $MR = 1,800 - 1.8q^2$, use definite integrals to find:

(i) TR when q is 10;
(ii) the change in TR when q increases from 10 to 20;
(iii) consumer surplus when q is 10;

Solution

(i) TR when q is 10 will be

$$\int_0^{10} MR\, dq = \int_0^{10} (1,800 - 1.8q^2)dq$$
$$= [1,800q - 0.6q^3]_0^{10}$$
$$= 18,000 - 600 = £17,400$$

(ii) The change in TR when q increases from 10 to 20 will be

$$\int_{10}^{20} MR\, dq = \int_{10}^{20} (1,800 - 1.8q^2)dq$$
$$= [1,800q - 0.6q^3]_{10}^{20}$$
$$= (36,000 - 4,800) - (18,000 - 600) = £13,800$$

(iii) Consumer surplus when q is 10 will be the definite integral of the demand function minus total revenue actually spent

$$\int_0^{10} (1,800 - 0.6q^2)dq = [1,800q - 0.2q^3]_0^{10} = 18,000 - 200 = £17,800$$

$TR = pq = 1,800q - 0.6q^3 = 18,000 - 600 = £17,400$
Thus consumer surplus $= £400$

QUESTIONS 12.5

1. Given the non-linear demand schedule $p = 600 - 6q^{0.5}$ and the corresponding marginal revenue function $MR = 600 - 9q^{0.5}$, use definite integrals to find:

   (a) total revenue when q is 2,500;
   (b) the change in total revenue when q increases from 2,025 to 2,500;
   (c) consumer surplus when q is 2,500 and price is £300;
   (d) the change in consumer surplus when q increases from 2,025 to 2,500 owing to a price fall from £330 to £300.

2. If a firm faces the marginal cost function

$$MC = 40 - 18q + 4.5q^2$$

   what would be the increase in total cost if output were increased from 30 to 40?

3. Specify your own function representing a marginal concept in economics, find the indefinite integral and the definite integral over a specified range of values and interpret the meaning of your answers.

13

Dynamics and difference equations

13.1 DYNAMIC ECONOMIC ANALYSIS

Most of the economic analysis you have learned so far has probably been what is known as 'comparative statics'. This entails the comparison of different (static) equilibrium situations. For example, supply and demand analysis predicts that, *ceteris paribus*, an increase in demand will result in an increase in both equilibrium price and quantity. The new equilibrium is compared with the old equilibrium, but no mention is made of the mechanism by which price and quantity adjust to their new equilibrium values. The branch of economics that looks at the way in which variables adjust between equilibrium values is known as 'dynamics', and this chapter gives an introduction to some simple dynamic economic models.

The ways in which markets adjust over time vary tremendously and so there is no one simple model that can explain the dynamic adjustment process. In some markets, such as commodity exchanges, prices are changed by the minute in response to changes in supply and demand and so adjustments to new equilibrium prices are almost instantaneous. In other markets the adjustment process may be a slow trial and error process over several years, in some cases so slow that price and quantity hardly ever reach their proper equilibrium values because supply and demand schedules shift before equilibrium has been reached.

One simple dynamic adjustment model that can be applied to some competitive markets, the cobweb model, is explained in Sections 13.2 and 13.3 below. The dynamic adjustment process in a basic Keynesian macroeconomic model is then considered in Section 13.4 and a model of price adjustment in a market with only two firms is analysed in Section 13.5. These different applications of dynamics will give you an idea of how adjustments can take place between equilibria and how mathematics can be used to calculate the values of variables at different points in time during the adjustment process. They are only very basic models, however, designed to give you an introduction to this branch of economics. The mathematics required to analyse more complex dynamic models goes beyond that covered in this text.

In this chapter, time is considered as a discrete variable and the dynamic adjustment process between equilibria is seen as a step-by-step process rather than one of continual adjustment. (The distinction between discrete and continuous variables was explained in Section 7.2.) This discrete definition of time enables us to calculate different values of the variables that are adjusting to new equilibrium levels

1 using a spreadsheet, and
2 using the mathematical concept of 'difference equations', explained in Section 13.3.

13.2 THE COBWEB: ITERATIVE SOLUTIONS

In some markets, particularly agricultural markets, supply cannot immediately expand to meet increased demand. Crops have to be planted and grown and livestock takes time to raise. Some manufactured products can also take quite a while to produce when orders suddenly increase. The cobweb model takes into account this delayed response on the supply side of a market by assuming that quantity supplied now (Q_t^s) depends on the ruling price in the previous time period (P_{t-1}), i.e. $Q_t^s = f(P_{t-1})$ where the subscripts denote the time period. Consumer demand for the same product (Q_t^d), however, is assumed to depend on the current price, i.e.

$$Q_t^d = f(P_t)$$

This is a reasonable picture of many agricultural markets. The quantity offered for sale this year depends on what was planted at the start of the growing season, which in turn depends on last year's price. Consumers look at current prices, though, when deciding what to buy.

The cobweb model also assumes that the market is perfectly competitive and that supply and demand are both linear schedules.

Before we go any further, it must be stressed that this model does *not* explain how price adjusts in all perfectly competitive markets, or even in all perfectly competitive agricultural markets. It is a simple model with some highly restrictive assumptions that can only explain how price adjusts in these particular circumstances. Some markets may have a more complex lag structure, e.g. $Q_t^s = f(P_{t-1}, P_{t-2}, P_{t-3})$, or may not have linear demand and supply. You should also not forget that intervention in agricultural markets, such as the EC Common Agricultural Policy, usually means that price is not competitively determined and hence the cobweb assumptions do not apply. Having said all this, the cobweb model can still give a fair idea of how price and quantity adjust in many markets with a delayed supply, even if precise predictions cannot be made because not all the assumptions hold.

The assumptions of the cobweb model set out above mean that the demand and supply functions can be specified in the format

$$Q_t^{\mathrm{d}} = a + bP_t$$

$$Q_t^{\mathrm{s}} = c + dP_{t-1}$$

where a, b, c and d are parameters specific to individual markets. P_t and Q_t are, of course, restricted to positive quantities.

Note that, as demand schedules slope down from left to right, the value of b is expected to be negative. As supply schedules usually cut the price axis at a positive value (and therefore the quantity axis at a negative value if the line were theoretically allowed to continue into negative quantities) the value of c will also usually be negative. Remember that these functions have Q as the dependent variable but in supply and demand analysis Q is usually measured along the horizontal axis.

Although desired quantity demanded only equals desired quantity supplied when a market is in equilibrium, it is always true that actual quantity bought equals quantity sold. In the cobweb model it is assumed that in any one time period producers supply a given amount (determined by the previous time period's price) and then price adjusts so that all the produce supplied is bought by consumers. In other words

$$Q_t^{\mathrm{d}} = Q_t^{\mathrm{s}}$$

Therefore

$$a + bP_t = c + dP_{t-1}$$

$$bP_t = c - a + dP_{t-1}$$

$$P_t = \frac{c-a}{b} + \frac{d}{b}P_{t-1} \tag{1}$$

This is what is known as a 'linear first-order difference equation'. A difference equation expresses the value of a variable in one time period as a function of its value in earlier periods; in this case

$$P_t = f(P_{t-1})$$

It is clearly a linear relationship as the terms $(c-a)/b$ and d/b will each take a single numerical value in an actual example. It is 'first order' because only a single lag on the previous time period is built into the model and the coefficient of P_{t-1} is a simple constant. From this difference equation we can derive an expression for P_t in terms of t, i.e. the difference equation

$$P_t = f(P_{t-1})$$

can be 'solved' to get the function

$$P_t = f(t)$$

Before going through the mathematics of difference equations, explained in the next section, let us first get a picture of how the cobweb price adjustment mechanism operates using a numerical example.

Example 13.1

In an agricultural market where the assumptions of the cobweb model apply, the demand and supply schedules are

$$Q_t^d = 400 - 20P_t$$

$$Q_t^s = -50 + 10P_{t-1}$$

A long-run equilibrium has been established for several years but then one year there is an unexpectedly good crop of 160. Explain how price will behave over the next few years following this 'shock' to the market.

(Note: In this example and in most other examples in this chapter, no specific units of measurement for P or Q are given in order to keep the analysis as simple as possible. In actual applications, of course, price will usually be measured in pounds and quantity in physical units, e.g. thousands of tonnes.)

Solution

In long-run equilibrium, price and quantity will remain unchanged each time period. This means that the long-run equilibrium price $P^* = P_t = P_{t-1}$ and the long-run equilibrium quantity $Q^* = Q_t^d = Q_t^s$. Therefore

$$Q^* = 400 - 20P^*$$

and

$$Q^* = -50 + 10P^*$$

Equating to solve for P^* and Q^* gives

$$400 - 20P^* = -50 + 10P^*$$

$$450 = 30P^*$$

$$15 = P^*$$

$$Q^* = 400 - 20P^* = 400 - 300 = 100$$

These values correspond to the point where the supply and demand schedules intersect in Figure 13.1. (Note that, as P is measured on the vertical axis, the inverses of the functions given in the question have been used to draw this diagram.)

If an unexpectedly good crop causes an amount of 160 to be supplied onto the market one year, then this means that the short-run supply schedule effectively becomes the vertical line S_0 in Figure 13.1. To sell this amount price has to be reduced to P_0, corresponding to the point A where S_0 cuts the demand schedule.

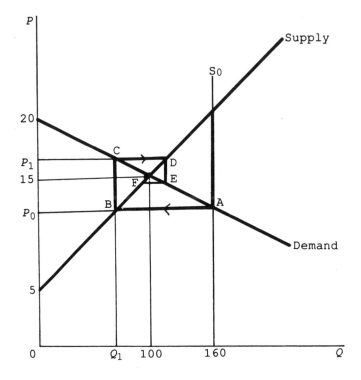

Fig 13.1

Producers will then plan production for the next time period on the assumption that P_0 is the ruling price. The amount supplied will be Q_1, corresponding to point B. However, when Q_1 is put onto the market it will sell for price P_1, corresponding to point C. Further adjustments in quantity and price are shown by points D, E, F etc. These trace out a cobweb pattern (hence the 'cobweb' name) which converges on the long-run equilibrium values of P and Q where the supply and demand schedules intersect.

Price will not always return towards its long-run equilibrium level, as we shall see later when some other examples are considered. However, first let us concentrate on finding the actual pattern of price adjustment which this question asks for.

Approximate values for the first few prices could be read off the graph in Figure 13.1, but as price converges towards the centre of the cobweb it gets difficult to read off the different values accurately. We shall therefore calculate the first few values of P manually so that you can become familiar with the mechanics of the cobweb model and then we shall set up a spreadsheet that can rapidly calculate patterns of price adjustment over a much longer period.

406

Dynamics and difference equations

Quantity supplied in each time period is calculated by simply entering the previously ruling price into the market's supply function

$$Q_t^s = -50 + 10P_{t-1}$$

but how is P calculated? There are two ways:

(i) from first principles, using the supply and demand schedules given in the question, and

(ii) using a difference equation, in the format (1) derived earlier.

First, method (i) is used.

The demand function

$$Q_t^d = 400 - 20P_t$$

can be rearranged to give the inverse function

$$P_t = 20 - 0.05Q_t^d$$

The model assumes that a fixed quantity arrives on the market each time period and then price adjusts until $Q_t^d = Q_t^s$. Thus, P_t can be found by inserting the current quantity supplied, Q_t^s, into the function

$$P_t = 20 - 0.05Q_t^s$$

since $Q_t^s = Q_t^d$ at the market clearing price.

Assuming that the initial disturbance to the system when Q^s rises to 160 occurs in time period 0, the values of P and Q over the next three time periods can be calculated as follows:

$$Q_0^s = 160 \qquad \text{(given value)}$$
$$P_0 = 20 - 0.05Q_0^s = 20 - 0.05(160) = 20 - 8 = 12$$

$$Q_1^s = -50 + 10P_0 = -50 + 10(12) = -50 + 120 = 70$$
$$P_1 = 20 - 0.05Q_1^s = 20 - 0.05(70) = 20 - 3.5 = 16.5$$

$$Q_2^s = -50 + 10P_1 = -50 + 10(16.5) = -50 + 165 = 115$$
$$P_2 = 20 - 0.05Q_2^s = 20 - 0.05(115) = 20 - 5.75 = 14.25$$

$$Q_3^s = -50 + 10P_2 = -50 + 10(14.25) = -50 + 142.5 = 92.5$$
$$P_3 = 20 - 0.05Q_3^s = 20 - 0.05(92.5) = 20 - 4.625 = 15.375$$

The pattern of price adjustment is therefore 12, 16.5, 14.25, 15.375 etc.

These values correspond to the pattern of price adjustment shown in Figure 13.1. Price initially falls below its long-run equilibrium value of 15 and then converges back towards this equilibrium, alternating above and

407

below it but with the magnitude of the difference becoming smaller each period.

The same pattern of price adjustment can be obtained by method (ii) using the difference equation

$$P_t = \frac{c-a}{b} + \frac{d}{b}P_{t-1} \tag{1}$$

and substituting in the given values of a, b, c and d to get

$$P_t = \frac{(-50)-400}{-20} + \frac{10}{-20}P_{t-1}$$
$$P_t = 22.5 - 0.5P_{t-1} \tag{2}$$

The original price P_0 still has to be derived by inserting the shock quantity 160 into the demand function, as already explained, which gives

$$P_0 = 20 - 0.05(160) = 12$$

Thus, using our difference equation (2),

$$P_1 = 22.5 - 0.5P_0 = 22.5 - 0.5(12) = 16.5$$
$$P_2 = 22.5 - 0.5P_1 = 22.5 - 0.5(16.5) = 14.25$$
$$P_3 = 22.5 - 0.5P_2 = 22.5 - 0.5(14.25) = 15.375$$

These prices are the same as those calculated by method (i), as expected.

Clearly it can be time consuming to calculate price over a large number of time periods, but fortunately a spreadsheet can be set up to perform the necessary calculations. One possible format for an appropriate Lotus 1–2–3 worksheet is set out in Table 13.1. This calculates price each period from first principles but if you are familiar with Lotus 1–2–3 you can construct your own spreadsheet based on the difference equation approach.

The spreadsheet in Table 13.1 is constructed as follows. Enter the following labels in the cells shown and right justify (RJ) where indicated:

COBWEB	cell A1	
MODEL	cell B1	
where	cell A3	
Qd = a + bPt	cell B3	
and Qs =	cell C3	(RJ)
c + dPt − 1	cell D3	
PARAMETER	cell A5	
VALUES:-	cell A6	
a =	cell B5	(RJ)
b =	cell B6	(RJ)
c =	cell D5	(RJ)
d =	cell D6	(RJ)

Initial	cell B7
shock qua	cell C7
ntity =	cell D7
EQUILIBRI	cell A9
UM PRICE =	cell B9
EQUIL	cell A10
QUANTITY =	cell B10
Time	cell A12 (RJ)
Price	cell B12 (RJ)
P Change	cell C12 (RJ)
Quantity	cell D12 (RJ)
STABLE	cell E9
OR	cell F9
UNSTABLE	cell E10
?	cell F10

Table 13.1

```
COBWEB    MODEL

where    Qd=a+bPt  and  Qs= c+dPt-1

PARAMETER    a =      400      c =     -50
VALUES:      b =      -20      d =      10
        Initial  shock quantity  =     160

EQUILIBRIUM PRICE=      15        STABLE   OR
EQUIL    QUANTITY=     100        UNSTABLE ?        STABLE
```

Time	Price	P Change	Quantity
0	12.0000		160.0000
1	16.5000	4.5000	70.0000
2	14.2500	-2.2500	115.0000
3	15.3750	1.1250	92.5000
4	14.8125	-0.5625	103.7500
5	15.0938	0.2813	98.1250
6	14.9531	-0.1406	100.9375
7	15.0234	0.0703	99.5313
8	14.9883	-0.0352	100.2344
9	15.0059	0.0176	99.8828
10	14.9971	-0.0088	100.0586
11	15.0015	0.0044	99.9707
12	14.9993	-0.0022	100.0146
13	15.0004	0.0011	99.9927
14	14.9998	-0.0005	100.0037
15	15.0001	0.0003	99.9982
16	15.0000	-0.0001	100.0009
17	15.0000	0.0001	99.9995
18	15.0000	0.0000	100.0002
19	15.0000	0.0000	99.9999
20	15.0000	0.0000	100.0001

Enter the given values for this problem:

(*a*)	400	cell C5
(*b*)	− 20	cell C6
(*c*)	− 50	cell E5
(*d*)	10	cell E6
(initial *Q*)	160	cell E7

To calculate the equilibrium price using the equation

$$P^* = \frac{c-a}{b-d}$$

enter the formula +(E5–C5)/(C6–E6) in cell C9.

To calculate the equilibrium quantity ($Q^* = a + bP^*$) enter the formula +C5+C6*C9 in cell C10.

For the time period use the /DF (Data Fill) command to enter numbers 0–20 in cells A13–A33.

The initial 'shock' quantity has been specified in cell E7 so to put this value in the first cell of the Quantity column enter the formula +E7 in cell D13.

Price in any time period can be determined by inserting the relevant quantity into the demand function.

$$Q_t^d = a + bP_t$$

which can be rearranged to give

$$P_t = \frac{Q_t - a}{b}$$

given that price always adjusts so that all quantity supplied in a given time period is sold, i.e.

$$Q_t^d = Q_t^s = Q_t$$

Thus the formula +(D13–$C$5)/$C$6 should be entered in cell B13 and then copied down the column.

Quantity for each year (apart from year 0) can now be determined from the supply function $Q_t^s = c + dP_{t-1}$ by entering the formula +$E$5+$E$6*B13 in cell D14 and then copying it down the column.

Your spreadsheet should now show a series of prices and quantities converging on the equilibrium values of 15 for price and 100 for quantity. The first few values can be checked against the manually calculated values and are, as expected, the same. Reformat all computed columns of figures to 4 decimal places to make the table easier to read. To bring home the point that each price adjustment is smaller than the previous one, the change in price can be calculated by entering the formula +B14–B13 in cell C14 and then copying this formula down the column.

Although the stability of this example is obvious from the way that price converges on its equilibrium value of 15, a stability check can be entered which may be useful when this spreadsheet is used for other examples. Assuming that b is always negative and d is positive, the market will be stable if $d/-b < 1$ and unstable (i.e. price will not converge back to its equilibrium) if $d/-b > 1$. (The reasons for this rule are explained later in Section 13.3.)

The Lotus 1–2–3 @IF command can be used to check stability using this rule by entering @IF(+E6/−C6<1, E9, E10) in cell G10. If $d/-b < 1$ then the contents of cell E9, i.e. the word 'STABLE', will appear in cell G10, and if $d/-b \geqslant 1$ then 'UNSTABLE' appears. (Note that if $d = -b$ then this particular test gives the result 'UNSTABLE' even though, in this special case, price neither converges nor diverges.)

Save this spreadsheet so that it can be used for other examples.

It was mentioned earlier that price will not always return to its long-run equilibrium level in markets where the cobweb model applies. To understand why this may be so consider Example 13.2 below.

Example 13.2

In a market where the assumptions of the cobweb model apply the demand and supply functions are

$$Q_t^d = 120 - 4P_t$$

$$Q_t^s = -80 + 16P_{t-1}$$

If in one time period the long-run equilibrium is disturbed by output unexpectedly rising to a level of 90, explain how price will adjust over the next few time periods.

Solution

The long-run equilibrium price can be determined from the formula

$$P^* = \frac{c-a}{b-d} = \frac{(-80)-120}{-4-16} = \frac{-200}{-20} = 10$$

Thus, the long-run equilibrium quantity is

$$Q^* = 120 - 4P^* = 120 - 4(10) = 80$$

You could use the spreadsheet developed for Example 13.1 above to trace out the subsequent pattern of price adjustment but if a few values are calculated manually it can be seen that calculations after period 2 are irrelevant.

Using the difference equation

411

$$P_t = \frac{c-a}{b} + \frac{d}{b} P_{t-1} \qquad \text{·(1)}$$

and substituting our known values, we get

$$P_t = \frac{(-80) - 120}{-4} + \frac{16}{-4} P_{t-1} = 50 - 4 P_{t-1} \qquad (2)$$

The initial price P_0 can be found by inserting the shock quantity 90 into the demand function. Thus

$$90 = 120 - 4P_0$$

$$4P_0 = 30$$

$$P_0 = 7.5$$

Putting this value into the difference equation (2) above we get

$$P_1 = 50 - 4P_0 = 50 - 4(7.5) = 20$$

$$P_2 = 50 - 4P_1 = 50 - 4(20) = -30$$

There is not much point in going any further with the calculations. Assuming that producers will not pay consumers to take goods off their hands, negative prices cannot exist. What has happened is that price has followed the path *ABCD* traced out in Figure 13.2.

The initial quantity 90 put onto the market causes price to drop to 7.5. Suppliers then reduce supply for the next period to

$$Q_1^s = -80 + 16P_0 = -80 + 16(7.5) = 40$$

This sells for price $P_1 = 20$ and so supply for the following period is increased to

$$Q_2^s = -80 + 16P_1 = -80 + 16(20) = 240$$

Consumers would only consume 120 even if price were zero (where the demand schedule hits the axis) and so, when this quantity of 240 is put onto the market, price will collapse to zero and there will still be unsold produce. Producers will not wish to supply anything for the next time period if they expect a price of zero and so no further production will take place in the market.

This is clearly an unstable market, but why is there a difference between this market and the stable market considered in Example 13.1? This question is. answered in Section 13.3 where the method of solution for difference equations is explained.

It must be remembered, though, that even if in theoretical models of unstable markets (such as Example 13.2 above) price 'explodes' and the market collapses, this may not happen in reality if

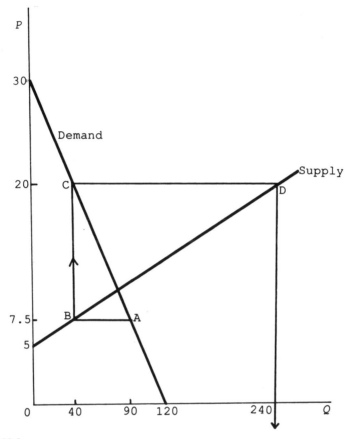

Figure 13.2

1 producers learn from experience and do not simply base production
   plans for the next period on the current price,
2 supply and demand schedules are not linear along their entire length or
3 government intervention takes place to support production.

Another example of an exploding market is Example 13.3 below, which is
solved using the spreadsheet developed for Example 13.1.

Example 13.3

In an agricultural market where the cobweb assumptions hold and

$$Q_t^d = 360 - 8P_t$$

$$Q_t^s = -120 + 12P_{t-1}$$

413

Basic mathematics for economists

a long-run equilibrium is disturbed by an unexpectedly good crop of 175 units. Use a spreadsheet to trace out the subsequent path of price adjustment and find out either

(i) how long price will take to return to within 1% of its equilibrium value or
(ii) how long it will take until the market collapses completely.

Solution

When the given parameters and shock quantity are entered, your spreadsheet should look like Table 13.2. This is clearly unstable as both the stability check and the price adjustments show. (Note that only eleven time periods are shown in Table 13.2 because the subsequent large values of P and Q computed in this unstable market would require excessively wide column widths.)

According to these figures, the market will continue to operate until the eighth time period following the initial shock. In period 9 nothing will be produced (mathematically the model gives a negative quantity, as shown in the spreadsheet) and the market collapses.

QUESTIONS 13.1

(In all these questions, assume that the assumptions of the cobweb model apply to each market.)

Table 13.2

COBWEB	MODEL					
where	$Qd=a+bPt$ and $Qs=c+dPt-1$					
PARAMETER	a =	360	c =	-120		
VALUES:	b =	-8	d =	12		
	Initial shock quantity		=	175		
EQUILIBRIUM PRICE=		24		STABLE	OR	
EQUIL QUANTITY=		168		UNSTABLE ?		UNSTABLE

Time	Price	P Change	Quantity
0	23.1250		175.0000
1	25.3125	2.1875	157.5000
2	22.0313	-3.2813	183.7500
3	26.9531	4.9219	144.3750
4	19.5703	-7.3828	203.4375
5	30.6445	11.0742	114.8438
6	14.0332	-16.6113	247.7344
7	38.9502	24.9170	48.3984
8	1.5747	-37.3755	347.4023
9	57.6379	56.0632	-101.1035
10	-26.4569	-84.0948	571.6553
11	99.6854	126.1423	-437.4829

414

1. The agricultural market whose demand and supply schedules are

$$Q_t^d = 240 - 20P_t$$

$$Q_t^s = -33\tfrac{1}{3} + 16\tfrac{2}{3}P_{t-1}$$

is initially in long-run equilibrium. Quantity then falls to 50% of its previous level as a result of an unexpectedly poor harvest. How many time periods will it take for price to return to within 1% of its long-run equilibrium level?

2. The demand and supply schedules

$$Q_t^d = 200 - 12.5P_t$$

$$Q_t^s = -60 + 20P_{t-1}$$

describe an unstable market. A shock reduction of quantity to 80 throws the system out of equilibrium. How long will it take for the market to collapse completely?

3. By tracing out the pattern of price adjustment after an initial shock that disturbs the previously ruling long-run equilibrium, say whether or not the following markets are stable.

(a) $Q_t^d = 150 - 1.5P_t$ $Q_t^s = -30 + 3P_{t-1}$

(b) $Q_t^d = 180 - 1.25P_t$ $Q_t^s = -20 + P_{t-1}$

13.3 THE COBWEB: DIFFERENCE EQUATION SOLUTIONS

The cobweb difference equation

$$P_t = \frac{c-a}{b} - \frac{d}{b}P_{t-1} \tag{1}$$

is an example of a linear first-order difference equation. The general format of a linear first-order difference equation is

$$x_t = \alpha + \beta x_{t-1}$$

where x is the variable which changes over time and α and β are constants.

In the cobweb difference equation, $\alpha = (c-a)/b$, $\beta = d/b$ and, of course, $x_t = P_t$.

Solving the cobweb difference equation means putting it into the format

$$P_t = f(t)$$

so that the value of P_t at any given time can be immediately calculated without the need to calculate all the preceding values of P_t.

There are two parts to the solution of linear first-order difference equations.

1 The equilibrium solution: in the cobweb case this is the long-run equilibrium price. As price is the same each time period in equilibrium, then
$P_t = P_{t-1}$. In general, the equilibrium solution is a constant value about
which adjustments in the variable in question take place over time.

2 The complementary solution: this tells us how the variable in question,
i.e. price in the cobweb model, varies from the equilibrium solution as
time changes. It is assumed that $P_t = Ak^t$ where A and k are constants.
(Note that in this formula, t denotes the power to which k is raised and
is not just a time superscript.)

This may seem a rather difficult concept to understand but what it boils down
to is that the solution to the cobweb model linear first-order difference equation contains two parts:

1 the long-run equilibrium price, and
2 a function that tells us how much price diverges from this equilibrium
level at different points in time.

The first part is straightforward. In the long run the equilibrium price P^* holds
in each time period and so

$$P^* = P_t = P_{t-1}$$

Substituting P^* into the difference equation

$$P_t = \frac{c-a}{b} + \frac{d}{b} P_{t-1} \qquad (1)$$

we get

$$P^* = \frac{c-a}{b} + \frac{d}{b} P^*$$

$$bP^* = c - a + dP^*$$

$$a - c = (d - b)P^*$$

$$\frac{a-c}{d-b} = P^*$$

This, of course, is the same equilibrium value of P that would be derived in
the single-time-period linear supply and demand model

$$Q^d = a + bP$$

$$Q^s = c + dP$$

To find the complementary solution, we return to the difference equation

$$P_t = \frac{c-a}{b} + \frac{d}{b} P_{t-1} \qquad (1)$$

416

but ignore the first term, which is a constant that does not vary over time, i.e. we look at the equation

$$P_t = \frac{d}{b} P_{t-1} \tag{2}$$

This may seem rather a strange procedure, but it works, as we shall see later when some numerical examples are tackled.

Using the method of finding the complementary solution referred to above, we assume that P_t depends on t according to the function

$$P_t = Ak^t$$

where A and k are some (as yet) unknown constants. This function applies to all values of t, which means that $P_{t-1} = Ak^{t-1}$. Substituting these formulations for P_t and P_{t-1} back into equation (2) above we get

$$Ak^t = \frac{d}{b} Ak^{t-1}$$

Dividing through by Ak^{t-1} gives

$$k = \frac{d}{b}$$

Thus the complementary solution is

$$P_t = A \left(\frac{d}{b} \right)^t$$

The value of A cannot be ascertained unless some specific value of P_t is known for a specific value of t. The complete solution to the cobweb difference equation therefore becomes

$$P_t = \text{equilibrium solution} + \text{complementary solution}$$

$$= \frac{a-c}{d-b} + A \left(\frac{d}{b} \right)^t$$

From this solution we can see that the stability of the model depends on the value of d/b. If A is a non-zero constant, then there are three possibilities

1 If $\left| \dfrac{d}{b} \right| < 1$, then $\left(\dfrac{d}{b} \right)^t \to 0$ as $t \to \infty$

This occurs in a stable market as the divergence of price from its equilibrium value gets smaller over time. (Note that it is the absolute value of d/b that we look at because b will normally be a negative quantity.)

2 If $\left| \dfrac{d}{b} \right| > 1$, then $\left(\dfrac{d}{b} \right)^t \to \infty$ as $t \to \infty$

417

This occurs in an unstable market. As time goes by price will diverge from its equilibrium level by greater and greater amounts after an initial disturbance.

3 If $\left|\dfrac{d}{b}\right| = 1$, then $\left|\left(\dfrac{d}{b}\right)^t\right| = 1$ as $t \to \infty$

Price will neither return to its equilibrium nor 'explode'. Normally, $b < 0$, and so $d/b < 0$, which in this case means that $d/b = -1$. Therefore, $(d/b)^t$ will oscillate between $+1$ and -1 depending on whether or not t is an even or odd number. Price will continually fluctuate between two levels (see Example 13.6 below).

We can now use the above difference equation solution to the general cobweb model to answer some specific numerical problems.

Example 13.4

Use the cobweb difference equation solution to answer the question in Example 13.1 above, i.e. what happens in the market where

$$Q_t^d = 400 - 20P_t$$

$$Q_t^s = -50 + 10P_{t-1}$$

if Q_t^s suddenly changes to 160?

Solution

Substituting the values $a = 400$, $b = -20$, $c = -50$ and $d = 10$ into the general cobweb difference equation solution

$$P_t = \frac{a-c}{d-b} + A\left(\frac{d}{b}\right)^t \qquad (1)$$

gives

$$P_t = \frac{400 - (-50)}{10 - (-20)} + A\left(\frac{10}{-20}\right)^t$$

$$= \frac{450}{30} + A(-0.5)^t$$

$$= 15 + A(-0.5)^t \qquad (2)$$

To find the value of A we then substitute in the known value of P_0 at time 0. The question tells us that the initial 'shock' output level Q_0 is 160 and so, as price adjusts until all output is sold, P_0 can be calculated from the demand schedule $Q_t^d = 400 - 20P_t$ by substituting this quantity. Thus

$$160 = 400 - 20P_0$$

$$20P_0 = 240$$

$$P_0 = 12$$

Therefore, the difference equation solution (2) for time period 0 will be

$$12 = 15 + A(-0.5)^0$$

$$12 = 15 + A$$

since $(-0.5)^0 = 1$, and therefore

$$A = -3$$

Thus the complete solution to the difference equation in this example is

$$P_t = 15 - 3(-0.5)^t$$

We can use this solution to calculate the first few values of P_t and compare with those we obtained when answering Example 13.1.

$$P_1 = 15 - 3(-0.5)^1 = 15 + 1.5 = 16.5$$

$$P_2 = 15 - 3(-0.5)^2 = 15 - 3(0.25) = 14.25$$

$$P_3 = 15 - 3(-0.5)^3 = 15 - 3(-0.125) = 15.375$$

As expected, these values are identical to those calculated by the iterative method.

In this particular example, price converges fairly quickly towards its long-run equilibrium level of 15. By time period 9, price will be

$$P_9 = 15 - 3(-0.5)^9$$

$$= 15 - 3(-0.0019531)$$

$$= 15 + 0.0058594$$

$$= 15.01 \text{ (to 2 dp)}$$

This is clearly a stable solution. In the difference equation

$$P_t = 15 - 3(0.5)^t$$

we can see that, as t gets larger, the value of $(-0.5)^t$ approaches zero. This satisfies the stability condition outlined above, that in the general cobweb model $|d/b| < 1$ ensures that price will return towards its equilibrium value.

Note that, because $-0.5 < 0$, the direction of the divergence from the equilibrium value alternates between time periods. This is because for any negative quantity $-x$, it will always be true that

$$x < 0, \ (-x)^2 > 0, \ (-x)^3 < 0, \ (-x)^4 > 0 \text{ etc.}$$

419

Thus for even-numbered time periods (in this example) price will be above its equilibrium value, and for odd-numbered time periods price will be below its equilibrium value.

Although in this example price converges towards its long-run equilibrium value, it would never actually reach it if price and quantity were divisible into infinitesimally small units. Theoretically, this example is a bit like the case of the 'hopping frog' back in Chapter 7 when infinite geometric series were examined. The distance from the equilibrium gets smaller and smaller each time period but never actually reaches zero. For practical purposes, which is what economics is concerned with, a reasonable cut-off point can be decided upon to define when a full return to equilibrium has been reached. In this numerical example the difference from the equilibrium is less than 0.01 by time period 9, which is for all intents and purposes a full return to equilibrium.

You could, of course, have worked all this out fairly quickly using a calculator, or even more speedily using a spreadsheet, without having to learn about difference equations. However, the point of this example was to explain the method of solution of difference equations, and to illustrate it with a simple problem where the answers could be checked against iterative solutions. In other cases, one may need to calculate values for more distant time periods. The method of solution of difference equations will also be useful for those of you who go on to study intermediate economic theory where some models, particularly in macroeconomics, are based on difference equations in an algebraic format which cannot be solved using a spreadsheet.

We shall now consider another cobweb example which is rather different from Example 13.4 in that

1 price does not return towards its equilibrium level and
2 the process of adjustment is more gradual over time.

Example 13.5

In a market where the assumptions of the cobweb model hold

$$Q_t^d = 200 - 8P_t$$

and

$$Q_t^s = -43 + 8.2P_{t-1}$$

The long-run equilibrium is disturbed when quantity suddenly changes to 90. What happens to price in the following time periods?

Solution

In long-run equilibrium

$$Q^* = Q_t^d = Q_t^s$$

and

$$P^* = P_t = P_{t-1}$$

Thus

$$200 - 8P^* = Q^* = -43 + 8.2P^*$$

$$243 = 16.2P^*$$

$$15 = P^*$$

This will be an unstable equilibrium as

$$\left| \frac{d}{b} \right| = \left| \frac{8.2}{-8} \right| = 1.025 > 1$$

The difference equation that describes the relationship between price in one period and the next will take the usual format

$$P_t = \frac{c-a}{b} + \frac{d}{b}P_{t-1} \tag{1}$$

where $a = 200$, $b = -8$, $c = -43$ and $d = 8.2$, giving

$$P_t = \frac{-43 - 200}{-8} + \frac{8.2}{-8}P_{t-1} = 30.375 - 1.025\,P_{t-1}$$

The solution to this difference equation will be in the form

$$P_t = \frac{a-c}{d-b} + A\left(\frac{d}{b}\right)^t$$

$$= \frac{200 - (-43)}{8.2 - (-8)} + A\left(\frac{8.2}{-8}\right)^t$$

$$= \frac{243}{16.2} + A(-1.025)^t$$

$$= 15 + A(-1.025)^t \tag{2}$$

The first part of this solution is of course the equilibrium value of P which has already been calculated above. To derive the value of A, we need to insert a known value of P_t at a given time period. In this example the quantity is given as 90 in period 0. To find the price that this quantity will sell for, this value is substituted into the demand function. Thus

$$Q_0^d = 90 = 200 - 8P_0$$

$$8P_0 = 110$$

$$P_0 = 13.75$$

Substituting this value into the difference equation solution (2) we get

$$P_0 = 13.75 = 15 + A(1.025)^0$$

$$13.75 = 15 + A$$

$$-1.25 = A$$

Note that, as in Example 13.1 above, the value of the parameter A in the difference equation solution is the difference between the equilibrium value of price and the value it initially takes when quantity is disturbed from its equilibrium level.

The complete solution to the difference equation in this example now becomes

$$P_t = 15 - 1.25(-1.025)^t$$

Using this formula to calculate the first few values of P_t (to 2 decimal places)

$$P_0 = 15 - 1.25(1.025)^0 = 13.75$$

$$P_1 = 15 + 1.25(1.025)^1 = 16.28$$

$$P_2 = 15 - 1.25(1.025)^2 = 13.69$$

$$P_3 = 15 + 1.25(1.025)^3 = 16.35$$

We can see that, although price is gradually moving away from its long-run equilibrium value of 15, it is a very slow process. By period 10, price is still above 13.00,

$$P_{10} = 15 - 1.25(1.025)^{10} = 13.40$$

and it takes until time period 102 before the market collapses when price becomes negative, as the figures below show:

$$P_{100} = 15 - 1.25(1.025)^{100} = 0.23$$

$$P_{101} = 15 + 1.25(1.025)^{101} = 30.14$$

$$P_{102} = 15 - 1.25(1.025)^{102} = -0.51$$

This example is not a particularly realistic picture of an agricultural market as many changes in supply and demand conditions would take place over a 100 year time period. However, it illustrates the usefulness of the difference equation solution in immediately computing values for distant time periods without first needing to compute all the preceding values.

The following example illustrates what happens when a market is neither stable nor unstable.

Example 13.6

The cobweb model assumptions hold in a market where

$$Q_t^d = 160 - 2P_t$$

and

$$Q_t^s = -20 + 2P_{t-1}$$

If the previously ruling long-run equilibrium is disturbed by an unexpectedly low output of 50 in one time period, what will happen to price in the following time periods?

Solution

Substituting the values $a = 160$, $b = -2$, $c = -20$ and $d = 2$ for this market into the general cobweb difference equation solution

$$P_t = \frac{a-c}{d-b} + A\left(\frac{d}{b}\right)^t \qquad (1)$$

gives

$$P_t = \frac{160 - (-20)}{2 - (-2)} + A\left(\frac{2}{-2}\right)^t$$

$$= \frac{180}{4} + A(-1)^t$$

$$= 45 + A(-1)^t \qquad (2)$$

To determine the value of A, first substitute the given value of 50 for Q_0 into the demand function so that

$$160 - 2P_0 = 50$$

$$110 = 2P_0$$

$$55 = P_0$$

Now substitute this value into (2) above for period 0 so that

$$P_0 = 55 = 45 + A(-1)^0$$

$$55 = 45 + A$$

$$10 = A$$

The complete solution to the difference equation for this example is therefore

$$P_t = 45 + 10(-1)^t$$

Using this formula to calculate the first few values of P_t, we see that

$$P_0 = 45 + 10(-1)^0 = 45 + 10 = 55$$

Basic mathematics for economists

$P_1 = 45 + 10(-1)^1 = 45 - 10 = 35$

$P_2 = 45 + 10(-1)^2 = 45 + 10 = 55$

$P_3 = 45 + 10(-1)^3 = 45 - 10 = 35$

$P_4 = 45 + 10(-1)^4 = 45 + 10 = 55$

etc.

Price therefore continually fluctuates between the prices of 35 and 55.

This is the third possibility in the stability conditions examined earlier. In this example

$$\left|\frac{d}{b}\right| = \left|\frac{2}{-2}\right| = |-1| = 1$$

Therefore, as $t \to \infty$, P_t neither converges on its equilibrium level nor explodes until the market collapses. This fluctuation between two price levels from year to year is sometimes observed in certain agricultural markets.

QUESTIONS 13.2

(Assume that the usual cobweb assumptions apply in these questions.)
1. In a market where

$$Q_t^d = 160 - 20P_t$$

and

$$Q_t^s = -80 + 40P_{t-1}$$

quantity unexpectedly drops from its equilibrium value to 75. Derive the difference equation which will calculate price in the time periods following this event.
2. If

$$Q_t^d = 180 - 0.9P_t$$

and

$$Q_t^s = -24 + 0.8P_{t-1}$$

say whether or not the long-run equilibrium price is stable and then use the difference equation method to calculate price in the thirtieth time period after a sudden one-off increase in quantity to 117.
3. Given the demand and supply schedules

$$Q_t^d = 3450 - 6P_t$$

and

$$Q_t^s = -729 + 4.5P_{t-1}$$

424

use difference equations to predict what price will be in the tenth time period after an unexpected drop in quantity to 354, assuming that the market was previously in long-run equilibrium.

13.4 THE LAGGED KEYNESIAN MACROECONOMIC MODEL

In the basic Keynesian model of the determination of national income, if foreign trade and government taxation and expenditure are excluded, the model reduces to the two equations of the accounting identity,

$$Y = C + I \tag{1}$$

and the consumption function

$$C = a + bY \tag{2}$$

where Y is national income (equal to total expenditure), C is consumption, I is investment, exogenously determined, a is a constant parameter and b is the marginal propensity to consume.

We substitute (2) into (1) to determine the equilibrium level of Y:

$$Y = a + bY + I$$
$$Y(1 - b) = a + I$$
$$Y = \frac{a + I}{1 - b}$$

This can be evaluated for given values of a, b and I.

If there is a disturbance from this equilibrium, e.g. exogenous investment I alters, then the adjustment to a new equilibrium will not be instantaneous. This is the basis of the well-known multiplier effect. An initial injection of expenditure will become income for another sector of the economy. A proportion of this will be passed on as a further round of expenditure, and so on until the 'ripple effect' dies away.

Because consumer expenditure may not adjust instantaneously to new levels of income, a lagged effect may be introduced into the consumption function (2) above. If it is assumed that consumers' expenditure in one time period depends on the income that they received in the previous time period, then the consumption function becomes

$$C_t = a + bY_{t-1} \tag{3}$$

where the subscripts denote the time period.

National income, however, will still be determined by the sum of all expenditure within the current time period. Therefore the accounting identity (1) can be written as

$$Y_t = C_t + I_t \tag{4}$$

425

From (3) and (4) we can derive a difference equation that explains how Y_t depends on Y_{t-1}. Substituting (3) into (4)

$$Y_t = (a + bY_{t-1}) + I_t$$
$$Y_t = bY_{t-1} + a + I_t \qquad (5)$$

This difference equation (5) can be solved using the method explained in Section 13.3 above. However, let us first illustrate how this lagged effect works using a numerical example.

Example 13.7

In a basic Keynesian macroeconomic model it is assumed that initially

$$Y_t = C_t + I_t$$

where $I_t = 134$ is exogenously determined, and

$$C_t = 40 + 0.6Y_{t-1}$$

The level of investment I_t then falls to 110 and remains at this level each time period. Trace out the pattern of adjustment to the new equilibrium value of Y, assuming that the model was initially in equilibrium.

Solution

Although this pattern of adjustment can best be viewed using a spreadsheet, let us first work out the first few steps of the process manually and relate them to the familiar 45° line income–expenditure graph (illustrated in Figure 13.3) often used to show how Y is determined in introductory economics texts.

If the system is initially in equilibrium then income in one time period is equal to expenditure in the previous time period, i.e.

$$Y_t = Y_{t-1} = Y^*$$

where Y^* is the equilibrium level of Y. Therefore, when the original value of I_t is inserted into the accounting identity the model becomes

$$Y^* = C_t + 134 \qquad (1)$$

$$C_t = 40 + 0.6Y^* \qquad (2)$$

By substitution of (2) into (1)

$$Y^* = (40 + 0.6Y^*) + 134$$

$$Y^*(1 - 0.6) = 40 + 134$$

$$0.4Y^* = 174$$

$$Y^* = 435$$

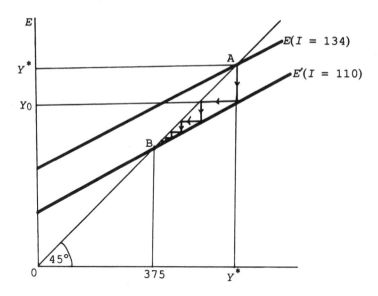

Figure 13.3

This is the equilibrium value of Y before the change in I.

Assume time period 0 is the one in which the drop in I to 110 occurs. Consumption in time period 0 will be based on income earned the previous time period, i.e. when Y was still at the old equilibrium level of 435. Thus

$$C_0 = 40 + 0.6(435) = 40 + 261 = 301$$

Therefore

$$Y_0 = C_0 + I_0 = 301 + 110 = 411$$

In the next time period, consumption will be based on Y_0; thus

$$Y_1 = C_1 + I_1$$
$$= (40 + 0.6Y_0) + 110$$
$$= 40 + 0.6(411) + 110$$
$$= 40 + 246.6 + 110 = 396.6$$

The value of Y for other time periods can be calculated in a similar fashion:

$$Y_2 = C_2 + I_2$$
$$= (40 + 0.6Y_1) + 110$$
$$= 40 + 0.6(396.6) + 110$$
$$= 387.96$$

427

$$Y_3 = C_3 + I_3$$
$$= (40 + 0.6Y_2) + 110$$
$$= 40 + 0.6(387.96) + 110$$
$$= 382.776$$

etc. It can be seen that in each time period Y decreases by smaller and smaller amounts as it readjusts towards the new equilibrium value. This new equilibrium value can easily be calculated using the same method as that used above to work out the initial equilibrium.

When $I = 110$ and $Y_t = Y_{t-1} = Y^*$ then

$$Y^* = C_t + I = C_t + 110$$
$$C_t = 40 + 0.6Y^*$$

By substitution

$$Y^* = (40 + 0.6Y^*) + 110$$
$$(1 - 0.6)Y^* = 150$$
$$Y^* = \frac{150}{0.4} = 375$$

This path of adjustment is illustrated in Figure 13.3 by the zigzag line with arrows which joins the old equilibrium at A with the new equilibrium at B. (Note that this diagram is not to scale and does not show actual values as these cannot be shown accurately for the numbers in this example. Its purpose is to show the direction and relative magnitude of the steps in the adjustment process.)

Unlike the cobweb model described earlier, the adjustment in this Keynesian model is always in the same direction. Instead of alternating either side of the final equilibrium, successive values of Y approach the equilibrium by smaller and smaller increments but always from the same direction. This is because the ratio in the complementary solution to the difference equation (explained below) is not negative as it was in the cobweb model. If the initial equilibrium had been below the new equilibrium then, of course, Y would have approached its new equilibrium from below instead of from above.

A Lotus 1–2–3 spreadsheet to illustrate further steps in the adjustment of Y in this model can be constructed as follows. Insert the labels shown below and right justify (RJ) where indicated.

LAGGED	cell A1	
KEYNESIAN	cell B1	
MODEL	cell C1	(RJ)
where	cell A3	

Dynamics and difference equations

Yt = Ct + It	cell B3	
Ct = a + bYt−1	cell C3	
GIVEN	cell B5	
VALUES:	cell C5	
OLD I =	cell D4	(RJ)
NEW I =	cell D5	(RJ)
a =	cell D6	(RJ)
b =	cell D7	(RJ)
INIT C =	cell D8	(RJ)
OLD EQ Y =	cell A7	
NEW EQ Y =	cell A8	
TIME	cell A10	
Y	cell C10	(RJ)
C	cell E10	(RJ)

Insert the given values in this example:

(OLD I)	134	cell E4
(NEW I)	110	cell E5
(a)	40	cell E6
(b)	0.6	cell E7

Use the Data Fill command to enter numbers 0–21 in cells A11–A32. To calculate the original and new equilibrium values of Y insert the formulae

$$+(\$E\$6+\$E\$4)/(1-\$E\$7) \text{ in cell B7}$$

and

$$+(\$E\$6+\$E\$5)/(1-\$E\$7) \text{ in cell B8}$$

The initial consumption level in period 0 is based on the old equilibrium level of Y so enter the formula

$$+\$E\$6+\$E\$7*\$B\$7 \text{ in cell E11}$$

and copy it into cell E8 to emphasize the point that this value of C is calculated differently from other values.

The level of Y in period 0 is this initial value of C plus the new I, so enter the formula

$$+E11+\$E\$5 \text{ in cell C11}$$

Copy this formula down the column, as current Y will always be current C plus the given value of I. The lagged effect of income on consumption from period 1 onwards can now be built into the model by entering the formula

$$+\$E\$6+\$E\$7*C11 \text{ in cell E12}$$

and then copying it down the column.

429

Your spreadsheet should now show the values of Y and C for periods 0–21. Reformat these columns of numbers to 3 decimal places to align the first few values. The spreadsheet should now look like Table 13.3, which clearly shows Y closing in on its new equilibrium as time goes by.

Table 13.3

LAGGED where	KEYNESIAN $Yt=Ct+It$	MODEL $Ct=a+bYt-1$			
	GIVEN	VALUES	OLD	I=	134
			NEW	I=	110
				a=	40
OLD EQ Y=	435			b=	0.6
NEW EQ Y=	375		INIT	C=	301
TIME		Y			C
0		411.000			301.000
1		396.600			286.600
2		387.960			277.960
3		382.776			272.776
4		379.666			269.666
5		377.799			267.799
6		376.680			266.680
7		376.008			266.008
8		375.605			265.605
9		375.363			265.363
10		375.218			265.218
11		375.131			265.131
12		375.078			265.078
13		375.047			265.047
14		375.028			265.028
15		375.017			265.017
16		375.010			265.010
17		375.006			265.006
18		375.004			265.004
19		375.002			265.002
20		375.001			265.001
21		375.001			265.001

Let us now return to the problem of how to solve the difference equation

$$Y_t = bY_{t-1} + a + I_t \tag{5}$$

in general terms. This general solution can then be applied to numerical problems, such as that set out in Example 13.7 above. By 'solving' this difference equation we mean putting it in the form

$$Y_t = f(t)$$

so that the value of Y_t can be determined for any given value of t.

Using the method explained in Example 13.4, we first need to find the equilibrium solution. This is when

$$Y_t = Y_{t-1} = Y^*$$

From the preceding discussion of the adjustment process in the numerical Example 13.7, we know that an equilibrium is likely to exist and that Y will return to its equilibrium eventually if any shock disturbs the system. (The necessary mathematical conditions are explained below for equation (8).)

In equilibrium, the equations

$$C_t = a + bY_{t-1} \tag{3}$$

and

$$Y_t = C_t + I_t \tag{4}$$

can therefore be written as

$$C_t = a + bY^*$$

$$Y^* = C_t + I_t$$

By substitution

$$Y^* = a + bY^* + I_t$$

$$(1 - b)Y^* = a + I_t$$

$$Y^* = \frac{a + I_t}{1 - b} \tag{6}$$

If the given values of a, b and I_t are put into (6) then the equilibrium value of Y is determined. This is the first part of the difference equation solution.

Returning to the difference equation (5) which we are trying to solve

$$Y_t = bY_{t-1} + a + I_t$$

If the two constant terms a and I_t are removed then this becomes

$$Y_t = bY_{t-1} \tag{7}$$

To find the complementary solution assume that this solution is in the format

$$Y_t = Ak^t$$

where A and k are unknown parameters. This means that

$$Y_{t-1} = Ak^{t-1}$$

Substituting into (7), this gives

$$Ak^t = bAk^{t-1}$$

$$k = b$$

Thus the complementary solution is

$$Y_t = Ab^t$$

The general solution to the difference equation is the sum of the equilibrium and complementary solutions. Hence

$$Y_t = \frac{a + I_t}{1 - b} + Ab^t \tag{8}$$

If t is increased, then the value of b^t will diminish as long as $|b| < 1$. This condition will be met since b is the marginal propensity to consume which has been estimated to lie between 0 and 1 in empirical studies.

The value of A can be determined if an initial value Y_0 is known. Substituting into (8), this gives

$$Y_0 = \frac{a + I_t}{1 - b} + Ab^0$$

Remembering that $b^0 = 1$, this means that

$$A = Y_0 - \frac{a + I_t}{1 - b} \tag{9}$$

Thus A is the value of the difference between the initial level of Y and its final equilibrium value.

Putting this result into (8) above, the general solution to our difference equation becomes

$$Y_t = \frac{a + I_t}{1 - b} + \left(Y_0 - \frac{a + I_t}{1 - b} \right) b^t \tag{10}$$

This may seem to be a rather cumbersome formula but it is straightforward to use. If you remember that

$$\frac{a + I_t}{1 - b} = Y^*$$

is the equilibrium value of Y and rewrite (10) as

$$Y_t = Y^* + (Y_0 - Y^*) b^t \tag{11}$$

you will find it easier to work with.

We can now check that this solution to the lagged Keynesian model difference equation works with the numerical Example 13.7 considered above.

This model initially assumed

$$Y_t = C_t + I_t$$

where $I_t = 134$ and

$$C_t = 40 + 0.6Y_{t-1}$$

which corresponded to an equilibrium of $Y = 435$.

When I_t was exogenously decreased to 110, the adjustment path towards the new equilibrium value of Y of 375 was worked out by an iterative method.

Now let us see what values for Y_t the solution to our difference equation will give. We have to be careful in determining the initial value of Y, i.e. Y_0. This is *not* exogenously determined as it depends on I_t and C_t. However, C_t *is* given as it depends on the previously existing level of Y which was 435 in time period 'minus one'. Therefore

$$C_0 = a + bY_{t-1} = 40 + 0.6(435) = 301$$

$$Y_0 = C_0 + I_0 = 301 + 110 = 411$$

This is the same initial value of Y as that calculated in Example 13.7.

The new equilibrium value of Y is

$$Y^* = \frac{a + I_t}{1 - b} = \frac{40 + 110}{1 - 0.6} = \frac{150}{0.4} = 375$$

The general solution to the difference equation

$$Y_t = Y^* + (Y_0 - Y^*)b^t \qquad (11)$$

therefore becomes, for this numerical example,

$$Y_t = 375 + (411 - 375)0.6^t$$

$$= 375 + 36(0.6)^t$$

The first few values of Y are thus

$$Y_1 = 375 + 36(0.6) = 375 + 21.6 = 396.6$$

$$Y_2 = 375 + 36(0.6)^2 = 375 + 12.96 = 387.96$$

$$Y_3 = 375 + 36(0.6)^3 = 375 + 7.776 = 382.776$$

These are exactly the same as the answers computed by the iterative method in Example 13.7 and also the same as those produced by the spreadsheet in Table 13.3, which is what one would expect.

One advantage of using the difference equation solution is that it is not necessary to calculate all the preceding values of Y to obtain its value in a given time period. For example, the value of Y in time period 9 can be calculated directly as

$$Y_9 = 375 + 36(0.6)^9 = 375 + 0.3628 = 375.3628$$

As t increases in value, eventually the value of $(0.6)^t$ becomes so small as to make the second term negligible.

In the above example we can say that for all intents and purposes Y has effectively reached its equilibrium value of 375 by the ninth time period, although theoretically Y would never actually reach 375 if infinitesimally small increments were allowed.

By now, many of you may be thinking that this difference equation method of computing the different values of Y in the adjustment process in a Keynesian macroeconomic model is extremely long-winded and it would be much quicker to compute the values by the iterative method, particularly if a spreadsheet can be used.

In many cases you may be right. However, you must remember that this chapter is only intended to give you an insight into the methods that can be used to trace out the time path of adjustment in dynamic economic models. The mathematical methods of solution explained here can be adapted to tackle more complex problems that cannot be illustrated on a spreadsheet. Also, economists need to set up mathematical formulations for functional relationships in order to estimate the parameters of these functions. Those of you who study econometrics after the first year of your course will discover that the algebraic solutions to difference equations can help in the setting up of models for testing certain dynamic economic relationships.

Now that the general solution to the lagged Keynesian macroeconomic model has been derived, it can be applied to other numerical examples and may even allow you to compute answers more quickly than by switching on your computer and setting up a spreadsheet.

Example 13.8

In the basic Keynesian model

$$Y_t = C_t + I_t$$

$$C_t = 650 + 0.5Y_{t-1}$$

there is initially an equilibrium, with I_t remaining at 300. Then I_t suddenly increases to 420. What will be the actual level of Y six time periods after this change?

Solution

The initial equilibrium level of Y is

$$Y^* = \frac{a + I_t}{1 - b} = \frac{650 + 300}{1 - 0.5} = \frac{950}{0.5} = 1,900$$

Therefore the value of C when the increase in I takes place is

$$C_0 = 650 + 0.5(1,900) = 650 + 950 = 1,600$$

and so the value of Y immediately after this shock is

$$Y_0 = C_0 + I_0 = 1,600 + 420 = 2,020$$

The new equilibrium level of Y is

$$Y^* = \frac{a + I_t}{1-b} = \frac{650 + 420}{1 - 0.5} = \frac{1,070}{0.5} = 2,140$$

Substituting these values into the general solution for the lagged Keynesian macroeconomic model difference equation for time period 6, we get

$$Y_t = Y^* + (Y_0 - Y^*)b^t$$
$$= 2,140 + (2,020 - 2,140)\,0.5^t$$
$$= 2,140 - 120(0.5)^6$$
$$= 2,140 - 1.875 = 2,138.125$$

Example 13.9

How many time periods will it take Y to reach 2,130 in the preceding example?

Solution

We know that $Y_t = 2,130$ and we wish to find t. Thus substituting into the solution

$$Y_t = Y^* + (Y_0 - Y^*)b^t$$

we get

$$2,130 = 2,140 + (2,020 - 2,140)\,0.5^t$$
$$-10 = -120(0.5)^t$$
$$0.08333 = (0.5)^t$$

To get t, put this into the log form

$$\log 0.08333 = t \log 0.5$$
$$\frac{\log 0.08333}{\log 0.5} = t$$
$$3.585 = t$$

using the [LOG] function on a calculator. Therefore Y will have exceeded 2,130 by the end of the fourth time period.

Only the most basic lagged Keynesian model has been considered so far in this section. Other possible formulations have been suggested for the ways in

Basic mathematics for economists

which past income levels can determine current expenditure. For example

$$C_t = a + bY_{t-2}$$

or

$$C_t = a + b_1 Y_{t-1} + b_2 Y_{t-2}$$

The latter example is known as a 'distributed lag' model. A general form of the distributed lag consumption function is

$$C_t = a + b_1 Y_{t-1} + b_2 Y_{t-2} + \ldots + b_n Y_{t-n}$$

The solutions of these more complex models require more advanced mathematical methods than have been explained in this basic mathematics text. Those students who continue with mathematical economics in the next stage of their course will probably learn about these methods of solution but at this stage it is sufficient that you understand the basic concept of difference equations and the solution of simple single-lag models. You should be able to adapt the spreadsheet set up in Table 13.3, however, to trace out the adjustment path of income in a distributed lag model with given parameters.

Example 13.10

Use a spreadsheet to estimate the value of Y_t for the twelve time periods after I is increased to 140, assuming that Y is determined by the distributed lag Keynesian model

$$Y_t = C_t + I_t$$
$$C_t = 320 + 0.5Y_{t-1} + 0.3Y_{t-2}$$

and that the system had previously been in equilibrium with I at 90.

Solution

This is a spreadsheet exercise that you can do yourself by making the necessary adjustments to the formulae that were used to set up the spreadsheet in Table 13.3 when tackling Example 13.7. Be careful in setting up the initial values, however, as C will depend on the old equilibrium level of Y up to period 1. Some of the initial values are calculated manually below for you to check against. The initial equilibrium level Y^* would have satisfied the equations

$$Y^* = C_t + 90 \tag{1}$$
$$C_t = 320 + 0.5Y^* + 0.3Y^* = 320 + 0.8Y^*$$

By substitution into (1)

436

Dynamics and difference equations

$$Y^* = 320 + 0.8Y^* + 90$$

$$0.2Y^* = 410$$

$$Y^* = 2{,}050 = Y_{t-1} = Y_{t-2}$$

Thus

$$C_0 = 320 + 0.5(2{,}050) + 0.3(2{,}050) = 1{,}960$$
$$Y_0 = C_0 + I_0 = 1{,}960 + 140 = 2{,}100$$
$$C_1 = 320 + 0.5(2{,}100) + 0.3(2{,}050) = 1{,}985$$
$$Y_1 = C_1 + I_1 = 1{,}985 + 140 = 2{,}125 \quad \text{etc.}$$

Your spreadsheet should show Y converging on 2,300.

QUESTIONS 13.3

1. A Keynesian macroeconomic model with a single-time-period lag on the consumption function described below is initially in equilibrium with the level of I_t given at 500.

$$Y_t = C_t + I_t$$
$$C_t = 750 + 0.5Y_{t-1}$$

   I_t is then increased to 650. Use difference equation analysis to find the value of Y in the fourth time period after this disturbance to the system. Will it then be within 1% of its new equilibrium level?

2. In the macroeconomic model

$$Y_t = C_t + I_t$$
$$C_t = 2{,}500 + 0.9Y_{t-1}$$

   there is initially an equilibrium with the level of I_t set at 1,100. Investment is then increased to a level of 1,500 where it remains for future time periods. Calculate what the level of Y will be in the fortieth time period after this investment increase.

3. In a basic Keynesian model with a government sector

$$Y_t = C_t + I_t + G_t$$

   where $I_t = 269$, $G_t = 310$ (exogenously determined Government expenditure), and

$$C_t = 80 + 0.8Y_{t-1}^D$$

   where Y_t^D is disposable (after-tax) income. Assume that all income is taxed at a rate of 25%. Government expenditure is then increased to 450 and kept at this level. What tax revenue can the government expect to raise five time periods after this initial rise in expenditure?

437

4. Use a spreadsheet to trace out the pattern of adjustment of Y towards its new equilibrium value in the model

$$Y_t = C_t + I_t$$

$$C_t = 310 + 0.7Y_{t-1}$$

if I is exogenously increased from 240 to 350 and then kept at this new level. Assume the system was initially in equilibrium. What is the value of C in the fourteenth time period after this increase in I?

13.5 DUOPOLY PRICE ADJUSTMENT

A duopoly is a market where there are only two sellers. It is difficult for economists to predict price and output in markets with only a few sellers (oligopoly) because firms' reactions to their rivals' actions can vary depending on the strategy they adopt. Firms may naively assume that rivals will not react to whatever pricing policy they themselves operate, they may try to outguess their rivals, or they may collude. You will learn more about these different models in your economics course. Here we will just examine how price may adjust over time in one of the simpler 'naive' models.

Both the Cournot and Bertrand models assume that firms do not think ahead and they set their output (Cournot) or their price (Bertrand) at the levels that will maximize profits assuming that their rivals' current output or price will not change. Without going into the details of the model, the predictions of the Bertrand model can be summarized in terms of the 'reactions functions' shown in Figure 13.4 for two duopolists X and Y. These show the price that will

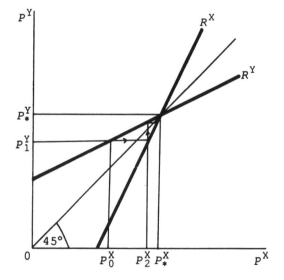

Figure 13.4

maximize one firm's profits given the value of the other firm's price read off the other axis. For example, X's reaction function R^X slopes up from right to left. If Y's price is higher, then X can get away with a higher price, but if Y lowers its price then X also has to reduce its price otherwise it will lose sales. The model assumes that both firms have identical cost structures and so the reaction functions are symmetrical, intersecting where they both cross the 45° line representing equal prices. The prediction is that prices will eventually settle at levels P_*^X and P_*^Y, which are equal. The path of adjustment from an initial price P_0^X is shown in Figure 13.4. In time period 1, Y reacts to P_0^X by setting price P_1^Y; then X sets price P_2^X in time period 2, and so on until P_*^X and P_*^Y are reached. Note that we are assuming that the firms take it in turns to adjust price and so each firm only sets a new price every other time period.

Let us now use our knowledge of difference equations to derive a function that will tell us what the price of one of the firms will be in any given time period with the aid of a numerical example.

Example 13.11

Two duopolists, firms X and Y, have the reaction functions

$$P_t^X = 45 + 0.8 P_{t-1}^Y$$

$$P_t^Y = 45 + 0.8 P_{t-1}^X$$

Assuming that the assumptions of the Bertrand model hold, derive a difference equation for P_t^X and calculate what P_t^X will be in time period 10 if firm X starts off by setting a price of 300.

Solution

Substituting Y's reaction function into X's, we get

$$P_t^X = 45 + 0.8\,(45 + 0.8 P_{t-2}^X)$$

$$P_t^X = 45 + 36 + 0.64 P_{t-2}^X$$

$$P_t^X = 81 + 0.64 P_{t-2}^X \tag{1}$$

Note that the value P_{t-2}^X appears because we have substituted in the reaction function of Y for period $t-1$ to correspond to the value of P_{t-1}^Y in X's reaction function, i.e. we have substituted in the equation

$$P_{t-1}^Y = 45 + 0.8 P_{t-2}^X$$

The equilibrium solution to the difference equation (1) will be where price no longer changes and

$$P_{t-1}^X = P_t^X = P_*^X$$

which is the equilibrium value of P_t^X. Substituting P_*^X into (1)

$$P_*^X = 81 + 0.64P_*^X$$

$$0.36P_*^X = 81$$

$$P_*^X = 225$$

To find the complementary solution we use the standard method of ignoring the constant term and assuming that

$$P_t^X = Ak^t$$

where A and k are constant parameters. Substituting into the difference equation

$$P_t^X = 81 + 0.64P_{t-2}^* \tag{1}$$

where the constant 81 is ignored, gives

$$Ak^t = 0.64Ak^{t-2}$$

Cancelling the common term Ak^{t-2}, we get

$$k^2 = 0.64$$

$$k = 0.8$$

The complementary solution is therefore

$$P_t^X = A(0.8)^t$$

Adding the complementary and equilibrium solutions, the general solution to difference equation (1) becomes

$$P_t^X = A(0.8)^t + 225$$

To find the value of A we need to know the specific value of P_t^X at some point in time. The question specifies that initially (i.e. in time period 0) P_t^X is 300 and so

$$P_0^X = 300 = A(0.8)^0 + 225$$

$$300 = A + 225$$

$$A = 75$$

Thus, as in the previous difference equation applications, A is the difference between the final equilibrium and the initial value of the variable in question, i.e. P_t^X in this example.

The full solution to the difference equation (1) is therefore

$$P_t^X = 75(0.8)^t + 225$$

This can be used to calculate the value of P_t^X *every alternate* time period. (Remember that in the intervening time periods X keeps price constant while Y adjusts price.) The price adjustment path will therefore be

$$P_0^X = 75(0.8)^0 + 225 = 75 + 225 = 300$$

$$P_2^X = 75(0.8)^2 + 225 = 48 + 225 = 273$$

$$P_4^X = 75(0.8)^4 + 225 = 30.72 + 225 = 255.72$$

etc.

The question asks what price will be in time period 10 and so this can be calculated as

$$P_{10}^X = 75(0.8)^{10} + 225 = 8.05 + 225 = 233.05$$

Example 13.12

Two firms X and Y in an oligopolistic market take a short-sighted view of their situation and set price on the basis of their rivals' price in the previous time period according to the reaction functions

$$P_t^X = 300 + 0.75 P_{t-1}^Y$$

$$P_t^Y = 300 + 0.75 P_{t-1}^X$$

Assume that each adjusts its price every other time period. The market is initially in equilibrium with $P_t^X = P_t^Y = 1{,}200$. Firm X then decides to try to improve its profits by raising price to 1,650. Taking into account the reactions to rivals' price changes described in the above functions, calculate what X's price will be in the eighth time period after its breakaway price rise.

Solution

$$P_t^X = 300 + 0.75 P_{t-1}^Y$$
$$= 300 + 0.75(300 + 0.75 P_{t-2}^X)$$
$$= 300 + 225 + 0.5625 P_{t-2}^X$$
$$= 525 + 0.5625 P_{t-2}^X \tag{1}$$

This difference equation is in the same format as that found in Example 13.11 above. Its solution will therefore also be in the same format, i.e.

$$P_t^X = Ak^t + P_*^X$$

441

Basic mathematics for economists

The equilibrium value P^X_* is given in the question as 1,200. This can easily be checked in our difference equation (1) because in equilibrium

$$P^X_t = P^X_{t-1} = P^X_*$$

and so

$$P^X_* = 525 + 0.5625 P^X_*$$

$$0.4375 P^X_* = 525$$

$$P^X_* = 1{,}200$$

To find the complementary solution, let $P^X_t = Ak^t$ and substitute into the difference equation (1) after dropping the constant 525. Thus

$$Ak^t = 0.5625 Ak^{t-2}$$

Cancelling Ak^{t-2} this gives

$$k^2 = 0.5625$$

$$k = 0.75$$

Adding the complementary and equilibrium solutions, the complete solution to the difference equation becomes

$$P^X_t = A(0.75)^t + 1{,}200$$

The value of A can be found when the initial $P^X_0 = 1{,}650$ is substituted. Thus

$$P^X_0 = 1{,}650 = A(0.75)^0 + 1{,}200$$

$$1{,}650 = A + 1{,}200$$

$$450 = A$$

The complete solution to the difference equation is therefore

$$P^X_t = 450(0.75)^t + 1{,}200$$

and so in the eighth period after the initial price rise

$$P^X_8 = 450(0.75)^8 + 1{,}200$$

$$= 45.05 + 1{,}200$$

$$= 1{,}245.05$$

QUESTIONS 13.4

1. Two duopolists X and Y react to each others' prices according to the functions

$$P^X_t = 240 + 0.9 P^Y_{t-1}$$

442

$$P_t^Y = 240 + 0.9 P_{t-1}^X$$

If firm X sets an initial price of 2,900, what will its price be twenty time periods later? Assume that each firm adjusts price every alternate time period.

2. In an oligopolistic market, the two firms X and Y have the following price reaction functions:

$$P_t^X = 800 + 0.6 P_{t-1}^Y$$

$$P_t^Y = 800 + 0.6 P_{t-1}^X$$

The usual assumptions of the Bertrand model apply and price is initially in equilibrium at a level of 2,000 for both firms. Firm X then decides to cut price to 1,500 to try to steal Y's market share. We know from the analysis of this model that X's price reduction will be short-lived and price will creep back towards its equilibrium level, but how short-lived? Calculate whether or not P^X will be back within 1% of its equilibrium value within six time periods. (Be careful how you calculate the value of A in the difference equation as this time the initial value is below the equilibrium value of P^X.)

3. In a duopoly where the assumptions of the Bertrand model hold, the two firms' reaction functions are

$$P_t^X = 95.54 + 0.83 P_{t-1}^Y$$

$$P_t^Y = 95.54 + 0.83 P_{t-1}^X$$

If firm X unexpectedly changes price to 499, derive the solution to the difference equation that determines P_t^X and use it to predict P_t^X in the twelfth time period after the initial change.

14

Exponential functions and continuous growth

14.1 CONTINUOUS GROWTH

In Chapter 7, growth and decay were treated as processes taking place over discrete time intervals. In order to analyse growth and decay as continuous processes it is first necessary to understand the concepts of exponential functions and natural logarithms.

You will also come across exponential and logarithmic functions elsewhere in economics as functional forms that explain some data patterns more accurately than would a simple linear function. They will be particularly useful for those of you who go on to study econometrics.

Although exponential and logarithmic functions can be used to tackle many conceptual problems in economics, particularly those concerning economic growth, it should be noted that the numerical answers to several of the basic problems in this chapter are identical to those that could be obtained using the discrete growth concepts already covered in Chapter 7. The reasons why this is so are explained in Section 14.5.

14.2 EXPONENTIAL FUNCTIONS

One definition of an exponential function is

$$y = A^x$$

where A is a constant and $A > 0$, $A \neq 1$. This is known as an exponential function to base A. It would obviously be very steep if A was any number substantially greater than 1.

Some values for exponential functions to different bases are given in Table 14.1. For all values of A it can be seen that

when $x = 0$ then $A^x = 1$,
when $x = 1$ then $A^x = A$.

These general rules for exponents were explained back in Chapter 2.

Exponential functions and continuous growth

Table 14.1 Selected values of exponential functions to different bases

	A^x				
x	$A = 0.1$	$A = 0.5$	$A = 1.1$	$A = 2$	$A = 10$
0	1	1	1	1	1
0.1	0.79	0.93	1.01	1.072	1.26
0.5	0.32	0.71	1.05	1.41	3.16
1	0.1	0.5	1.1	2	10
2	0.01	0.25	1.21	4	100
5	0.00001	0.031	1.61	32	100,000
-0.1	1.26	1.072	0.99	0.93	0.79
-0.5	3.16	1.41	0.95	0.71	0.32
-1	10	2	0.91	0.5	0.1
-2	100	4	0.83	0.25	0.01
-5	100,000	32	0.62	0.03	0.00001

From Table 14.1 we can also see that the values of 2^x and 10^x for positive values of x are identical to the values of 0.5^x and 0.1^x for corresponding negative values of x. Again, this is as one would expect from the general rules for dealing with exponents. For example,

$$0.5^1 = 2^{-1}$$

and

$$(0.5)^{-1} = \frac{1}{0.5} = 2$$

In mathematics there is a special number which when used as a base for an exponential function yields several useful results. This number is

$$2.7182818 \text{ (to 7 dp)}$$

and is usually represented by the letter 'e'. You should be able to get this number on your calculator by entering 1 and then using the $[e^x]$ function key.

Other values of e^x can also be found using the $[e^x]$ function key on a calculator. The usual procedure is to enter the number (x) and then press the $[e^x]$ function key. There also exist tables of exponential values which were used by students before calculators with exponential function keys became available. To check that you can accurately find values of e^x, use your calculator to obtain the following exponential values:

$$e^{0.05} = 1.0512711$$

$$e^{0.5} = 1.6487213$$

$$e^4 = 54.59815$$

$$e^{-2.624} = 0.0725122$$

If you do not get these values ask your tutor for assistance.

In economics, exponential functions to this base are particularly useful for analysing growth rates of different variables. This number, e, is also used as a base for natural logarithms, explained in the next section. Although strictly speaking the term 'exponential function' can be applied to any function of the form $y = A^x$ for positive values of A, and the special function $y = e^x$ should be known as the 'natural exponential function', it is usual to refer to functions of the form $y = e^x$ simply as 'exponential functions'.

To understand how the rather awkward number e is derived, we return to the methods used for calculating the value of an investment developed in Chapter 7. You will recall that the final value (F) of an initial investment (A) deposited for t time periods at an interest rate of i can be calculated from the formula

$$F = A(1 + i)^t$$

If the interest rate is assumed to be 100% then $i = 1$ and the formula becomes

$$F = A(1 + 1)^t = A(2)^t$$

Let us also assume that the sum invested $A = 1$. If interest is paid at the end of each year then after 1 year

$$F_1 = (1 + 1)^1 = 2$$

If interest is paid monthly at the annual rate divided by 12 this will give a larger final return because interest credited at the end of the first month will itself earn further interest in the second and subsequent months, interest credited at the end of the second month will be reinvested in the third month, and so on. (This point was made when the relationship between monthly rates of interest and the APR was explained.) Again, assuming an annual rate of interest of 100% ($i = 1$) and an initial investment of 1, the final sum after 12 months at an interest rate of $1/12$ will be

$$F_{12} = \left(1 + \frac{1}{12}\right)^{12} = 2.6130353$$

If interest was to be credited daily at the rate $1/365$ then the final sum would be

$$F_{365} = \left(1 + \frac{1}{365}\right)^{365} = 2.7145677$$

If interest was to be credited by the hour at a rate of $\frac{1}{8,760}$ (given that there are 8,760 hours in a 365 day year) then the final sum would be

$$F_{8,760} = \left(1 + \frac{1}{8,760}\right)^{8,760} = 2.7181209$$

Exponential functions and continuous growth

From the above calculations we can see that the more frequently that interest is credited the closer the value of the final sum accumulated gets to the value of e of 2.7182818. . . .

In fact e is the number that one would get if interest was credited instantaneously throughout the time period. In other words,

$$e = \left(1 + \frac{1}{n}\right)^n$$

where $n \to \infty$.

In terms of interest payments, what the above result means is that an annual interest rate of 100% credited instantaneously is equivalent to an APR of 271.83% (to 2 decimal places). Where the interest rate is less than 100% but is still paid instantaneously, the ratio of the APR to the nominal annual rate is not the same as e, as is explained in Section 14.5.

Although it may be unrealistic to assume that a bank pays interest instantaneously so that a sum invested grows continually every second, the almost continuous daily payment of interest is not uncommon. It is also not unreasonable to assume continuous growth of other variables relevant to economics, e.g. population, the amount of natural materials mined. Other variables may continuously decline in value over time, e.g. the stock of a non-renewable natural resource. Instead of considering the growth of an investment at a given annual interest rate i, let us consider the continuous growth of any economic variable at an annual rate of r. (If we use the notation r for rate of continuous growth and i for annual interest rate, it will help distinguish the two concepts and allow comparisons.)

For any economic variable that grows continuously, assume that A is the initial value, r is the rate of growth (annual), n is the number of times per year that increments in growth are accumulated, t is the number of years of growth and y is the final value. This means that

$$y = A\left(1 + \frac{r}{n}\right)^{nt}$$

where $n \to \infty$ for continuous growth. To reduce this to a simpler formulation, multiply top and bottom of the exponent by r so that

$$y = A\left(1 + \frac{r}{n}\right)^{nt} = A\left(1 + \frac{r}{n}\right)^{\left(\frac{n}{r}\right)rt}$$

This can be written as

$$y = A\left(1 + \frac{1}{m}\right)^{mrt} \tag{1}$$

where $m = n/r$. As the growth becomes continuous and $n \to \infty$ then $n/r \to \infty$, i.e. $m \to \infty$. Therefore,

447

$$\left(1+\frac{1}{m}\right)^{m} \to e \quad \text{as} \quad m \to \infty$$

Substituting this result back into (1) above gives

$$y = Ae^{rt}$$

This formula can be used to find the final value of any variable growing continuously at a known annual rate from a given original value.

Example 14.1

Population in a Third World nation is growing continuously at a rate of 3%. If the population is now 4.5 million, what will it be in 15 years' time?

Solution

The final value of the population (in millions) is found by using the formula $y = Ae^{rt}$ and substituting the given numbers: initial value $A = 4.5$; rate of growth $r = 3\% = 0.03$; number of time periods $t = 15$, giving

$$y = 4.5e^{0.03(15)} = 4.5e^{0.45} = 4.5 \times 1.5683122 = 7.0574048$$

Thus the predicted final population is 7,057,405.

Example 14.2

An economy is forecast to grow continuously at a rate of 2.5%. If GNP is currently £56 billion, what will the forecast for GNP be at the end of the third quarter the year after next?

Solution

In this example

$$t = 1.75 \text{ years}$$
$$r = 2.5\% = 0.025$$
$$A = 56 \ (\text{£ billion})$$

Therefore,

$$y = Ae^{rt} = 56e^{0.025(1.75)} = 56e^{0.04375} = 58.504384$$

Thus the forecast for GNP is £58,504,384,000.

So far we have only considered positive growth, but the exponential function can also be used to analyse continuous decay.

We have already shown that

$$y = Ae^{rt}$$

Dividing both sides by e^{rt} gives

$$\frac{y}{e^{rt}} = A$$

This can be rewritten as

$$A = ye^{-rt}$$

This formula tells us what initial sum A would grow to a value of y after t years of continuous growth at rate r. It also tells us what would happen if the sum y was allowed to decline continuously at rate r for t years. In this case, r is known as the rate of decay.

Example 14.3

A river flow through a hydroelectric dam is 18 million gallons a day. The river flow is shrinking continuously at an annual rate of 4%. What will the river flow be in 6 years' time?

Solution

In this example we need to find A, and we know $y = 18$, $r = 4\% = 0.04$ and $t = 6$. Thus

$$A = ye^{-rt} = 18e^{-0.04(6)} = 18e^{-0.24} = 14.16$$

Therefore, the river flow will shrink to 14.16 million gallons per day.

To help clarify this concept of a continuous rate of growth we can use calculus to derive the rate of change of an exponential function. Assume that the variable y changes over time according to the function $y = Ae^{rt}$. The rate of change of y with respect to t will therefore be the derivative dy/dt. It is not a straightforward exercise to differentiate this function, however. For the time being let us accept the result that if $y = e^t$ then $dy/dt = e^t$, i.e. the differential of an exponential function is the function itself. (This result is explained in Section 14.4.) Thus, using the chain rule, when $y = Ae^{rt}$ then

$$\frac{dy}{dt} = rAe^{rt}$$

This derivative tells us how much y increases with respect to an infinitesimally small increment in t, but in problems of growth we are interested in the *proportional* increase in y with respect to its original value. The *rate* of growth is therefore

$$\frac{dy/dt}{y} = \frac{rAe^{rt}}{Ae^{rt}} = r$$

Even though r is the instantaneous rate of growth at any given moment in time, it must be expressed with reference to a time interval, usually a year in economic applications, e.g. 4.5% per annum. It is rather like saying that the slope of a curve is, say, 1.78 at point X. A slope of 1.78 means that height increases by 1.78 for every 1 unit increase along the horizontal axis, but at a single point on a curve there is no actual movement along the axis.

Example 14.4

Owing to improved technology and efficiency in production an empirical study found the output of product Q at any moment in time to be determined by the function

$$Q = 40e^{0.03t}$$

where t is the number of years from the base year in the empirical study and Q is the output per year in thousands of tonnes. What is the growth rate of production?

Solution

The growth rate can simply be read off from the function

$$Q = 40e^{0.03t}$$

as $0.03 = 3\%$.

QUESTIONS 14.1

1. A country's population is currently 32 million and is growing continuously at an annual equivalent rate of 3.5%. What will the population be in 20 years' time if this rate of growth persists?
2. A company launched a successful new product last year. The current weekly sales level is 56,000 units. If sales are expected to grow continually at an annual equivalent rate of 12.5% what will be the expected level of sales 36 weeks from now? (Assume that 1 year is exactly 52 weeks.)
3. Current stocks of mineral M are 250 million tonnes. If stocks are continually being used up at an annual equivalent rate of 9%, what amount of M will remain after 30 years?
4. A renewable natural resource R will allow an estimated maximum consumption rate of 200 million units per annum. Current annual usage is

65 million units. If the annual level of usage grows continually at an annual equivalent rate of 7.5% will there be sufficient R to satisfy annual demand after (a) 5 years, (b) 10 years, (c) 15 years, (d) 20 years?

5. Stocks of resource R are shrinking continually at an annual equivalent rate of 8.5%. How much will remain in 30 years' time if current stocks are 725,000 units?

6. A statistician estimates that a country's population N is growing continuously and can be determined by the function

$$N = 3,620,000e^{0.02t}$$

where t is the number of years after 1950. What is the population growth rate? Will population reach 10 million by the year 2000?

14.3 NATURAL LOGARITHMS

In Chapter 2 we saw how logarithms to base 10 were defined and utilized in mathematical problems. You will recall that the logarithm of a number to base X is the power to which X must be raised in order to equal that number.

For many purposes in mathematics it is useful to work with logarithms to the base e. These are known as 'natural logarithms', and the usual notation is ln (as opposed to log for logarithms to base 10).

As with values of the exponential function, natural logarithms can be found from a calculator using the [LN] function key, or from printed tables. Check that you can derive the following values from your calculator:

$$\ln 1 = 0$$
$$\ln 2.6 = 0.9555114$$
$$\ln 8,463 = 9.043459$$

The rules for using natural logarithms are the same as for logarithms to any other base. For example, to multiply two numbers, their logarithms are added. But how do you then transform the sum of the logarithms back to a number, i.e. what is the 'antilog' of a natural logarithm?

Consider the exponential function $y = e^x$. The natural logarithm of y will be x, i.e. $\ln y = x$. If we only know the value of $\ln y$ and wish to find y then, given that $\ln y = x$, it must be true that

$$y = e^x = e^{\ln y}$$

Therefore y can be found from $\ln y$ by finding the exponential of $\ln y$.

Example 14.5

Multiply 5,623.76 by 441.873 using natural logarithms.

Basic mathematics for economists

Solution

$$\ln 5{,}623.76 = 8.6347558 +$$
$$\ln 441.873 = \underline{6.0910225}$$
$$14.725778 \quad \text{(to 6 dp)}$$

To transform this logarithm back to its corresponding number we find
$e^{14.725778} = 2{,}484{,}987.7$

This answer can be verified by carrying out a straightforward multiplication on your calculator. You would not, of course, actually use natural logarithms for basic numeric multiplication problems and this example is only given to illustrate how natural logarithms operate.

One use of natural logarithms is to help determine rates of growth.

Example 14.6

The consumption of natural resource R has risen from 38 million tonnes (per annum) to 68.4 million tonnes over the last 12 years. If it is assumed that growth has been continuous at the same rate what is this rate of growth?

Solution

If growth is continuous then the final consumption level of R will be determined by the exponential function:

$$R = R_0 e^{rt} \tag{1}$$

where r is the rate of growth and R_0 is the initial consumption level. Substituting the known values into (1) gives

$$68.4 = 38 e^{12r}$$
$$1.8 = e^{12r} \tag{2}$$

In (2), $12r$ is the power to which e must be raised to equal 1.8. Therefore,

$$\ln 1.8 = 12r$$
$$r = \frac{\ln 1.8}{12} = 0.0489822$$

Thus consumption has risen at an annual equivalent rate of 4.9%.

Natural logarithms can also be used to work out rates of decay.

Example 14.7

The annual catch of fish from a specific sea area is declining continually at a constant rate r. Ten years ago the total catch was 940 tonnes and this year the total catch is 784 tonnes. What is the rate of decline?

Exponential functions and continuous growth

Solution

If the decline is continuous then the catch C at any point in time will be determined by the function

$$C = C_0 e^{-rt}$$

Substituting the known values into this function

$$784 = 940 e^{-10r}$$
$$0.8340426 = e^{-10r}$$
$$\ln 0.8340426 = -10r$$
$$-0.1814799 = -10r$$
$$0.0181471 = r$$

Therefore the rate of decline is 1.8%.

Rates of growth and decline can also be determined over short time periods of less than a year.

Example 14.8

Consumption of mineral M is known to be increasing continually at a constant rate per annum. The daily rate of consumption was 46.4 tonnes on 1 March and had risen to 47.2 tonnes by 1 May. What is the annual equivalent growth rate for consumption of this mineral?

Solution

From 1 March to 1 May is 61 days. Thus

$$47.2 = 46.4 e^{r(61/365)}$$

$$\ln\left(\frac{47.2}{46.4}\right) = r\left(\frac{61}{365}\right)$$

$$r = \ln\left(\frac{47.2}{46.4}\right) \times \frac{365}{61} = 0.1022864$$

Therefore the annual equivalent growth rate is 10.2%.

QUESTIONS 14.2

1. In an advanced industrial economy, population is observed to have grown from 50 to 55 million over the last 20 years. What is the annual rate of growth?

2. If the average quantity of petrol used per week by a typical private motorist has increased from 16.1 litres to 24.2 litres over the last 20 years, what has been the average annual growth rate in petrol consumption assuming that this increase in petrol consumption has been continuous?

   If, over the same time period, petrol consumption for a typical private car has fallen from 8.75 litres per 100 km to 6.56 litres per 100 km, what has been the average annual growth rate in the distance covered each week by a typical motorist?

3. World reserves of mineral M are observed to have declined from 830 million tonnes to 675 million tonnes over the last 25 years. Assuming this decline to have been continuous, calculate the annual rate of decline and then predict what reserves will be left in 10 years' time.

4. An economy's GNP grows from £5,682 million to £5,727 million during the first quarter of a new government's term of office. If this growth rate persisted through its entire term of office of 4 years, what would GNP be at the time of the next election?

5. If the number of a protected species of animal in a reserve increased continually from 600 in 1980 to 1,450 in 1990, what was the average annual equivalent growth rate?

14.4 DIFFERENTIATION OF LOGARITHMIC AND EXPONENTIAL FUNCTIONS

We have already used the result that if $y = e^t$ then $dy/dt = e^t$. It can be derived if we accept as given that if

$$y = \ln x$$

then

$$\frac{dy}{dx} = \frac{1}{x}$$

This result can be proved mathematically but the proof is rather complex. As there is no need for you to understand it, the result is just stated in this basic mathematics text. Note that this also implies that $\int (1/x)\, dx = \ln x$. This was the exceptional case in integration not dealt with in Chapter 12.

Returning to the exponential function $y = e^t$, if natural logarithms of both sides are taken, then $\ln y = t$. We then use the rule for differentiating natural logarithmic functions stated above (noting that the term y is used differently in this function). Thus, given $t = \ln y$,

$$\frac{dt}{dy} = \frac{1}{y} \tag{1}$$

The inverse function rule in calculus states that

454

Exponential functions and continuous growth

$$\frac{dy}{dt} = \frac{1}{dt/dy} \qquad (2)$$

Therefore, substituting (1) into (2) gives

$$\frac{dy}{dt} = \frac{1}{1/y} = y$$

and so when $y = e^t$ then

$$\frac{dy}{dt} = e^t$$

which is the result we wished to prove.

14.5 DISCRETE AND CONTINUOUS GROWTH COMPARED

In mathematics and economics the use of exponential functions to analyse the concept of continuous growth has led to some important results. However, for many simple numerical problems it may be just as easy to work with the concept of discrete growth. Note, however, that a given increase in the value of a variable will correspond to an instantaneous growth rate that is lower than the discrete growth rate.

Example 14.9

The annual usage of a mineral M grows from 460 tonnes to 2,990 tonnes in 10 years. Find the rate of growth of annual usage and estimate usage after a further 5 years assuming (i) continuous growth and (ii) discrete growth.

Solution

(i) If growth is continuous then

$$2,990 = 460e^{10r}$$
$$6.5 = e^{10r}$$
$$\ln 6.5 = 10r$$
$$r = \frac{\ln 6.5}{10} = 0.1871802 = 18.7\%$$

Using the more accurate figure for r of ln 6.5/10, usage in year 15 will be

$$460e^{15r} = 460e^{1.5\ln 6.5}$$
$$= 7,623 \quad \text{(to nearest whole number)}$$

(ii) Assuming discrete growth (at rate i)

$$2,990 = 460(1 + i)^{10}$$

455

$$6.5 = (1+i)^{10}$$
$$1.2058446 = 1+i$$
$$i = 0.2058446 = 20.6\%$$

Usage in year 15 will be

$$460(1+i)^{15} = 460(1.2058446)^{15} = 7,623$$

The above example shows that an instantaneous growth rate of 18.7% is equivalent to a discrete growth rate of 20.6%. Although these rates are different they will still predict the same final outcome.

The discrete growth concept can also quite easily be applied to problems involving fractions of a year. An analogy is the situation where someone closes a building society account three-quarters of the way through a year. If interest is normally credited annually at rate i then the interest credited when the account is closed will be at the rate $0.75i$.

Example 14.10

An investment grows from £64,000 to £72,000 in 2 years. Find the annual rate of growth and predict its value at the end of a further $1\frac{1}{4}$ years assuming (i) continuous growth and (ii) discrete growth.

Solution

(i) Assuming continuous growth

$$72,000 = 64,000e^{2r}$$
$$1.125 = e^{2r}$$
$$\ln 1.125 = 2r$$
$$r = 0.5(\ln 1.125)$$
$$= 0.0588915$$
$$= 5.9\%$$

In $3\frac{1}{4}$ years from the original start date, the value of the investment will be

$$64,000e^{3.25r} = 64,000e^{3.25(0.0588915)}$$
$$= £77,500.20$$

(ii) If growth is discrete

$$72,000 = 64,000(1+i)^2$$
$$1.125 = (1+i)^2$$

Exponential functions and continuous growth

$$\sqrt{(1.125)} = 1 + i$$

$$i = \sqrt{(1.125)} - 1$$

$$= 0.0606602 = 6.1\%$$

In $3\frac{1}{4}$ years the investment will be

$$64{,}000(1 + i)^{3.25} = 77{,}500.198$$

$$= £77{,}500.20$$

A direct comparison of the continuous growth rate r that would accumulate a given sum A to the same final sum F over 1 year as would a discrete growth rate i can be found as follows:

Continuous growth $F = Ae^r$
Discrete growth $F = A(1 + i)$

Therefore

$$Ae^r = A(1 + i)$$

$$e^r = (1 + i)$$

$$r = \ln(1 + i) \qquad\qquad (1)$$

From (1) we can also write

$$i = e^r - 1 \qquad\qquad (2)$$

Example 14.11

(i) Find the continuous growth rate that would correspond to a discrete growth rate of
   (a) 0   (b) 0.1   (c) 0.5   (d) 1   (e) e − 1
(ii) Find the discrete growth rate that would correspond to continuous growth rates (a), (b), (c) and (d) above and (e) ln 2. Give all answers to 2 significant decimal places.

Solution

(i) Using the formula $r = \ln(1 + i)$ the answers are as follows:

   (a) $i = 0$ $r = 0$
   (b) $i = 0.1$ $r = 0.095$
   (c) $i = 0.5$ $r = 0.41$
   (d) $i = 1$ $r = 0.69$
   (e) $i = e - 1$ $r = 1$

(ii) Using the formula $i = e^r - 1$ the answers are

   (a) $i = 0$ $r = 0$

457

(b) $i = 0.11$ $r = 0.1$
(c) $i = 0.65$ $r = 0.5$
(d) $i = 1.72$ $r = 1$
(e) $i = 1$ $r = \ln 2$

To sum up, the point of this section is that, although growth may be continuous, the solution of simple numerical problems can be arrived at assuming discrete growth and using the methods learned in Chapter 7. This may be particularly helpful to students who have difficulty in understanding how to use exponential functions.

QUESTIONS 14.3

1. An economy's annual usage of resource R has increased from 3,560 units to 5,823 units over the last 15 years. Find the annual growth rate in the amount used and predict what usage will be in 10 years' time assuming (a) that growth is instantaneous and (b) that growth is discrete on an annual basis.
2. Over a 12 month period what instantaneous growth rate is equivalent to a discrete growth rate of 6%?
3. What discrete annual growth rate is equivalent to an instantaneous growth rate of 6% persisting over 12 months?
4. What interest rate would you prefer to be used to add interest to your savings: 8% applied on a continuous basis or 9% applied once a year?

Answers

CHAPTER 2

2.1	1. 3,555	2. 865	3. 92,920	4. 23
2.2	1. 919	2. 225	3. 164	4. 627
	5. 440	6. 101		
2.3	1. 840	2. 17	3. 172	4. 122
	5. £13,800	6. 598	7. £176	

2.4 1. $\dfrac{73}{168}$ 2. $\dfrac{101}{252}$ 3. $4\dfrac{4}{5}$ 4. $\dfrac{19}{30}$

5. $2\dfrac{13}{21}$ 6. $4\dfrac{37}{60}$ 7. $14\dfrac{1}{12}$ 8. $3\dfrac{12}{13}$

9. $18\dfrac{39}{40}$ 10. $1\dfrac{1}{2}$

2.5 1. (a) $\dfrac{1}{3}$ (b) $\dfrac{5}{7}$ (c) $1\dfrac{2}{5}$ (d) 3 (e) 11

2. 1   3. (a) 1 (b) 5

4. 15, $4\dfrac{1}{3}$, $2\dfrac{1}{5}$, $1\dfrac{2}{7}$, $\dfrac{7}{9}$, $\dfrac{5}{11}$, $\dfrac{3}{13}$, $\dfrac{1}{15}$

2.6 1. 36.914   2. 751.4   3. 435.1096
4. 36,082   5. 0.09675   6. 610   7. 300
8. (a) 0.1   (b) 0.001   (c) 0.000001
9. (a) 0.452   (b) 2.431   (c) 0.075   (d) 0.002

2.7 1. -2 2. -24 3. -33 4. 0.45
5. 0.35   6. -117 7. -330 8. 3600

9. $\dfrac{-157}{140}$ 10. $\dfrac{17}{16}$

2.8 1. 0.25   2. 123   3. 6   4. 64
5. 11.641754   6. 531,441   7. 0.0015328
8. 36   9. -618.47021 10. 25.000655

2.9 1. 25   2. 2   3. 0.2   4. 7
5. 2.4494897   6. 96   7. 10
8. 5.2780316   9. 0.03423   10. 87.977857

2.10 1. 270,818.98   2. 220.9478
3. 2.8563×10^9 4. 1.5728×10^8

5. 1.2683   6. 16,552,877

7. $\log Q = \log A + \alpha \log K + \beta \log L + \gamma \log R$

8. 93.696376   9. 4.38228   10. 5.1331868

CHAPTER 3

3.1   1. (a) $0.01x$ (b) $0.5x$ (c) $0.5x$ 2. $0.01rx + 0.5wx + 0.5mx$

3. (a) $\dfrac{x}{40}$ (b) $\dfrac{xp}{40}$ 4. (a) $0.25x$ lb (b) $0.5x$ lb (c) $x(0.25m + 0.5p)$

5. $0.25w + 0.25$ 6. Own example   7. $10.5x + 6y$ 8. $3q - 6000$

3.2   1. 456   2. 77.312   3. $r + z$, 9%

4. Own example   5. 1.0940895   6. £465.58   7. £2,100

8. (a) $333.33 + 0.75(M - 333.33)$ (b) £1,320.83

3.3   1. $30x + 4$ 2. $24x - 18y + 7xy - 12$

3. $6x + 5y - 650$ 4. $9H - 120$

3.4   1. $6x^2 - 24x$ 2. $x^2 + 4x + 9$

3. $2x^2 + 6x + xy + 3y$ 4. $42x^2 - 16y^2 - 34xy + 6y$

5. $33x + 2y - 20y^2 + 62xy - 21$

6. $120 + 2x + 54y + 40z - x^2 + 6y^2 + xy + 4xz + 8yz$

7. $200q - 2q^2$ 8. $13x + 11y$

9. $8x^2 + 60x + 76$ 10. $4,000 + 150x$

3.5   1. $(x+4)^2$ 2. $(x-3y)^2$ 3. Does not factorize

4. Does not factorize   5. Own example

3.6   1. $3x + 7 - 20x^{-1}$ 2. $x + 9$ 3. $4y + x + 12$

4. $200x^{-1} + 21$ 5. $179x$

6. $2(x+3) + 4 - x - 3 - x + 2 = 9$

7. Own example

3.7   1. 4   2. $\dfrac{1}{11}$ 3. 7   4. 14

5. 82   6. 20 p   7. 33%   8. 40p   9. £1,444.44

10. 16 feet 11. 26

3.8   1. $\dfrac{1}{n}\sum\limits_{i=1}^{n} H_i$, 173.7 cm   2. 35   3. 60

4. $\sum\limits_{i=1}^{n} 6,000(0.9)^{i-1}$, 16,260 tonnes

5. (a) $\dfrac{1}{n}\sum\limits_{i=1}^{n} R_i$, £4,425 (b) $\dfrac{1}{3}\sum\limits_{n-3}^{n-1} R_i$, £4,933

6. 13.25%, 8.2%

3.9   1. (a) $\leqslant$ (b) $<$ (c) $\geqslant$ (d) $>$

2. (a) $>$ (b) $\geqslant$ (c) $>$ (d) $>$

3. (a) $Q_1 < Q_2$ (b) $Q_1 = Q_2$ (c) $Q_1 > Q_2$

4. $P_2 > P_1$

CHAPTER 4

4.1   1. (a) Quantity demanded depends on the price of tea, average expenditure etc.

(b) Q_t dependent, all others independent.

Answers

(c) $Q_t = 99 - 6P_t - 0.5Y_t + 0.8A + 1.2N + 1.4P_C$

(suggested number assumes tea is an inferior good)

2. (a) 202 (b) 7 (c) 6, $x \geqslant 0$

3. Yes; no

4.2   1. °F = 32 + 1.8°C   2. $P = 2,400 - 2Q$

3. It is not monotonic, e.g. TR = 200 when $q = 5$ or 10

4. $(0.025X - 25)^2$; no   5. Own example

4.3   (Answers to 1 to 5 give intercepts on axes)

1. $x = -12, y = 6$ 2. $x = 3\frac{1}{3}, y = -40$

3. $P = 60, Q = 300$ 4. $P = 150, Q = 750$

5. $K = 24, L = 40$

6. Goes through origin and ($Q = 10$, TC = 80)

7. Goes through ($Q = 0$, TC = 200) and ($Q = 10$, TC = 250)

8. Horizontal line at TFC = 75   9. Own example

10. (a) and (d); both slope upwards and have a positive intercept on the P axis

4.4   1. $Q = 90 - 5P$; 50; $Q \geqslant 0, P \geqslant 0$

2. $C = 30 + 0.75Y$ 3. By £20 to £100

4. $P = 12 - 0.015Q$ 5. £6,440

4.5   1. 3.75, 0.75, 0.375, -0.75 2. $P = 12, Q = 40$; £4.50; 10

3. (a) $\frac{2}{3}$ (b) 3   4. (c) (i) (a) (ii) (b)   5. (a), (d)

6. APC $= 400Y^{-1} + 0.5 > 0.5 =$ MPC assuming $Y > 0$

7. (a) 0.263 (b) 0.714 (c) 1.667

8. Own example

4.6   1. -1.5 (a) becomes -1 (b) becomes -1 (c) no change   (d) no change

2. $K = 100, L = 160, P_K = £8, P_L = £5$

3. Cost £520 > budget; P_L reduced by £10 to £30

4. (a) -10 (b) -1 (c) -0.1 (d) -0.025 (e) 0

5. No change

6. Height £120, base 12, slope $-10 = -$(wage)

7. Own example

4.7   1. Sketchgraphs   2. Sketchgraphs   3. Steeper

4. Like $y = x^{-1}$; £260   5. Own example

4.8   1. Sketchgraphs   2. Own example

3. $\pi = 20x - 400 - 0.2x^3$; inverted U always < 0

4. (a) $40 = 3250q^{-1}$ (b) Original firms' π per unit = £27.50 but new firms' AC = £170 > price

4.9   1. (a) $16L^{-1}$ (b) 0.16 (c) constant

2. (a) $57,243.34L^{-1.5}$ (b) 57.243 (c) constant

3. (a) $322.54L^{-1}$ (b) 3.2254 (c) increasing

4. (a) $3,125L^{-1.25}$ (b) 9.882 (c) increasing

5. (a) $23,415,916L^{-1.6667}$ (b) 10,868.71

(c) decreasing

6. (a) $4,093.062L^{-1.7714}$ (b) 1.173 (c) decreasing

4.10   1. MR $= 33.33 - 0.00667Q$ for $Q \geqslant 500$

2. MR $= 76 - 0.222Q$ for $Q \geqslant 22.5$

3. MR $= 80 - 0.555Q$ for $Q \geqslant 562.5$

461

4. MC = 30 + 0.0714Q for $Q \geqslant 56$
5. MC = 56 + 0.1333Q for $Q \geqslant 30$
6. MC = 3 + 0.01695Q for $Q \geqslant 59$

CHAPTER 5

5.1   1. $q = 40$, $p = 6$ 2. $x = 67$, $y = 17$ (approximately)
       3. No solution exists
5.2   1. $q = 118$, $p = 256$ 2. (a) $q = 80$, $p = 370$
       (b) q falls to 78, p rises to 376   3. Own example
       4. (a) 40   (b) rises to 50   5. $x = 2.102$, $y = 62.25$
5.3   1. $A = 24$, $B = 12$ 2. 200   3. $x = 190$, $y = 60$
5.4   1. $x = 30$, $y = 60$ 2. $A = 6$, $B = 36$ 3. $x = 25$, $y = 20$
5.5   1. $x = 24$, $y = 14.4$, $z = 19.2$
       2. $x = 4$, $y = 6$, $z = 4$ 3. $A = 6$, $B = 22$, $C = 2$
       4. $x = 17$, $y = 4$, $z = 18$
       5. $A = 82.5$, $B = 35$, $C = 6$, $D = 9$
5.6   1. $q = 500$, $p = 275$ 2. $K = 17.5$, $L = 16$, $R = 10$
       3. (a) p rises from £8 to £10   (b) p rises to £9
       4. $Y = £3,750$ m; government deficit £150 m
       5. $Y = £1,625$ m; balance of payments deficit £15 m
       6. $L = 80$, $w = 52$
5.7   1. $q_1 = 60$, $q_2 = 80$, $p_1 = £10$, $p_2 = £8$
       2. $q_1 = 180$, $q_2 = 200$, $p = £39$
       3. $q_1 = 1,728$, $q_2 = 780$, $p = £190.70$
       4. $q_1 = 40$, $q_2 = 50$, $p_1 = £6$, $p_2 = £4$
       5. $p_1 = £8.75$, $q_1 = 60$, $p_2 = £6.10$, $q_2 = 550$
       6. £81 for extra 65 units   7. £7.50 for extra 25 units
       8. $q_1 = 1,510$, $q_2 = 1,540$, $q_A = 800$, $q_B = 2,250$, $p_A = £500$, $p_B = £625$
       9. $q_1 = 160$, $q_2 = 600$, $q_A = 293\frac{1}{3}$, $q_B = 266\frac{2}{3}$, $q_C = 200$, $p_A = £95$,
          $p_B = £80$, $p_C = £60$
       10. $q_1 = 48$, $q_2 = 39$, $p_1 = £12$, $p_2 = £8.87$
       11. (a) 190 units   (b) £175 for extra 75 units
       12. $q_1 = 106$, $q_2 = 172$, $q_3 = 132$, $p = £19.80$
5A.1   1. 8.4 of A, 4.64 of B (tonnes);   (a) no change   (b) no B, 12.16 of A
       2. $A = 13$, $B = 27$ 3. 12 of A, 5 of B
       4. 22.5 of A, 7.5 of B   5. 6 of A, 32 of B   6. Own example
       7. 13.64 of A, 21.82 of B; £7092; surplus 2.72 of R, 22.72 of mix additive
       8. Produce 15 of A, 21 of B   9. 30 of A, B = 0
       10. Objective function parallel to first constraint
       11. 24,000 shares in X, 18,000 shares in Y, return £8,640
       12. Own example
5A.2   1. $C = 70$ when $A = 1$, $B = 1.5$, slack in $x = 30$
       2. $A = 3$, $B = 0$
       3. $Q = 2.5$, $R = 1.5$; excesses 62.5 mg of B, 2.25 mg of C

Answers

4. 10 of A, 5 of B; space for 50 extra loads of X
5. Zero R, 15 tonnes of T; G exceeds by 45 kg
6. 100 of A, 40 of B   7. Own example

5A.3   1. 2 of A, 1 of B   2. 7.5 of X and 15 of Y (tonnes)
3. 8   4. Own example

CHAPTER 6

6.1   1. 2 or 3   2. 10 or 60   3. When $x = 2$
4. 0.5   5. 9
6.2   1. 10 or -12.5 2. £16.35
3. (a) 1.01 or 98.99   (b) 11.27 or 88.73   (c) no solution exists
4. Own example
6.3   1. $x = 15, y = 15$ or $x = -3, y = 249$ 2. $x = 1.75, y = 3.15$ or
$x = -1.53, y = 20.97$ 3. 16.4   4. $q_1 = 3.2, q_2 = 4.8,$
$p_1 = £136, p_2 = £96$ 5. $p_1 = £15, q_1 = 80, p_2 = £8.50, q_2 = 70$
6.4   1. 52   2. 106.9

CHAPTER 7

7.1   1. £4,630.50   2. £314.70   3. £17,623.16
4. £744.71   5. £40,441.40   6. £5,030.03
7. £43,747.41; 12.68%   8. £501,159.74; 7.44%   9. (a) APR 11.35%
10. £2,083.61; 19.25%
7.2   1. £6,301.69   2. £355.89   3. No, $A = £9,106.27$
4. £6,851.65   5. (a) £9,638.58   (b) £11,579.83 (c) £13,318.15
6. 5   7. 5.27 years   8. 12.1 years
9. 5.45 years   10. 3.42 years   11. 10.7%
12. 9.5%   13. 7.5%   14. 0.8%
15. 10.3%   16. 8.4%   17. (b) as PV = £5,269.85
7.3   1. (a) £90.75   (b) $-£100.07$ (c) $-£474.01$ (d) £622.86   (e) £1,936.87
(f) £877.33   (g) £791.25   (h) £992.16
2. B, PV = £6,569.10   3. (a) All viable   (b) A best, NPV = £6,824.68
4. Yes, NPV = £7,433.56   5. Yes, NPV = £4,363.45
6. (a) Yes, NPV = £610.02   (b) no, NPV = $-£522.30$
7. B, NPV = £856.48
7.4   1. $r_A = 20\%, r_B = 41.6\%, r_C = 20\%, r_D = 20\%$; B consistently best, but
others have same IRR with different NPV ranking
2. (a) A, $r_A = 21.25\%, r_B = 20.42\%$ (b) B, $NPV_B = £2,698.94,$
$NPV_A = £2,291.34$ 3. IRR = 16.93%
7.5   1. (a) 2.5, 781.25, 50,857.3 (b) 3, 121.5, 14,762 (c) 1.4, 10.756, 139.6
(d) 0.8, 19.66, 267.8   (e) 0.75, 0.57, 9.06
2. 5,741 (to nearest whole unit)
3. A, £1,149.32; B, £2,980.91; C, £45,216.47
4. Yes, NPV = £3,774.71   5. £4,149.20
7.6   1. (a) $k = 1.5$, not convergent   (b) $k = 0.8$, converges on 600   (c) $k = -1.5$,
not convergent   (d) $k = 1/3$ converges on 54   (e) converges on 961.54
(f) not convergent
2. £3,076.92   3. Yes, NPV = £10,714.29
4. (a) £240,000   (b) £120,000   (c) £80,000   (d) £60,000   5. £155.56

463

7.7 1. £152.59   2. £197.38   3. £191.46
4. £794.66   5. (a) 14.02% (b) 26.08% (c) 23.86% (d) 14.71%
6. Loan is marginally better deal (PV of payments = £6,348.33 + £1,734 deposit = £8,082.33, less than cash price by £12.67)

7.8 1. 6.82 years   2. After 15.21 years   3. 4%
4. Yes, sum of infinite GP = 1,300 millions tonnes   5. 4.85%

Chapter 8

8.1 1. $36x^2$ 2. 192   3. 21.6
4. $260x^4$ 5. Own example
8.2 1. $3x^2 + 60$ 2. 250   3. $-4x^{-2} - 4$
4. 1   5. $0.2x^{-3} + 0.6x^{-0.7}$ 6. Own example
8.3 1. $120 - 6q$, 20   2. 25   3. 14,400
4. £200   5. Own example
8.4 1. 7.5   2. $12q^2 - 40q + 60$
3. (a) $1.5q^2 - 6q + 25$ (b) $0.5q^2 - 3q + 25 + 20q^{-1}$ (c) $q - 3 - 20q^{-2}$
4. MC constant at 0.8   5. Own example
8.5 1. 4   2. (a) 80 (b) 158.33 (c) 40 or 120   3. 6
8.6 1. $50 - \frac{2}{3}q$ 2. 900   3. $24 - 1.2q^2$
8.7 1. 0.8   2. Proof   3. 0.16667   4. 1
8.8 1. £77.50   2. Own example
3. Rise, maximum TY when $t = £39$
8.9 1. (a) 0.8 (b) 4,400 (c) 5 (d) 120 (e) Yes, both 940

Chapter 9

9.1 1. 62.5   2. 150   3. (a) 500 (b) 600 (c) 300   4. 50
9.2 1. 1,200, max.   2. 25, max.   3. 4,096, max.
4. 4, not max.
9.3 1. 6, min.   2. 14.4956, min.   3. 0, min.
4. 3, not min.   5. No stationary point exists
9.4 1. (a) MC = $2q^2 - 28q + 222$, min. when $q = 7$, MC = 124
(b) AVC = $\frac{2}{3}q^2 - 14q + 222$, min. when $q = 10.5$, AVC = 148.5
(c) AFC = $50q^{-1}$, min. when $q \to \infty$, AFC $\to 0$
(d) TR = $200q - 2q^2$, max. when $q = 50$, TR = 5,000
(e) MR = $200 - 4q$, no turning point, end-point max. when $q = 0$
(f) $\pi = -\frac{2}{3}q^3 + 12q^2 - 22q - 50$, max. when $q = 11$, $\pi = 272.67$ (π min. when $q = 1$, $\pi = -60\frac{2}{3}$)
2. Own example   3. (a) 16 (b) 8 (c) 12
4. No turning point but end-point min. when $q = 0$
5. No turning point but end-point min. when $q = 0$
6. Max. when $x = 63.33$, no minimum
9.5 1. π max. when $q = 4$ (theoretical min. when $q = -1.67$ not realistic)
2. (a) Max. when $q = 10$ (b) no min. exists
3. π max. when $q = 12.67$, gives $\pi = -48.8$
4. 5,075 when $q = 10$ 5. 27.6 when $q = 37$
9.6 1. 15 orders of 400   2. 560
3. 480 every 4.5 months   4. 140

Answers

Chapter 10

10.1 1. (a) $3 + 8x$, $16 + 4z$ (b) $42x^2z^2$, $28x^3z$ (c) $4z + 6x^{-3}z^3$, $4x - 9x^{-2}z^2$

2. $MP_L = 4.8K^{0.4}L^{-0.6}$, falls as L increases

3. $MP_K = 12K^{-0.7}L^{0.3}R^{0.4}$, $MP_L = 12K^{0.3}L^{-0.7}R^{0.4}$, $MP_R = 16K^{0.3}L^{0.3}R^{-0.6}$

4. $MP_L = 0.7$, does not decline as L increases 5. No 6. $12x_j^2$

10.2 1. (a) 0.228 (b) falls to 0.224 (c) inferior as $\partial q/\partial m < 0$ (d) elasticity with respect to $p_s = 0.379$ and so a 1% increase in both prices would cause a percentage rise in q of $0.379 - 0.228 = 0.151\%$

2. (a) Yes, MU_A and MU_B will rise at first but then fall; (b) no, MU_A falls but MU_B continually rises, therefore law not obeyed; (c) yes, both MU_A and MU_B continually fall.

3. No, MU will never reach zero for finite values of A or B.

4. 3,738.46; balance of payments changes from 4.23 deficit to 68.85 surplus.

5. $25 + 0.6q_1^2 + 2.4q_1q_2$ 6. 0.45; 1.81818; 55

10.3 1. $-2K^{0.6}L^{1.5}$, $2.4K^{0.4}L^{0.5}$

2. $Q_{LL} = 6.4$, MP_L function has constant slope; $Q_{LK} = 35 + 2.8K$, position of MP_L will rise as K rises; $Q_{KK} = 2.8L$, MP_K has constant slope, actual value varies with L; $Q_{KL} = 35 + 2.8K$, increase in L will increase MP_K, effect depends on level of K.

3. $TC_{11} = 0.008q_3^2$, $TC_{22} = 0$, $TC_{33} = 9q_2 + 0.008q^2$
$TC_{12} = 1.2q_3 = TC_{21}$, $TC_{23} = 9 + 1.2q_1$
$TC_{31} = 0.016q_1q_3 + 1.2q_2 = TC_{13}$

10.4 1. $q_1 = 12.46$; $q_2 = 36.55$ 2. $p_1 = 97.60$, $p_2 = 101.81$ 3. $q_1 = 0$, $q_2 = 501.55$ (mathematical answer gives $q_1 = -1,292.24$, $q_2 = 1,701.767$ so rework without market 1)

4. £575.81 when $q_1 = 47.86$ and $q_2 = 39.01$ 5. $q_1 = 266.67$, $q_2 = 333.33$
6. $q_1 = 1,580.2$, $q_2 = 1,719.8$ 7. $K = 2,644.2$, $L = 3,718.5$
8. £29,869.47 when $K = 1,493.47$ and $L = 2,489.12$
9. Because max. $\pi = £18,137.95$ when $K = 2,176.5$ and $L = 2,015.22$
10. $K = 10,149.1$, $L = 9,743.1$

10.5 1. (a) $12K^{-0.4}L^{0.4}dK + 8K^{0.6}L^{-0.6}dL$
(b) $14.4K^{-0.7}L^{0.2}R^{0.4}dK + 9.6^{0.3}L^{-0.8}R^{0.4}dL + 19.2K^{0.3}L^{0.2}R^{-0.6}dR$
(c) $(4.8K^{-0.2} + 1.6KL^2)dK + (3.5L^{-0.3} + 1.6K^2L)dL$

2. (a) Yes (b) no, surplus (c) no, surplus

3. $40x^{-0.6}z^{-0.45} + 12x^{0.4}z^{-0.7}$ 4. $\dfrac{\partial Q_A}{\partial P_A} + \dfrac{\partial Q_A}{\partial M}\dfrac{dM}{dP_A}$

Chapter 11

11.1 1. $K = 12.6$, $L = 21$ 2. $K = 500$, $L = 2,500$
3. $A = 6$, $B = 4$ 4. 141.42 when $K = 25$, $L = 50$
5. Own example 6. (a) $K = 1,000$, $L = 50$ (b) $K = 400$, $L = 20$
7. 1,950 when $K = 60$, $L = 120$ 8. $L = 241$, $K = 201$, TC = £3,617

11.2 See answers to 11.1

Answers

11.3   1. See answers to 11.1   2. $L = 38.8$, $K = 20.7$, TC = £3,104.50
       3. C_1 = £480,621, C_2 = £213,609   4. $L = 19.04$, $K = 8.18$
       TC = £1,145.30

11.4   1. $x = 30$, $y = 30$, $z = 90$ 2. 840.6 when $K = 34.94$, $L = 19.74$, $R = 13.97$
       3. $x = 50$, $y = 100$, $z = 150$ 4. 79,602.1 when $x = 300$, $y = 300$, $z = 1,875$
       5. $K = 26.7$, $L = 33.3$, $R = 8.9$, $M = 55.6$ 6. Own example
       7. $L = 60$, $K = 45$, $R = 40$

Chapter 12

12.1   1. 9   2. Answer given
       3. $3M(1 + i)^2$
       4. $0.6x(3 + 0.6x^2)^{-0.5}$ 5. $0.5(6 + x)^{-0.5}$
       6. $\text{MRP}_L = 60L^{-0.5} - 8$, $L = 16$ 7. 169 units
       8. £8   9. 0.000868

12.2   1. $(6x + 7)^{-0.5}(39x^2 + 36.4x - 5.7)$
       2. 12   3. $76.5L^{-0.5}(0.5K^{0.8} + 3L^{0.5})^{-0.4}$
       4. 312.5   5. £190   6. Own example
       7. (a) $-0.05(60 - 0.1q)^{-0.5}$ (b) rate of change of slope =
       $-0.025(60 - 0.1q)^{-1.5} < 0$ when $q < 600$ (c) 400

12.3   1. $(24 + 6.4x - 4.5^{1.5} - 3x^{2.5})(8 - 6x^{1.5})^{-1.5}$
       2. $(18,000 + 360q)(25 + q)^{-1.5}$ 3. -0.113
       4. $q = 1,333\frac{1}{3}$, $d^2\text{TR}/dq^2 = -0.00367$ 5. $L = 4.8$, $H = 7.2$
       6. Adapt proof in text for MC and AC to AVC = $\text{TVC}(q)^{-1}$

12.4   1. (a) $12.5x^2$ (b) $5x + 0.6x^2 + 0.05x^3$ (c) $24x^5 - 15x^4$
       (d) $42x + 18x^{-1}$ (e) $60x^{1.5} + 220x^{-0.2}$
       2. (a) $4q + 0.05q^2$ (b) $42q - 9q^2 + 2q^3$ (c) $35q + 0.3q^3$
       (d) $62q - 8q^2 + 0.5q^3$ (e) $185q - 12q^2 + 0.3q^4$

12.5   1. (a) £750,000 (b) £81,750 (c) £1,000,000 (d) £149,500
       2. £49,600   3. Own example

Chapter 13

13.1   1. 20   2. No production in period 4
       3. (a) Unstable (b) stable

13.2   1. $P_t = 4 + 0.25(-2)^t$ 2. Stable, 118.54   3. 404.64

13.3   1. 2,790.625; yes   2. 39,946.789
       3. 503.44   4. 1,848.259

13.4   1. 2,460.79   2. No, 1,976.67 < 1,980
       3. $P_t^X = 562 - 63(0.83)^t$, 555.27

Chapter 14

14.1   1. 64.44 million   2. 61,062 units   3. 16.8 million tonnes
       4. Usage in million units: (a) 94.6, yes (b) 137.6, yes (c) 200.2, no
       (d) 291.31, no
       5. 56,609 units   6. 2%; 9.84 million; no

Answers

14.2   1. 0.48%   2. 2.04%; 3.48%
        3. 0.83%, 621.43 million tonnes   4. £6,446.39 million
        5. 8.8%

14.3   1. (a) 3.28%, 8,084   (b) 3.33%, 8,084
        2. 5.83%   3. 6.18%
        4. 9% discrete (equivalent to 8.61% continuous)

Index

absolute value 22, 62; and convergency conditions 233, 417–18
accounting identity 75
accumulated value of an investment 189–94
addition 9–10, 20; of algebraic terms 39–41; in equations 54, 162; with inequalities 62
adjustment between equilibria 402–3
agriculture and time lags in supply 403, 405–6, 413
algebra: and economics 4, 361; revision of basics 35–64
annual percentage rate (APR) 192–3, 238–42
annuities 228–31, 235; perpetual 232–5
antilogs 30–1, 451
arc elasticity 16–19
area 44; derived by integration 396–400
arithmetic 9–34
average 59
average cost 52–3, 94–5, 264; minimum point 280–1, 386–7
average fixed cost 52–3, 93–4
average propensity to consume 84
average variable cost 52–3, 94–5; minimum point 264
axes 72, 404

balance of payments 304–5
base for logarithms 29, 451
brackets 12–13, 43, 46
break-even chart 114–16, 283–4
break-even point 116, 179
budget constraint 85–9, 124, 160; and maximum output 55, 181–5, 348–53
business 3

calculators 3–5, 25, 28–9, 192, 210
calculus 251–76
cancellation 229

cardinal utility 300
cartels 128, 136
cartesian axes 71
cause and effect 68
ceteris paribus 295, 298, 309, 336, 402
chain rule 370–6, 378; and partial differentiation 374–6
Cobb–Douglas production function 33, 102–4, 353; convexity to origin 295–6; derivation of marginal product 295–6; returns to scale 332–3; product exhaustion 332–6
cobweb model 403; iterative solution 403–15; difference equation solution 415–24; spreadsheet solution 408–11
coefficients 80, 118, see also parameters
collusion 134, 438
common ratio: and convergency 233–4; of geometric series 225
comparative statics 402
competing products 311–12
complementary solution to difference equations 416–18, 431–2, 440
complement prices 298–300, 306
composite functions 94–9
compound interest 189–91
computers 3, 5–6, 165
constant in functions 91, 94; when differentiated 255, 293
constant of integration 390, 393
constant returns to scale 103–4, 333–5
constrained maximization: by substitution 338–44; in linear programming 144–53; using Lagrange multiplier 346–55, 361–6
constrained minimization: by substitution 344–5; in linear programming 156–64; using Lagrange multiplier 355–61, 366–8

constrained optimization 338–69; and linear programming 144

constraint: budget 85–9; in linear programming 145–6; in substitution method 339; Lagrangian 346

consumer surplus 399–401

consumer utility 300–3

consumption function 73, 77–9

continuous functions 186–8

continuous growth 444–5

convergent series 232–5

convexity to origin 353–5

coordinates 71

corner solutions 284–5

cost minimization: subject to output constraint 344–5; subject to quality constraints 156–61

costs, *see* average, marginal and total cost

Cournot model 438

cross multiplying 365

cross partial derivatives 310, 313

cubic functions 91, 181–5, 263–4

cubic values 24

curved functions 90–4

data availability 3

decay 444, 449–50, 452–3

decimals 19–22, 189

decline 186, 243–50

decreasing returns to scale 104, 323, 333–5

definite integrals 395–401

degree of a polynomial 95

demand for labour 372–5

demand function 53, 69, 74–7; and marginal revenue 259–60, 267, 397–9; and supply equilibrium 110–11, 402–25; and total revenue 96, 110, 258–9; elasticity 16–19, 82–3, 269–71, 298–9, 375–6; signs of parameters 82, 404

denominator 13, 51

dependent variables 59

deposit accounts 186–7

derivatives: first-order 251–6; partial 292–6; second-order 280–1, 309; second-order partial 280–1, 309; total 336; used to predict changes 329–30

difference equations 404; and cobweb model 415–24; and Keynesian model 425–6, 430–7; and Bertrand duopoly 438–42

differentials 329–35; total 330–1

differential calculus 251

differentiation: basic rules 251–5; from first principles 256–7; of logarithmic and exponential functions 449–50, 454–5; partial 292–337; to derive MC from TC 262–6; to derive MR from TR 257–60; using chain rule 370–6; using product rule 382–9; using quotient rule 382–9

diminishing marginal productivity 295–6

diminishing marginal utility 301–3

discounting factors 205, 210–12, 233; changing 222–3

discrete functions 186–8, 403, 444

distributed lags 436

disturbances to equilibrium 405, 408

divergent series 232–5

division 9, 20, 54; in equations 54; of algebraic expressions 50–1

domain 66, 71, 284

doubling value of an investment 199

duopoly price adjustment 438–42

dynamics 402–43

e: derivation of 445–7; as base for natural logarithms 451

econometrics 2, 434

economic growth 448

elasticity: arc 16–19; income 299–300; point 82–3, 269–71, 298–9, 375–6

elimination of unknowns 110–11, 116, 118, 120

empirical data 33, 68

end point solutions 284–5

equality constraints 163–4

equality sign 52

equating to solve simultaneous equations 113–16

equations: cubic 181–5, 263–4; Lagrange 346; polynomial 95, 180–5; quadratic 180–5; simple linear solution 52–3; simultaneous 110–43

equilibrium: adjustment mechanism between 402–43; price and quantity 110–12

Euler's theorem 332–5

evaluation 38–9

exchange rate 38, 53–4

exhaustion of total product 332–6

exogenous variables 65; changes in 425

expenditure on inputs 85
exploding price 412
exponential functions 444–50; and growth rates 447–50; differentiation of 449–50, 454–5
exponents 25, 444–5; fractional 27; negative 25
export multiplier 303
expressions: algebraic 35–7; evaluation of 38–9; factorization 46–50; simplification 39–51
extraction rates 187, 243–50

factorization 46–9, 172–4
feasible area 145, 156
final value of investment: when continuous growth 447–8; when discrete growth 190–4
first degree price discrimination 132–5
first order derivatives 309
first order difference equations 404
fitting linear functions 76–9
fixed costs, *see* total and average fixed cost
fixed inputs 292
fractions 13–15
free goods 301
functional form 33
functions 53, 65–109; Cobb–Douglas production 33, 102–4, 295–6, 332–3, 353; cubic 181–5, 263–4; inverse 67–9, 74, 267–9; linear 73–9, 109; polynomial 95, 180–5; power 90–3; quadratic 166–85; two independent variables, graphing 100–4

general form of a function 65
geometric series 224–235; sum of 226–7; infinite 232–3, 420
government budget 304–5
government expenditure multiplier 303
gradient, *see* slope
graduations on axes 72, 396
graphs 70–109, 111–12, 167–72
group working 7
growth: continuous 444–58; discrete 186, 243–50; discrete and continuous compared 455–8
guesswork 3

Hicks 349–50
height of functions 100–2, 315
homogeneous functions 103

horizontal axis 70–2, 74, 81–3
horizontal functions 95
horizontal lines and zero slope 98, 277–83
horizontal summation: of MC functions 107–9; of MR functions 104–7
hours of work/leisure 387–9

identities 53
income effect 349–52
income elasticity 299–300
income-expenditure 45° line 426–8
increasing retuns to scale 104, 333–5
independent variables 59, 67
indifference curves 349–52
indirect functional relationships 336–7
individual labour supply 387–9
inequalities 61–4, 145, 165
inferior goods 349–52
infinite geometric series 232–3, 420
infinitesimally small changes 256–8, 269–70
infinity 71, 232–3
inflation 59–60
inflexion points 279, 283
initial term in geometric series 225–6, 229, 233
inputs: optimal combination 129, 339–40, 344–5; optimal order size 288–91; requirements in linear programming 119, 144; variable and fixed 292
integers 14, 49
integrals 289–90; definite 395–401
integration 389–401; of $1/x$ 454
intercepts 75, 86, 91, 134, 260, 404
interest 188–94, 199–201; annual rates 192–3, 238–42, 446–7; changes in 194, 222–3; compound 189–91; continuous and discrete compared 447, 455–8; determination of 199–201; instantaneously credited 447; monthly rates 193, 238–42; simple 188–9
internal rate of return 202, 215–24; deficiencies of 218–223; Lotus 1–2–3 command 216–18
intersection points 111
inventory control 288–91
inverse function rule 299, 454
inverse functions 67–9, 74, 81–3, 267–9
investment 189; appraisal 202–24; final value 45, 190–4, 228, 447–8,

470

456–7; initial amount 195–6; internal rate of return 202, 215–24; IRR and NPV methods compared 218–23; net present value 202–13, 228–30, 236; spread sheet solution 206–11, 216–17, 222–3; time period 196; time to double 199
isoquants 100–2, 296, 331, 338–9, 353; and output maximization 338–9

joint cost functions 306, 316–18

Keynesian macroeconomic model 125, 274–6, 303–5; lagged 425–37; spreadsheet solution 428–30

labour: individual supply 387–9; profit maximizing level to employ 372–5
Lagrange multiplier: maximization with two variables 346–52; minimization with two variables 355–60; second order conditions 352–5; with more than two variables 361–8
lags: in supply/production 403; in consumption function 425;
leisure hours 387–9
linear demand and supply schedules: and cobweb 403, 413
linear equations 53; solution of single 53–6; simultaneous 110–65
linear functions 70–5; fitting equations to 75–9
linear programming 144–65; constrained maximization 144–53; constrained minimization 156–64; equality constraints 163–4; mixed constraints 163–5; multiple solutions 150–2; solutions on axes 152–3; more than two variables 165
loan repayments 236–42; spreadsheet calculation 240–2
local maxima/minima 282–3
logarithms: base 10 29–34; in investment time period calculation 196–9; natural 435, 451–5
long run 177, 405
Lotus 1–2–3, *see* spreadsheets

macroeconomic model 117, 125, 303–5; dynamic adjustment 425–37
managerial economics 144
marginal cost 107–9, 306; by differentiation 262–6; cutting AVC

264; cutting minimum AC 386–7; horizontal summation 107–9, 135–41, 177–9; integral of 392–4, 397–400; intersection with MR 285–7; of labour 128
marginal product 124, 309–11, 331–2, 374; by partial differentiation 292–6
marginal propensity to consume 84, 274, 303
marginal propensity to import 303
marginal rate of technical substitution 331–2
marginal revenue 129; by differentiation 256–62, 384; for linear demand functions 129, 259; horizontal summation 105–7, 128–32; integral of 392–4, 397–400; interrelated demand functions 306–7, 397–9
marginal revenue product 128, 372–5
marginal utility 300–3, 343–4
market clearing price 407
maximization: by differentiation 277–80, 378–82; linear programming, of profit 265–7, 285–8, 321–3; of tax yield 273; when two independent variables 315–27
methodology 3
minimization: by differentiation 281–2; linear programming 156–64; when two independent variables 315–27
minimum point: of AC 386–7; of AVC 264
mixed constraints 163
money income 88
monopoly 68
monopsony 128
monotonic functions 67–8, 74
monthly interest rates 193, 238–42, 446
mortgage repayments 238–9
multiple operations 10–11
multiplication 9–10, 20; of algebraic expressions 41–5, 54; pairs of brackets 43–5
multi-plant monopoly 107–9, 135–41, 319–21, 323–5
multiplier: export 303; government expenditure 303; investment 274–6, 303–5, 425
multi-product firms 144; related costs 306, 316–18

national income determination 125, 274–6, 303–5, 425–37

natural exponential function 446
natural logarithms 451–5
natural resources, *see* resource depletion rates
necessary conditions for maximization 278
negative area below axis 397–9
negative slope 80–1, 86–7
negative numbers 22–3, 139, 174; and inequality signs 61–2
net present value 202–13, 228–30, 236; Lotus 1–2–3 command 206–8; of geometric series 228–31, 233–5
non-linear functions 33, 90–4, 173
non-negativity restictions 66, 139, 145, 152, 157, 284
non-renewable natural resources 447
normal goods 349–52
numerator 13, 51

objective functions 146, 353–5
oligopoly 438
opportunity cost 88–9, 188
optimal input determination 124–5, 339–40
optimization 1; constrained 338–9; unconstrained 277–91, 315–27
order to perform calculations 3–4, 10–11
ordinal utility 300
oscillating prices 418, 422–4
output, *see* total product
output maximization subject to budget constraint: by substitution 338–43; using Lagrange multiplier 346–8; using spreadsheet 181–3

parallel functions 112–13
parameters 2, 76–7
partial derivatives 292–6; cross 310, 313–14; second order 313–14
partial differentiation 292–337
percentages 21, 189
perfect competition 177–9, 321–3; and cobweb model 403
perfect price discrimination 132–5
perpetual annuities 232–5
per unit taxes 126, 271–4
plotting: linear functions 72–3; quadratic functions 167–72; polynomials 184; using spreadsheets 170–2

point elasticity: linear demand schedule 82–3; non-linear demand schedule 269–71, 298–9, 375–6
pollution levels 359–60
polynomial functions 95, 180–5; multiple solutions 220–1; solution by spreadsheet 181–4
population and demand 299–300
population growth 244–5
positive value restrictions 66, 145
power functions 90–4
powers 23–9, 32, 63–4, 90–3, 192; fractional 27–9, 93; negative 25, 93
predictions of models 1, 53, 76–9, 329–30
present value 196, 202, *see also* NPV
price determination: in competitive markets 110, 177–8; in dynamic models 403–24
price discrimination 104–7, 128–35; first degree (perfect) 132–5; second degree 132–5; third degree 104–7, 128–35, 139–41
price index 59–60
price ratio and slope of budget line 86–9
prime numbers 49
production functions 33, 65–6, 100–4, 295–6, 332–3, 353
product rule 377–82
profit function 145–7, 287
profit maximization: by differentiation 265–7; linear programming 144; second order conditions 285–8; unconstrained 321–3

quadratic equations 166–85; factorization 172–3; graphical solution 167–72; no solution to 168–70, 173, 175; simultaneous 176–9
quadratic formula 174–5
quadratic functions 166–72, 263; slope of 90–4, 251–2
qualitative predictions 1
quality constraints 156–7
quantitative predictions 1
quotient rule 382–9

range 66, 71, 107
ranking of investment projects: IRR criterion 215–16; NPV criterion 204
rate of change 251, 309

rate of change of slope 279–85
reaction functions 438–9
real price changes 60
rectangular hyperbola 101
regression analysis 33, 53
reorder costs 288–9
representation 35–6
repayments on loans 236–42
resource allocation 2, 338
resource depletion rates 243–50,
     452–3, 455–6
respecifying functions 267–9
returns to scale 103–4; and product
     exhaustion 333–5
revenue, *see* total revenue
risk 203
roots 27, 54
rounding errors 4, 15
row operations 118–20

sales revenue, *see* total revenue
second degree price discrimination
     132–5
second-order conditions 278–85, 315,
     325, 352–5; for a maximum 277–80,
     315–16; for a minimum 281–2,
     315–16; Lagrangian 352–5; using
     product rule 379–82; using quotient
     rule 385; when two independent
     variables 315–27
second-order derivatives 280, 309
second-order partial derivatives 297,
     309–14
sequences, *see under* geometric series
shocks to equilibrium 405, 408
short run 177, 295, 374
short run supply schedule 405
simple interest 188–9
simplex method 165
simplification 1, 39–52
simultaneous linear equations 110–43,
     147–9, 362; equating to same
     variable 113–16; graphical solution
     111–13; more than two unknowns
     120–3; row operations 118–20;
     substitution method 116–18
simultaneous quadratic equations
     176–9
slack 147
slope 79–83; and point elasticity
     269–70; budget constraint 86; by
     differentiation 251–3; demand and
     MR functions 259–60; non-linear

functions 96–8; rate of change of
     279–85; zero and maximization
     277–9; zero and minimization 281–2
social science of economics 1
solving simple equations 52–6
specific form functions 66
spreadsheets 6, 170–2, 181–5, 206–11,
     216–17, 222–3, 240–2, 246–7, 261–2,
     408–11, 428–30
square roots 27, 175
square values 23–4, 44
stability 411, 417–20
stationary points 278–9, 315, 350–1,
     361
statistics 2, 53
storage costs 288–9
straight lines, *see* linear functions
substitutes 298–9, 318–19
substitution 116–18, 338–45
substitution effect 349–52
subtraction 9, 20; of algebraic terms
     39–41; in equations 54; with
     inequalities 62
sufficient conditions for maximization
     278–85
summation: finite geometric series
     226–7; infinite geometric series 232–5
summation sign $\sum$ 57–60, 297
supply and demand equilibrium
     110–11, 402–25; short run 405
supply schedule 110, 177–8

tangents 80, 97–8
tax: income 42–3; per unit 126, 271–4
tax yield maximization 271–4, 381–2
technological advances and efficiency
     450
third degree price discrimination
     104–7, 128–35, 139–41
time periods 188, 196–9; adjustment
     between, *see* dynamics
total cost 35, 52, 85, 114, 181–3; as
     cubic function 263–4; in linear
     programming 158; relationship with
     MC 262–4
total derivatives 336
total differentials 329–36
total fixed cost 52, 391–2
total variable cost 52; by integrating
     marginal cost 391–2, 396–7
total product 292–3
total revenue 96–7, 144, 166;
     maximization 256–62, 277–8; by

integrating marginal revenue 392–4, 397–400
troughs of function 316
turning points 279–83
two independent variable functions 100–4
two part pricing 132–3

unconstrained optimization 277–91; with two independent variables 315–27
uneconomic region 296
unknown variables 76–7, 110–11, 116, 120, 166–7
units of measurement 72, 77, 396
unstable equilibrium 411–14, 417–20
U-shaped cost curves 179, 386–7
utility 300–3, 343–4

value of the marginal product 332–5
variable cost, *see* total and average variable cost
variable inputs 292
variables: dependent 59; independent 59, 67; unknown 101–11, 116, 120, 166
vertical axis 70–2, 74, 81–3
viability of investment projects: IRR criterion 215–16; NPV criterion 203–4

wage rates 372–5, 387–9
water resources 449
weak inequalities 61–4, 144, 156

welfare measure 399
working hours 387–9

zero slope 98, 277–83, 292, 295

SYMBOLS AND TERMINOLOGY

$\lvert x \rvert$	absolute value	62
[ ]	definite integral	395
dy/dx	derivative	251
e^x	exponential function	445
f_1	first-order derivative	297
$f(x)$	function	65
>	greater than inequality	61
$\geqq$	greater than or equal to weak inequality	61
$\equiv$	identity	53
∞	infinity	71
$\int$	integration sign	390
λ	Lagrange multiplier	346
$\mathscr{L}$	Lagrangian equation	346
<	less than inequality	61
$\leqq$	less than or equal to weak inequality	61
log	logarithm (base 10)	29
ln	natural logarithm	451
$\partial y/\partial x$	partial derivative	292
f_{12}	second-order partial derivative	297
Δx	small change in x	329
$\sqrt{}$	square root	27
Σ	summation	57